AF835806

RAY AND WAVE CHAOS IN OCEAN ACOUSTICS

Chaos in Waveguides

Series on Complexity, Nonlinearity and Chaos

ISSN 2010-0019

Series Editor: Albert C.J. Luo
(Southern Illinois University Edwardsville, USA)

Aims and Scope

The books in this series will focus on the recent developments, findings and progress on fundamental theories and principles, analytical and symbolic approaches, computational techniques in nonlinear physical science and nonlinear mathematics.

Topics of interest in Complexity, Nonlinearity and Chaos include but not limited to:

- New findings and discoveries in nonlinear physics and mathematics,
- Complexity and mathematical structures in nonlinear physics,
- Nonlinear phenomena and observations in nature,
- Computational methods and simulations in complex systems,
- New theories, and principles and mathematical methods,
- Stability, bifurcation, chaos and fractals in nonlinear physical science.

Vol. 1 Ray and Wave Chaos in Ocean Acoustics: Chaos in Waveguides
D. Makarov, S. Prants, A. Virovlyansky & G. Zaslavsky

CNC Series on Complexity, Nonlinearity and Chaos – Vol. 1

RAY AND WAVE CHAOS IN OCEAN ACOUSTICS

Chaos in Waveguides

Denis Makarov, Sergey Prants
Pacific Oceanological Institute, Vladivostok, Russia

Anatoly Virovlyansky
Institute of Applied Physics, Nizhny Novgorod, Russia

George Zaslavsky
New York University, New York, USA

World Scientific

NEW JERSEY • LONDON • SINGAPORE • BEIJING • SHANGHAI • HONG KONG • TAIPEI • CHENNAI

Published by

World Scientific Publishing Co. Pte. Ltd.
5 Toh Tuck Link, Singapore 596224
USA office: 27 Warren Street, Suite 401-402, Hackensack, NJ 07601
UK office: 57 Shelton Street, Covent Garden, London WC2H 9HE

British Library Cataloguing-in-Publication Data
A catalogue record for this book is available from the British Library.

Series on Complexity, Nonlinearity, and Chaos — Vol. 1
RAY AND WAVE CHAOS IN OCEAN ACOUSTICS
Chaos in Waveguides

Copyright © 2010 by World Scientific Publishing Co. Pte. Ltd.

All rights reserved. This book, or parts thereof, may not be reproduced in any form or by any means, electronic or mechanical, including photocopying, recording or any information storage and retrieval system now known or to be invented, without written permission from the Publisher.

For photocopying of material in this volume, please pay a copying fee through the Copyright Clearance Center, Inc., 222 Rosewood Drive, Danvers, MA 01923, USA. In this case permission to photocopy is not required from the publisher.

ISBN-13 978-981-4273-17-6
ISBN-10 981-4273-17-1

Printed in Singapore by Mainland Press Pte Ltd

To the memory of George Zaslavsky

George Zaslavsky (May 31, 1935–November 25, 2008)

This is the last book by George Zaslavsky. He planned it towards the end of 2007 when he already knew about his mortal disease. We began writing this book in three cities, New York, Nizhny Novgorod, and Vladivostok. His strong will and organizational skills was the force behind the completion of the book after he died in New York on 25 November, 2008.

George was born in Odessa in 1935 in a family of an artillery officer who dragged his cannon in World War II, and survived. George received his education at the University of Odessa and moved in 1957 to Novosibirsk when a golden age of Siberian physics had dawned. In 1965, George joined the Institute of Nuclear Physics where he became interested in nonlinear problems of accelerator and plasma physics. Roald Sagdeev and Boris Chirikov kindled his interest in the theory of dynamical chaos. In 1968, George with his former wife N. Filonenko introduced a separatrix map that became one of the major tools in theoretical study of Hamiltonian chaos. The work "Stochastical instability of nonlinear oscillations" by G. Zaslavsky and B. Chirikov, published in Physics Uspekhi in 1971, was the first review paper that "opened the eyes" of many physicists to the power of the theory of dynamical systems and modern ergodic theory. It was realized that very complicated behavior is possible in dynamical systems with only a few degrees of freedom. This complexity cannot be adequately described in terms of individual trajectories, and requires statistical methods. Typical Hamiltonian systems are not integrable but chaotic, and such chaos is not homogeneous. For the same values of the control parameters, there coexist regions in phase space with regular and chaotic motion. The results obtained in the 60th were summarized in the book *Statistical Irreversibility in Nonlinear Systems* (Nauka, Moscow, 1970).

The late 1960s was a hard time for George. He was forced to leave the Institute of Nuclear Physics in Novosibirsk for signing letters in defense of some Soviet dissidents. George got a position at the Institute of Physics in Krasnoyarsk, not far away from Novosibirsk. There, he founded a laboratory engaged in the theory of nonlinear processes which exists up to now. In Krasnoyarsk, George became interested in the theory of quantum chaos. The first rigorous theory of quantum resonance was developed in 1977 in collaboration with Gennady Berman. They introduced the important notion of quantum break time (the Ehrenfest time) after which quantum evolution begins to deviate from a semiclassical one. The results obtained in Krasnoyarsk were summarized in the book *Chaos in Dynamical Systems* (Nauka, Moscow and Harwood, Amsterdam, 1985).

In 1981, George and Sadrilla Abdullaev published the first paper on chaotic instability of sound rays in idealized underwater waveguides. The first results of their studies on this topic were summarized in a review paper published in Physics Uspekhi in 1991. It was one of his main focus of research up to the last days.

In 1984, R. Sagdeev invited George to the Institute of Space Research in Moscow. There he worked on the theory of degenerate and almost degenerate Hamiltonian systems, anomalous chaotic transport, plasma physics, and theory of chaos in waveguides. The book *Nonlinear Physics: from the Pendulum to Turbulence and Chaos* (Nauka, Moscow and Harwood, New York, 1988), written with R. Sagdeev, is now a classical textbook for everybody who studies chaos theory. When studying interaction of a charged particle with a wave packet, George with colleagues from the institute discovered that stochastic layers of different separatrices in degenerated Hamiltonian systems may merge producing a stochastic web. Unlike the famous Arnold diffusion in non-degenerated Hamiltonian systems, that appears only if the number of degrees of freedom exceeds 2, diffusion in the Zaslavsky webs is possible at one and half degrees of freedom. This diffusion is a rather universal phenomenon and its speed is much greater than that of Arnold diffusion. Beautiful symmetries of the Zaslavsky webs and their properties in different branches of physics have been described in the book *Weak Chaos and Quasi-Regular Structures* (Nauka, Moscow, 1991 and Cambridge University Press, Cambridge, 1991) coauthored with R. Sagdeev, D. Usikov and A. Chernikov. A funny review on this publication was published in a Cambridge student newspaper. The book appeared during the time of disintegration of the Soviet Union and the reviewer wrote: "Once again those Russians with their chaos even if it is a weak one!".

In 1991, George emigrated to the USA and became a Professor of Physics and Mathematics at the Department of the New York University and at the Courant Institute of Mathematical Sciences. The last 17 years of his life were devoted to principal problems of Hamiltonian chaos connected with anomalous kinetics and fractional dynamics, foundations of statistical mechanics, chaotic advection, quantum chaos, and long-range propagation of acoustic waves in the ocean. During his New York period, George published two important books on the Hamiltonian chaos: *Physics of Chaos in Hamiltonian Systems* (Imperial College Press, London, 1998) and *Hamiltonian Chaos and Fractional Dynamics* (Oxford University Press, NY, 2005).

George was a very creative scientist and a very good teacher whose former students and collaborators are now working in America, Europe, and Asia. He authored and coauthored 9 books and more than 300 papers in journals. Many of his works are widely cited. George worked hard all his life. He loved music, theater, literature and was an expert on good wine and food. Only a few people knew that he loved to paint. In the last years, he spent every summer in Provence (France) working, writing books and papers and painting in water-colors; the album with his water-colors was issued in 2009 in Moscow.

George Zaslavsky was one of the key persons in the theory of dynamical chaos and made many important contributions to a variety of other topics. His books and papers have influenced very much the advancement of modern nonlinear science.

We wish to dedicate this book to the memory of George Zaslavsky.

Denis Makarov
Sergey Prants
Anatoly Virovlyansky

Preface

Sound propagation in the ocean over hundreds and thousands kilometers is possible due to the existence of a natural refractive waveguide called the underwater sound channel (USC) [Flatté *et al.* (1979); Brekhovskikh and Lysanov (1991); Jensen *et al.* (1994)]. The vertical sound speed profile in a deep ocean usually has a minimum at a depth of about 1 km. Therefore, part of the sound energy is captured within the water bulk which prevents it from the interaction with the lossy bottom. Experiments on the long range sound propagation are usually performed at carrier frequencies of order 100 Hz. The dissipation of sound waves in sea water at these frequencies is comparatively small, only a few dB per 1000 km, and signals can be registered with sufficiently high signal-to-noise ratios even at ranges exceeding 10 000 km [Munk *et al.* (1994)].

The theoretical description of sound fields in the USC is based on the standard wave equation [Ahluwalia and Keller (1977); Brekhovskikh and Lysanov (1991); Jensen *et al.* (1994); Brekhovskikh and Godin (1999)]. The problem is linear, and usually it is treated using the same methods which are traditionally employed for analyzing wave fields in other waveguide media. Among them, the method of geometrical optics, or ray approach, is of particular importance [Flatté *et al.* (1979); Kravtsov and Orlov (1990); Brekhovskikh and Lysanov (1991); Brekhovskikh and Godin (1999)]. Since the time when the underwater acoustics became a quantitative science (it happened during World War II), this method has been one of the most efficient and frequently applied tools for a study of sound fields in the ocean. Significant advantages of the ray approach are that it provides a deep insight into the basic mechanisms of signal propagation in a complicated environment and is well suited for deriving analytical estimates.

Extensive theoretical and experimental studies of long range sound propagation in the ocean have been carried out for sixty years. Results obtained in this field are presented in many publications (see, e.g., monographs [De Santo (1979); Brekhovskikh (1980); Brekhovskikh and Lysanov (1991); Jensen *et al.* (1994)] and references therein). Already in the middle 1980s this topic was considered well understood. However, in the last two decades it turned out that there exists a factor, earlier not taken into consideration, which to a significant extent determines the structure of the wave field at long ranges.

This is the phenomenon of ray chaos whose significance was realized from the beginning of the 1980s [Abdullaev and Zaslavsky (1981, 1983, 1984); Palmer *et al.* (1988); Abdullaev and Zaslavsky (1991); Smith *et al.* (1992)]. In the ray approximation, the underwater sound propagation can be modelled by a Hamiltonian system representing a nonlinear oscillator driven by a weak nonstationary external perturbation. A range-independent background sound speed profile plays the role of an unperturbed potential on which a range-dependent perturbation of the sound speed along the waveguide, that can be caused by internal waves, mesoscale eddies, ocean fronts or something else, is superimposed. In the first papers on ray motion in range-dependent waveguides [Abdullaev and Zaslavsky (1981, 1983, 1984)], chaotic instability of rays has been found in idealized models. The point is that the ray path in a deterministic inhomogeneous medium behaves like a random process. Chaotic rays are very unstable. The initially close paths diverge exponentially with range r so that the difference between their vertical coordinates grows as $\exp(\nu r)$, where ν is the Lyapunov exponent. A typical value of the Lyapunov exponent in a deep sea varies from $1/300\ \text{km}^{-1}$ to $1/100\ \text{km}^{-1}$ [Beron-Vera *et al.* (2003)]. At ranges of order 1000 km, or at megameter ranges in terms of ocean acoustics, chaos is well developed and should be taken into account when describing the sound fields. Now the study of ray chaos and its manifestations at a finite wavelength, the so-called wave chaos, is a branch of the theory of long range sound propagation in the ocean [Worcester and Spindel (2005)]. However, the results obtained in this field are scattered over many journal publications and may be unknown even to specialists working in related fields. The objective of this book is to discuss the methods and concepts used in the theory of ray and wave chaos and present a comprehensive survey of the state of the art in this branch of ocean acoustics. Applications of the general approaches are illustrated by considering a number of particular problems which were solved with the participation of the authors.

The phenomena of ray and wave chaos have the well-known prototypes in mechanics called the dynamical and quantum chaos, respectively [Zaslavsky (1981); Gutzwiller (1990); Reichl (1992)]. The point is that the ray trajectory in an inhomogeneous waveguide is governed by the same Hamilton equations as a nonlinear oscillator driven by a nonstationary external force. Typically, the oscillator exhibits a quasi-random behavior [Sagdeev *et al.* (1988); Lichtenberg and Lieberman (1992)]. Analysis of the phase space of this oscillator is a classical problem in the theory of dynamical chaos. The theory of quantum chaos deals with mechanical systems whose classical analogues demonstrate chaotic motion. The nonlinear oscillator subjected to a nonstationary external force falls in this category. Its wave function obeys the Schrödinger equation that has the same form as the standard parabolic equation describing the sound field in the USC.

The above analogy with the nonlinear oscillator allows one to use the methods of dynamical and quantum chaos for a study of acoustic fields. However, despite the coincidence of the basic equations, formulations of the problems in mechanics and acoustics can be very different. For example, the parameter of a sound ray most extensively studied in underwater acoustics is a ray travel time, that is, the arrival time of a sound pulse coming to the receiver through a particular ray path [Simmen *et al.* (1997); Brown and Viechnicki (1998); Smirnov *et al.* (2001); Virovlyansky (2003); Beron-Vera *et al.* (2003); Makarov *et al.* (2004); Smirnov *et al.* (2005b)]. Ray travel times are used as the main observables in different schemes of acoustic monitoring of the ocean temperature fields [Munk and Wunsch (1979); Spiesberger and Metzger (1991); Worcester *et al.* (1999); Dushaw (1999)]. An analogue of this characteristic in mechanics is Hamilton's principal function, or mechanical action, i.e., a characteristic of dynamical systems that is typically not measured experimentally and therefore has not received much attention in the studies of dynamical chaos.

In a deep ocean, it is generally believed that the sound speed inhomogeneities induced by internal waves are the dominant cause of the acoustic fluctuations at long ranges [Flatté *et al.* (1979)]. An environmental model widely used in the studies of chaotic ray motion is a superposition of a smooth range-independent, or slowly varying with range, sound speed field and a weak range-dependent perturbation induced by internal waves. First works on ray chaos in ocean acoustics have considered perturbations with periodic range dependencies. Although such waveguide models look rather artificial, they demonstrate a number of general properties of the wave field observed in more realistic environments. Interest in range-periodic waveguides is caused by the fact that they can be examined by direct application of the methods borrowed from the theory of dynamical and quantum chaos, such, for example, as the method of Poincaré map. Later on, more realistic waveguide models, in which the sound speed perturbation represented the realization of a random field, were explored. In this book models of both types will be considered.

We are interested in long-range propagation of sound in the ocean over distances of the order of hundreds or even thousands of kilometers. We begin Chapter 1 with a brief description of a guided sound propagation in the ocean in Sec. 1.1.

In Sec. 1.2, we briefly describe the main features of the environmental model used throughout the book. The basic equations determining the sound field are also presented. The linear wave equation is the main one of them. In the case of a monochromatic wave field, it reduces to the Helmholtz equation. Our analysis of the ray and wave chaos in this book is based on the standard parabolic equation which follows from the Helmholtz equation in small-angle approximation.

The geometrical optics approach is introduced in Sec. 1.3. The ray dynamics is described in the scope of Hamiltonian formalism expressed in terms of position-momentum and action-angle variables. The main attention is paid to the action-angle variables which are rarely used in ocean acoustics and therefore may be unknown to the readers.

Section 1.4 is devoted to a discussion of the ray travel times in a range-independent waveguide. Although in this case all the rays are regular, their travel times exhibit some general properties which will be observed in the presence of a perturbation giving rise to ray chaos, as well. We consider a distribution of ray arrivals in the time–depth plane. This is the so-called timefront, a characteristic of the ray travel time distribution traditionally used in underwater acoustics. It is shown that there exist surprisingly simple analytical estimates relating the difference between the ray travel times and the action variables of ray trajectories. These estimates give a quantitative description of the timefront structure and its evolution with range.

The ocean is a complicated acoustic medium with a number of inhomogeneities of different scales. The main factors of the ocean variability influencing sound propagation, namely, internal and Rossby waves, fronts, currents, streamers, and eddies, are briefly discussed in Sec. 1.5. The acoustic tomography of the ocean means extracting information about the state of the ocean from measured acoustic data. The Munk–Wunsch scheme of a reconstruction of the ocean state is described in Sec. 1.6.

Section 1.7 is devoted to a review of important experiments on long-range sound propagation. The first one was conducted in 1960 with explosive charges when the sounds were detected by hydrophones at the distance of about 20 000 km. Modern experiments use sources generating coherent acoustic signals. Some results of the Heard Island Feasibility Test (1991) on long-range sound propagation with a base of 18 000 km in deep waters of the Atlantic, Pacific, and Indian Oceans are discussed. We review as well experiments with acoustic sources mounted into the downsloped bottom in shallow waters in Kaneohe Bay (Oahu island) from 1983 to 1989 with a base of 3709 km. Other experiments with a smaller base were conducted in 1999 and 2006 with bottom-mounted sources in shallow waters of the Sea of Japan nearby cape Shultz. A significant role in studies of ray and wave chaos was played by the Acoustic Engineering Test conducted in 1994. The Alternate Source Test in 1996 provided an opportunity to compare wave-field properties at different signal frequencies: 28, 75, and 84 Hz. We end the chapter with a brief review of aims and some results of the international project, known as Acoustic Thermometry of Ocean Climate, that combined remote acoustic and altimetric measurements along with direct temperature measurements to estimate changes in oceanic heat storage.

Chapter 2 is devoted entirely to the analysis of ray chaos in the USC with periodic range dependence. This highly idealized environmental model allows one to realize the basic mechanismes responsible for the emergence of ray chaos. We begin this chapter with a brief review of dynamics and chaos in Hamiltonian systems and introduce the key concepts of invariant tori, stochastic layers, chaotic mixing, local instability, stickiness, normal and anomalous diffusion and transport in the phase space (Sec. 2.1). In Sec. 2.2 the characteristic of instability of rays in an USC, the finite-range Lyapunov exponent, is introduced. The essential feature of range-periodic waveguides is the so-called ray-medium resonance analyzed in Sec. 2.3. Ray-medium resonance is the analogue of nonlinear resonance in Hamiltonian mechanics. The phase portrait of the ray system in such a waveguide, constructed using the Poincare map, has the same form as that of the nonlinear oscillator driven by a periodic external force. There can be two scenarios of the onset of chaos. If a sound-speed inhomogeneity varies smoothly with depth, then remarkable ray chaos emerges due to overlapping of neighbouring ray-medium resonances. This scenario is studied in Sec. 2.4. Overlapping results in the emergence of a chaotic sea in the phase space, with islands of stability within. In this case, ray instability is governed by the so-called stability parameter which is determined in the form of a background sound-speed profile.

In Sec. 2.5 we consider ray motion in a more complicated case where a weak range-periodic perturbation of sound speed is a rapidly oscillating function of depth. Such a perturbation mimics a fine structure produced by high modes of internal waves. Under these conditions, ray chaos arises due to scattering on vertical resonances whose properties are studied in Sec. 2.5. Instability is governed by the ratio of vertical and horizontal length scales of a sound-speed perturbation. We study interrelation between two scenarios of the onset of chaos and show that they are closely related to each other, except for the case when vertical resonance causes bifurcations of periodic ray orbits.

In Sec. 2.6 we consider the ray travel times. The emphasis is on the analysis of the relationship between the ray travel time and the topology (shape) of the ray trajectory characterized by the ray identifier $\pm M$, where $\pm$ determines the sign of the launch angle and M is the number of ray turning points. It appears that the rays with equal identifiers have close travel times. Therefore, the chaotic rays, arriving at a fixed observation point with equal identifiers, form compact clusters. Arrivals of all the rays with a given identifier are located within a compact area of the timefront. Some of these areas are isolated in the sense that they do not overlap with the areas corresponding to different identifiers. A specific property of range-periodic waveguide models is the coexistence of regular and chaotic rays. Ray tracing demonstrates that this property may manifest itself in the appearance of a gap in the timefront. Besides the gap, a focusing of ray travel times was found to exist within a comparatively small temporal interval preceding the gap. In the numerical simulation of sound pulse propagation, this phenomenon manifests itself in the appearance of a bright spot in the distribution of acoustic energy in the time–depth plane. It was shown that the focusing effect is a signature of the stickiness, i.e., the presence of such parts of a chaotic trajectory where the latter exhibits an almost regular behavior.

In Chap. 3 we discuss the wave chaos phenomenon, that is, the manifestation of chaotic ray motion at a finite wavelength, in a waveguide with a periodic range dependence. We begin this chapter by discussing a close connection between the wave chaos and quantum chaos. Many insights and methods of quantum chaos can be and have been applied in underwater acoustics. In Sec. 3.1, a few key concepts of the quantum chaos theory, that seem to be important in the underwater sound propagation, are discussed: the Ehrenfest break distance, fidelity or overlap of wave fields, dynamical localization, and the semiclassical propagator.

In Sec. 3.2 we introduce the modal representation of the wave field and examine the ray-mode relations. A ray-based approach for the description of normal mode amplitudes in a range-dependent waveguide is derived. In the scope of this approach, the mode amplitude is expressed through parameters of the ray paths. It is shown that the normal mode is formed by contributions from rays whose action variables at a given observation range satisfy the quantization rule. This approach is convenient for analysis of manifestations of the ray dynamics (both, regular and chaotic) in range variations of the mode amplitudes. In this section, it is also shown that the description of the wave field in a range periodic waveguide can be greatly simplified using the Floquet theory. Then the field is decomposed into a sum of terms with regular and relatively simple range dependence which we call the Floquet modes. They are complete analogues of the Floquet states in quantum mechanics (quantum states with fixed quasi-energies).

Section 3.3 deals with manifestations of the ray chaos in the modal structure of the wave field. At first, we consider the decomposition of the wave field into the normal modes of an unperturbed (range-independent) waveguide. It is shown that the normal mode dynamics is to a significant extent determined by the mode-medium resonance. This phenomenon — its prototype in quantum mechanics is called the nonlinear quantum resonance — is interpreted as a manifestation of the ray-medium resonance at a finite wavelength. We also consider an alternative representation of the wave field as a decomposition into the sum of the Floquet modes. We use the Wigner representation of the wave field to link the Floquet modes and the ray distribution in the phase space. It is shown that the characteristic features of the phase portrait of the ray system, namely stable islands and a chaotic sea, manifest themselves in the Wigner functions of the Floquet modes. A similar effect was observed in the theory of quantum chaos. We also consider an analogue of another effect from the theory of quantum chaos — the scarring, or an increase of the wave intensity in the vicinity of an unstable periodic ray path.

Typically, horizontal gradients of the variability in the deep ocean are $10^2 \div 10^3$ times weaker than the vertical ones which are caused mainly by high modes of internal waves. The applicability condition of the ray approximation is not met for the low-frequency sound signals propagating over long distances in the environment with a vertical variability even if the latter is small. So we arrive at a conflict between the ray and wave description: the vertical resonance produces strong ray chaos as it is shown in Sec. 2.5, whereas a small-scale vertical perturbation should be ineffective in the wave-field propagation. In Sec. 3.4, we analyze this conflict from the viewpoint of the modal representation of a wavefield. Another way we follow is the analysis of the Floquet modes. We track how discrepancies between ray and wave descriptions increase with decreasing of the vertical scale of a sound-speed perturbation.

Chapter 4 is devoted to studying the wave fields in a realistic environmental model with sound speed fluctuations induced by random internal waves. The wave fields in such a medium are studied from two different and complementary viewpoints by the theory of wave propagation in random media and the theory of ray and wave chaos. The principal difference between these two approaches is explained in the introductory part.

In Sec. 4.1, we construct a statistical description of the chaotic ray dynamics in the environmental model under consideration. The use of the action-angle variables greatly simplifies the analysis. It turns out that the range dependence of the action variable can be modeled by a random Wiener process representing the simplest mathematical model of diffusion. Then the angle variable is modeled by an integral of this process. In this approximation (we call it the Wiener process approximation), surprisingly simple analytical estimates for statistical characteristics of the ray parameters are derived. It is shown how this result can be applied for estimating the sound intensity smoothed over the depth with a sufficiently large smoothing scale. We also describe a rather general method to find regions of stability in the phase space of Hamiltonian systems driven by a weak noise with an arbitrary spectrum. This method is based on a specific Poincaré map. Physical manifestations of these regions of stability are coherent clusters — bunches of sound rays propagating coherently over long distances in an underwater waveguide through a randomly fluctuating ocean.

Section 4.2 is devoted to studying the ray travel times. It is demonstrated that some general properties observed in the range-independent and range-periodic waveguides remain valid in the presence of random internal waves. The most important of them is the effect of clusterization discussed in Sec. 2.6. This phenomenon sheds new light on the surprising stability of the early portion of the arrival pattern observed in both numerical simulations and field experiments. The quantitative explanation of this effect is obtained by combining an approximate analytical relation for the difference between travel times of chaotic and regular rays and stochastic ray theory derived in Sec. 4.1.

In Sec. 4.3, we explore the influence of the random perturbation on the modal structure of the wave field. This problem is solved by combining two results: (i) relations expressing mode amplitudes through the parameters of ray paths from Sec. 3.2 and (ii) stochastic ray theory from Sec. 4.1. For a monochromatic wave field, a simple analytical estimate is obtained for a coarse-grained distribution of acoustic energy between normal modes. Significant attention is paid to a study of the mode pulses, that is, sound pulses carried by individual modes. Analytical estimates for the spread of a mode pulse and the bias of its mean travel time in the presence of internal waves are derived.

In Sec. 4.4, we consider a wave beam radiated by a vertical antenna placed in an USC. We begin with deriving a ray-based formalism for the description of the beam propagation. First, this approach is used to find a beam whose mean width in a range-independent waveguide is less than that of any other beam with the same initial width. Then the ray-based approach, combined with the ray-mode relations, is applied for studying the modal structure of the beam in the fluctuating environment. The spread of the modal spectrum due to wave scattering at the sound speed fluctuations is estimated.

In Sec. 4.5, we study the delays of the sound pulses caused by a synoptic eddy (an example of the mesoscale inhomogeneity) in the presence of random internal waves. These delays are used as input parameters in the method of ocean acoustic tomography for reconstructing the temperature field variation caused by the eddy. The solution of the inverse problem is considerably hindered by the fact that in the case of signal registration by a point receiver, only the sound pulses propagating along steep rays can be resolved. We show that the use of a vertical antenna allows one to loosen this restriction. An appropriate space–time processing procedure is proposed. The procedure is based on the presence of such characteristics of the ray arrival distribution in the time–depth plane that remain stable even under conditions of ray chaos.

In Chap. 5 we give a brief description of those basic concepts and methods in the Hamilton chaos theory that have been used in this book. It is a kind of Glossary with a large number of illustrations, facilitating the understanding of abstract theoretical concepts, and short articles ranged in the alphabetic order. We hope that simple explanations of the important mathematical issues make reading easier. Throughout the text of the book, we give references to the Glossary in parenthesis (see the corresponding article in the Glossary) just after the term or the notion which are needed to be explained in more detail. Technical details of derivation of some equations and their solutions can be found in the appendicies.

In Appendix A, we describe a few analytic models of the sound-speed profile in an USC including the Munk profile, commonly used for modeling variations in the velocity of sound in real environment, and more artificial profiles — linear, bilinear, and biexponential ones used to provide analytic derivation of the action and angle variables. Details of derivation of ray equations and solution to the transport equation are given in Appendices B and C. In Appendix D, a stationary phase technique is described.

The book is intended for postgraduate students and researches, physicists, oceanographers, and applied mathematicians, who are interested in ocean acoustics and in the applications of ideas and methods of ray and wave chaos to other disciplines.

It is our great pleasure to thank our co-workers who participated in obtaining many of the results discussed in this book: S. Abdullaev, M.G. Brown, J.A. Colosi, M. Edelman, A. Iomin, A.Yu. Kazarova, L.E. Konkov, L.Ya. Lyubavin, J. Simmen, I.P. Smirnov, E.V. Sosedko, B. Sundaram, F.D. Tappert, S. Tomsovic, M.Yu. Uleysky, M.A. Wolfson. We gratefully acknowledge the friutful discussions with V.A. Bulanov, V.I. Klyatskin, Yu.N. Morgunov, L.A. Ostrovsky and A.I. Neishtadt. We thank V.A. Akulichev, the Pacific Oceanological Institute (Vladivostok), the Institute of Applied Physics (Nizhny Novgorod), Department of Physics and the Courant Institute of Mathematical Sciences (New York) for continuous and active support. We would like to express a special word of thanks to Michael Uleysky for his help in preparing the final draft of this book and valuable comments.

Finally, we wish to thank US National Science Foundation, Office of Naval Research, Russian Foundation for Basic Research, and Russian Academy of Sciences for support of our research in underwater acoustics.

Contents

Preface ix

1. Ray and Wave Propagation 1

1.1 Underwater Sound Channel 2
1.2 Basic Equations 5
1.2.1 Helmholtz equation 7
1.2.2 Parabolic equation 8
1.3 Geometrical Optics Approximations and Optical-Mechanical Analogy. The Hamiltonian Formalism 10
1.3.1 Eikonal and transport equations 11
1.3.2 Momentum-position variables 12
1.3.3 Action-angle variables in a range-independent waveguide 15
1.3.4 Action-angle variables in a range-dependent waveguide 19
1.3.5 Geometrical optics for the Helmholtz equation . . 22
1.4 Ray Travel Times 26
1.4.1 Timefront 27
1.4.2 Travel time as a function of starting momentum . 33
1.5 Range-Dependent Environments 36
1.5.1 Internal waves 37
1.5.2 Rossby waves 40
1.5.3 Currents 41
1.5.4 Eddies 41
1.6 Acoustic Ocean Tomography 42
1.7 Experiments on Long-Range Sound Propagation 45
1.7.1 The Heard Island Feasibility Test 45

1.7.2 Experiments with bottom-mounted sources 51
1.7.3 Acoustic Engineering Test 57
1.7.4 Alternate Source Test 60
1.7.5 Acoustic Thermometry of Ocean Climate 63
1.8 Summary 65

2. Ray Chaos 67

2.1 Hamiltonian Chaos 67
2.1.1 Dynamics of Hamiltonian systems 68
2.1.2 Statistical description of Hamiltonian chaos 72
2.2 Lyapunov Instability 78
2.3 Ray-Medium Resonance 83
2.4 Overlapping of Resonances 87
2.5 Vertical Resonance 98
2.5.1 Adiabatic approximation 100
2.5.2 Passage through a resonance 105
2.5.3 Vertical resonance versus ray-medium resonance . 107
2.5.4 Resonance-induced chaotic layer 112
2.5.5 Influence of vertical resonance on a timefront of a received pulse 117
2.6 Manifestation of Regular and Chaotic Ray Motion in Distributions of Ray Travel Times 119
2.6.1 Perturbed waveguide 119
2.6.2 Timefront 122
2.6.3 Amplitude of a pulse signal in plane "travel time–depth" 125
2.6.4 Gap between travel times of chaotic and regular rays 126
2.6.5 "Focusing" of ray travel times 130
2.6.6 Role of stickiness and chaotic jets in "focusing" of ray travel times 132
2.6.7 Smoothed intensity of pulse signal 134
2.7 Summary 136

3. Wave Chaos 139

3.1 The Problem of Wave Chaos 140
3.1.1 Ehrenfest time 141
3.1.2 Semiclassical propagator 142

3.1.3 Fidelity or overlap of wave fields 145
3.1.4 Dynamical localization 147
3.2 Normal Modes 150
3.2.1 Range-independent waveguide 150
3.2.2 Range-dependent waveguide 158
3.2.3 Normal modes corresponding to the Helmholtz equation 160
3.2.4 Ray-based description of mode amplitudes 163
3.2.5 Floquet modes of a periodic waveguide 168
3.3 Mode Coupling Under Chaotic Conditions 170
3.3.1 Spatial mode-medium resonance 171
3.3.2 Mode medium resonance and Floquet modes . . 177
3.3.3 Manifestations of stable islands and chaotic sea in structures of the Floquet modes 181
3.3.4 Scarring 186
3.4 Influence of Fine-Scale Inhomogeneities on Wave Dynamics 193
3.4.1 Introductory remarks 193
3.4.2 Mode coupling in the presence of a fine structure 194
3.4.3 Influence of vertical oscillations of a sound-speed perturbation on Floquet modes 198
3.5 Summary 214

4. Chaotic Phenomena in Random Environment 217

4.1 Ray Chaos in a Random Medium 217
4.1.1 Statistical description of chaotic ray dynamics . . 218
4.1.2 Environmental model 220
4.1.3 Fokker-Planck equation for action 222
4.1.4 Evaluation of diffusivity B 225
4.1.5 Wiener process approximation 227
4.1.6 Probability density functions of ray parameters . 232
4.1.7 Coarse-grained Wigner function 237
4.1.8 Specific Poincaré map for a randomly-inhomogeneous waveguide 241
4.2 Travel Times of Chaotic Rays 251
4.2.1 Timefront in a randomly inhomogeneous waveguide 251
4.2.2 Travel time differences 254
4.2.3 Statistics of ray travel times 259

4.3 Modal Structure of the Sound Field in a Waveguide with Random Inhomogeneities 269
4.3.1 Redistribution of acoustic energy between normal modes under conditions of ray chaos 269
4.3.2 Transient wave field 274
4.4 Wave Beam in an Ocean Acoustic Waveguide 287
4.4.1 Ray-based description of a wave beam 288
4.4.2 Wave beam and its modal structure in the presence of random internal waves 301
4.5 Arrival Times of Sound Pulses in the Presence of Internal Waves and a Mesoscale Inhomogeneity 306
4.5.1 Variation of the timefront in the presence of synoptic eddy 307
4.5.2 The use of a vertical receiving array for measuring travel time delays 310
4.6 Summary 315

5. Glossary of Some Concepts and Notations in Hamiltonian Chaos Theory 319

Appendix A Models of a Waveguide 351

A.1 Canonical Munk Profile 351
A.2 Linear Profile 354
A.3 Bilinear Profile 355
A.4 Biexponential Profile 358

Appendix B Derivation of Ray Equations 363

Appendix C Solution of the Transport Equation 365

Appendix D Stationary Phase Technique 369

Bibliography 371

Index 385

Chapter 1

Ray and Wave Propagation

In this chapter we review important preliminary material. In Sec. 1.1, we briefly describe a mathematical model of a deep ocean acoustic waveguide used throughout the book. Main equations governing the sound wave field are presented in Sec. 1.2. It is pointed out that the analysis is based on the linear wave equation. In the case of a monochromatic wave field, it reduces to the Helmholtz equation. In the small-angle approximation, the latter further simplifies to the standard parabolic equation, which will be our main tool for a study of the long range sound propagation in the ocean.

In Sec. 1.3, we consider solutions to the parabolic and Helmholtz equations obtained in the geometrical optics approximation. The Hamiltonian formalism for the examination of ray dynamics is introduced. Significant attention is paid to a detailed description of the action-angle variables which are widely used in the study of dynamical and quantum chaos, but are rarely applied in solving waveguide problems.

Section 1.4 focuses on the ray travel times, i.e., arrival times of sound pulses coming to an observation point through different ray paths. These quantities play an important role in the underwater acoustics: in many schemes of acoustic monitoring of ocean structure, ray travel times are main observables used to reconstruct variations in the environment. In the present chapter, we study general properties of these characteristics of the ray paths in a range-independent (unperturbed) USC. Ray travel times in more realistic waveguide models will be explored in the next chapters.

In Sec. 1.5, we consider different medium inhomogeneities affecting the long-range sound propagation. Main attention is paid to the internal waves. In underwater acoustics they are regarded as the main factor giving rise to ray and wave chaos. We briefly describe other important factors influencing the ocean variability, namely, Rossby waves, fronts, currents, streamers, and eddies.

Medium inhomogeneities discussed in Sec. 1.5 have an influence on the travel times of sound pulses. This fact allows one to extract information on the environment from measured acoustic data. This may be done using the method of the ocean acoustic tomography discussed in Sec. 1.6.

The last section of this chapter (Sec. 1.7) reviews experiments on long-range sound propagation. We start with considering two experiments in which sound propagated over especially large distances, which are close to 20 000 km. Then we proceed with a description of several experiments which enlighten many important properties of long-range sound propagation, associated with range variability of the ocean. The chapter is ended with a brief review of some results on the acoustic thermometry of the ocean.

1.1 Underwater Sound Channel

From the view point of underwater acoustics, the ocean is a very complicated waveguide medium. It contains inhomogeneities of different types which affect the sound propagation, such as surface and internal waves, currents, mesoscale eddies, frontal zones, turbulence, etc. [De Santo (1979); Brekhovskikh and Lysanov (1991)]. Detailed analysis of all these factors and their influence on the long range sound propagation is a difficult problem which is beyond the scope of this book. We will consider a comparatively simple environmental model which, nevertheless, properly describes the main features of the long-range sound propagation in deep water waveguides typical of low and moderate latitudes in the Pacific, Indian, and Atlantic oceans. Similar models are widely used by many researchers in the theoretical studies of acoustic wave fields at megameter ranges [Flatté *et al.* (1979); Brekhovskikh and Lysanov (1991)].

The environmental model adopted in this book is based on two main assumptions.

The motions of oceanic waves and water masses are rather slow compared with the speed of sound in seawater. Practically, any inhomogeneity of the sound speed field remains frozen during the transmit time of the acoustic waves. Therefore, in the first approximation, the ocean can be treated as essentially a frozen medium. The medium parameters formally will be considered time-independent in the equations governing the sound wave field.

The second approximation, which is also widely used in the ocean acoustics, consists in neglecting the horizontal refraction of sound waves. The point is that the cross-range gradients of the refractive index are typically two orders of magnitude smaller than the vertical gradients [Flatté *et al.* (1979); Brekhovskikh and Lysanov (1991); Simmen *et al.* (1997)]. Therefore, the sound propagation is well described as two-dimensional (having no out-of-plane scattering). This allows one to use an axially symmetric (two-dimensional) environmental model, in which the sound speed c is a function of two coordinates, range r and depth z. We assume that the z axis is directed downward and the water surface is located at $z = 0$.

In the deep ocean there is typically a minimum of the sound speed c at a certain depth z_{a}. Up to this depth the sound speed decreases with increasing z mainly due to decreasing the water temperature. Below z_{a} it increases due to hydrostatic pressure increases. The presence of the minimum of sound speed leads to the formation of *an underwater sound channel* (USC) [Ewing *et al.* (1948); Brekhovskikh (1949); Rosenberg (1949)]. The depth z_{a}, where c is minimal, is referred to as *an USC axis.* According to the Snell's law, sound tends to propagate toward regions of lower speed, and the sound speed profile $c(z)$ refracts waves toward the USC axis. Being trapped in the USC, the sound waves do not reach the lossy bottom and, therefore, they may propagate over long distances with relatively small attenuation.

The speed of sound c in the ocean ranges approximately from 1450 to 1540 m/s. A typical sound-speed profile in the deep ocean is depicted in Fig. 1.1(a). Namely the form of the profile $c(z)$ defines conditions of the sound propagation. With one type of the profile, sound may propagate over thousands of kilometers, whereas with another one, sound of the same frequency and power propagates only for a few tens of kilometers. Variations of the sound speed in the upper ocean are caused by variations in temperature and salinity. The temperature and salinity change weakly in the deep ocean, and c varies there due to the increasing hydrostatic pressure.

The depth of the USC axis depends mainly on how long the ocean is heated by sunlight and, therefore, is different for different geographical regions. In the tropics it falls down to 2000 m and rises to 100–500 m at the high latitudes [Brekhovskikh (1980); Brekhovskikh and Lysanov (1991)]. Seasonal ocean variability can shift the axis by tens or even hundreds meters. A depth dependence of the sound speed may have more than one minimum. If there are two of them, *a double underwater sound channel* occurs (Fig. 1.1(b)). It may arise due to the presence of warm and saline waters below the channel axis. Such channels occur in the Nothern part of the Atlantic [Brekhovskikh (1980); Boyles (1984)].

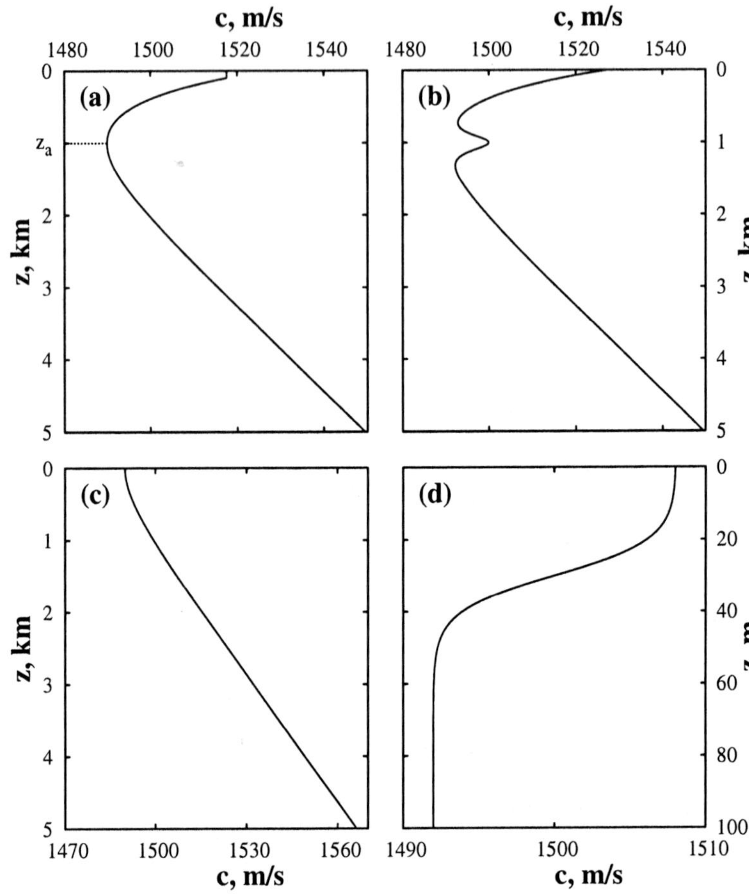

Fig. 1.1 (a) Typical sound-speed profile in the deep ocean. (b) Sound-speed profile in a double underwater sound channel. (c) Sound-speed profile in a near-surface USC. (d) Sound-speed profile in a near-bottom USC.

Near-surface USCs occur if the axis z_a is nearby the surface, and the sound speed increases continuously up to the bottom or to some depth (see Fig. 1.1(c)). The near-surface channels are typical in the Arctic and Antarctic cold seas and in some seas in the tropics where the upper mixed layer with practically constant values of temperature and salinity over the depth may occur due to wind mixing or a convection. *Near-bottom* USC typically occurs in shallow-water regions. In a shallow sea the sound speed may decrease with the increasing depth because the water is warmed practically to the bottom and salinity increases due to evaporation (see Fig. 1.1(d)).

In such a waveguide all sound rays reflect off the bottom and, therefore, the wave field rapidly decays with range.

1.2 Basic Equations

In the simplest range-independent environmental model, the sound speed field in each cross section is determined by the same function $\bar{c}(z)$. This model properly describes the capture of the sound energy within the water bulk and predicts some general properties of the waveguide propagation. A more realistic model of the sound speed field has the form

$$c(r,z) = \bar{c}(z) + \delta c(r,z), \tag{1.1}$$

where $\delta c(r,z)$ is a range-dependent perturbation superimposed on the background sound speed profile $\bar{c}(z)$.

The index of refraction is defined as

$$n(r,z) = \frac{c_0}{c(r,z)}, \tag{1.2}$$

where c_0 is the sound speed in a reference (homogeneous) medium. For the later convenience, we will assume that c_0 satisfies the condition

$$|c(r,z) - c_0| \ll c_0. \tag{1.3}$$

Then the index of refraction is close to unity. In the ocean acoustic waveguides, the variation of $c(r,z)$ typically does not exceed 2-3% and, therefore, the value of c_0 satisfying the above condition always exists.

Since $|\delta c| \ll c_0$, the refractive index, like the sound speed, can be presented in the form

$$n(r,z) = \bar{n}(z) + \delta n\,(r,z)\,, \tag{1.4}$$

where

$$\bar{n}(z) = c_0/\bar{c}(z) \tag{1.5}$$

is an unperturbed (background) constituent and

$$\delta n(r,z) \simeq -\frac{\delta c(r,z)}{c_0} \tag{1.6}$$

is a range-dependent perturbation. An example of a typical smooth background profile is shown in Fig. 1.2.

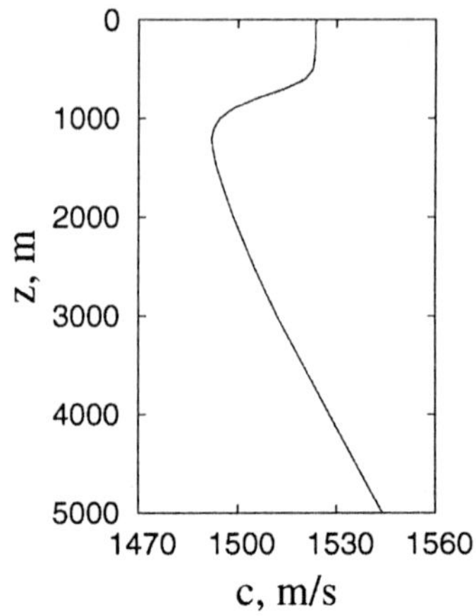

Fig. 1.2 A typical sound-speed profile in the deep ocean.

The perturbation $\delta c(r,z)$ can model different inhomogeneities of the water bulk. In this book, we will focus mainly on the sound speed fluctuations induced by random internal waves. Their influence accumulates with the range and is considered the main factor that causes the chaotic motion of the sound ray paths at megameter ranges. A detailed description of this perturbation will be given in Sec. 1.5 where we will review environmental models used in the theory of long range sound propagation in the ocean. Besides the internal wave induced variations in the sound speed, we will consider in Sec. 1.5 stronger perturbations with greater spatial scales.

At ranges of the order of hundreds or thousands of kilometers, the sound fields are formed by waves propagating at small grazing angles. These waves survive at so long ranges because they are captured in the waveguide and this prevents them from interaction with a lossy bottom. In this book we are concerned only with such waves. In other words, in our environmental model, the sound propagates in a fluid half-space $z > 0$ with an absolutely reflecting soft upper boundary $z = 0$. The finiteness of the ocean depth is accounted for implicitly by introducing an upper limit $\chi_{\max}$ for the grazing angles χ of waves considered in our analytical estimates and numerical simulations. According to the Snell law (see Eq. (1.92) below), the ray grazing angle χ takes its maximum value at the sound channel axis. The lowest point of the ray path $z_{\max}$ is defined by the relation

$$\frac{\cos\chi_{\max}}{c(z_{\rm a})} = \frac{1}{c(z_{\max})}. \tag{1.7}$$

A typical ocean depth at low and moderate latitudes is 5–6 km. The values of $\chi_{\max}$ for the rays with lower turning points at these depths are 14°–16°.

1.2.1 *Helmholtz equation*

In the theory of long range sound propagation in the ocean, it is assumed that the sound pressure $\tilde{v}(r, z, t)$ is governed by the linear wave equation [Jensen *et al.* (1994)]

$$\frac{1}{c^2}\frac{\partial^2 \tilde{v}}{\partial t^2} - \Delta\tilde{v} = 0, \tag{1.8}$$

where Δ is the Laplace operator. Sometimes, the sound field is characterized by the sound velocity potential ψ ($\nabla\psi$ is the velocity of fluid particles in the sound wave) which is governed by a linear wave equation identical to Eq. (1.8). As was noted above, we exploit the approximation of frozen inhomogeneities, that is, we assume that the sound speed field c varies in space, but not in time.

In the idealized two-dimensional environmental model used in this book wave equation (1.8) takes the form

$$\frac{1}{c^2}\frac{\partial^2 \tilde{v}}{\partial t^2} - \frac{1}{r}\frac{\partial}{\partial r}r\frac{\partial \tilde{v}}{\partial r} - \frac{\partial^2 \tilde{v}}{\partial z^2} = 0. \tag{1.9}$$

Consider a monochromatic wave field at a carrier (angular) frequency Ω. The time dependence of the field complex amplitude is expressed by a factor $e^{-i\Omega t}$. Substituting $\tilde{v}(\vec{r}, t) = v(r, z)e^{-i\Omega t}$ in Eq. (1.9), we arrive at the Helmholtz equation [Brekhovskikh and Lysanov (1991)]

$$\frac{1}{r}\frac{\partial}{\partial r}r\frac{\partial v}{\partial r} + \frac{\partial^2 v}{\partial z^2} + k_0^2 n^2(r, z)v = 0, \tag{1.10}$$

where

$$k_0 = \Omega/c_0 \tag{1.11}$$

is a wavenumber in the reference medium.

This equation is to be solved with appropriate boundary conditions. The wave field vanishes at the free water surface, that is, $v(r, 0) = 0$. A relevant definition of the bottom boundary condition is a more complicated issue. As was mentioned above, we take into account only the waves captured into the underwater sound channel. Therefore, we can use the idealized condition $v(r, z \to \infty) = 0$. Another bottom boundary condition we also use in this monograph is $\partial v(r, z = h)/\partial z = 0$. This is the Neumann condition implying that the bottom is a horizontal rigid wall at the depth h. In the limit of large r, the wave field should satisfy the radiation condition [Jensen *et al.* (1994)].

Equation (1.10) describes the wave field outside the source. The latter can be defined by introducing an appropriate right-hand side of this equation. In the case of a point source set at range $r = 0$ and depth $z = z_s$, the wave field is determined by the forced Helmholtz equation

$$\frac{1}{r}\frac{\partial}{\partial r}r\frac{\partial v}{\partial r} + \frac{\partial^2 v}{\partial z^2} + k_0^2 n^2(r,z)v = -2\frac{\delta(r)}{r}\delta\left(z - z_s\right). \tag{1.12}$$

In the immediate vicinity of the source the solution of Eq. (1.12) is approximately described by the same relation as in free space with constant refractive index n:

$$v = \frac{e^{ik_0 nR}}{R}, \quad R = \sqrt{r^2 + (z - z_s)^2}. \tag{1.13}$$

The transient wave field can be synthesized of solutions to the Helmholtz equation at different frequencies Ω as

$$\tilde{v}(r,z,t) = \int d\Omega\ s(\Omega) v(r,z,\Omega) e^{-i\Omega t}, \tag{1.14}$$

where the factor $s(\Omega)$ is the spectrum of a radiated signal $\tilde{s}(t)$

$$s(\Omega) = \frac{1}{2\pi}\int dt\ \tilde{s}(t) e^{i\Omega t}. \tag{1.15}$$

1.2.2 *Parabolic equation*

As was noted above, only the waves captured in the underwater sound channel and propagating at sufficiently small grazing angles survive at long ranges. These waves can approximately be described in the so-called parabolic (or small-angle) approximation. In this approximation, the complex amplitude of the wave field is represented in the form

$$v(r,z) = r^{-1/2} u(r,z)\, e^{ik_0 r}, \tag{1.16}$$

where $u(r,z)$ is an envelope function. The latter approximately satisfies the so-called standard parabolic equation

$$2ik_0\frac{\partial u}{\partial r} + \frac{\partial^2 u}{\partial z^2} - 2k_0^2\left[U(z) + V(r,z)\right] u = 0, \tag{1.17}$$

where

$$U(z) = \frac{1}{2}\left(1 - \bar{n}^2(z)\right) \simeq \frac{\Delta c(z)}{c_0} \tag{1.18}$$

and

$$V(r,z) = \left[\bar{n}^2(z) - n^2(r,z)\right]/2 \simeq \frac{\delta c(r,z)}{c_0}, \tag{1.19}$$

where $\Delta c(z) = \bar{c}(z) - c_0$. This equation is widely used in the theory of wave propagation. First, it was introduced to describe propagation of an electromagnetic wave along the Earth's surface [Leontovich (1944); Leontovich and Fock (1946)]. The wide application of the parabolic equation in the ocean acoustics began since the 1970s [Tappert (1977); Jensen *et al.* (1994)]. It is well known that the parabolic equation properly describes the main features of the sound speed fields in underwater sound channels.

Here we can use the same boundary conditions as in the case of the Helmholtz equation. The source can be determined by introducing the wave field at $r = 0$. For a point source set at a depth z_{s}, we have

$$u(0,z) = \delta(z - z_s). \tag{1.20}$$

For very small r, the solution has the same form as in a free space with $c(r,z) = c_0$:

$$u(r,z) = \sqrt{\frac{k_0}{2\pi i r}} \exp\left(ik_0 \frac{(z - z_s)^2}{2r}\right). \tag{1.21}$$

In the parabolic equation approximation, Eq. (1.14) for the transient wave field transforms to

$$\tilde{v}(r,z,t) = \frac{1}{\sqrt{r}} \int d\Omega\ s(\Omega) u(r,z,\Omega) e^{-i\Omega(t - r/c_0)}. \tag{1.22}$$

Besides the standard parabolic equation (1.17), there are the so-called wide-angle parabolic equations which properly describe propagation of waves in a broader range of grazing angles [Jensen *et al.* (1994)]. Unfortunately, the structures of these equations are too complicated for obtaining analytical results. Therefore, the wide-angle parabolic equations are not considered in this book.

The analysis of sound fields in this book is performed on the basis of the standard parabolic equation (1.17). This is done for the following reasons. First, Eq. (1.17) formally coincides with the Schrödinger equation in quantum mechanics. In this analogy, the range r plays the role of time, z is the coordinate of a mechanical particle of unit mass, k_0^{-1} is the Planck constant, $U + V$ is the potential energy. This fact simplifies application of the methods of dynamical and quantum chaos for a study of ray and wave chaos. Second, the parabolic equation, unlike the Helmholtz equation, can be comparatively easily solved numerically [Jensen *et al.* (1994)]. Third, the use of the parabolic equation can simplify the derivation of analytical estimates.

The last reason, however, is not significant. Most analytical results presented below can easily be generalized to the case where the wave field is governed by the Helmholtz equation. We do not discuss these generalizations, but in order to help the reader, the main formulas for the ray and mode representations of the wave field are represented in two forms, corresponding to the parabolic and Helmholtz equations.

1.3 Geometrical Optics Approximations and Optical-Mechanical Analogy. The Hamiltonian Formalism

In the limit of a small wavelength, the parabolic and Helmholtz equations can be solved using the geometrical optics approximation. A necessary condition of its applicability is that the wave length $\lambda_0 = 2\pi/k_0$ is small compared with the characteristic scale of the medium inhomogeneity $L_{\rm inh}$. Within the framework of this method, the wave field is expressed through parameters of the ray paths determined by the ray equations similar to the Hamilton equations in classical mechanics. The optical-mechanical analogy allows one to use the same Hamiltonian formalism for the description of ray dynamics and dynamics of a mechanical particle. This section provides some background material on this formalism.

We assume that this material, to a certain extent, is known to the reader. Our objective is to introduce the notation and recall the basic notions of the Hamiltonian formalism expressed in terms of momentum-position and action-angle canonical variables. The main attention is paid to the action-angle variables which are not often used in the wave propagation theory and may be unknown to many specialists working in this field. One of our goals in this book is to demonstrate that these variables can be an efficient tool in investigating the waveguide propagation.

1.3.1 *Eikonal and transport equations*

In the scope of the geometrical optics method, a solution to the parabolic equation (1.17) is sought in the form

$$u = Ae^{ik_0 S}, \tag{1.23}$$

where A and S are two new unknown functions called the amplitude and eikonal, respectively. It was noted above that in the analogy between the quantum mechanics and wave theory, k_0^{-1} plays the role of the Plack constant. Therefore, the geometrical optics approximation is the analogue of the semiclassical approximation. Wavenumber k_0 is a large parameter in our problem.[1]

Substituting Eq. (1.23) in Eq. (1.17) and collecting terms of like order in k_0, we obtain the eikonal equation

$$\frac{\partial S}{\partial r} + \frac{1}{2}\left(\frac{\partial S}{\partial z}\right)^2 + U + V = 0 \tag{1.24}$$

and the transport equation

$$\frac{\partial A}{\partial r} + \frac{\partial A}{\partial z}\frac{\partial S}{\partial z} + \frac{1}{2}A\frac{\partial^2 S}{\partial z^2} = 0. \tag{1.25}$$

Note that Eq. (1.24) has the form of the Hamilton-Jacobi equation in classical mechanics [Landau and Lifshitz (1976)]. Correspondingly, the eikonal S is the analogue of Hamilton's principal function or the action integral. In acoustics, this quantity has a simple and important physical meaning that becomes clear after the insertion of Eq. (1.23) into Eq. (1.22). Since the eikonal S is frequency-independent, making use of Eq. (1.15) we find

$$\tilde{v}(r, z, t) = \frac{1}{\sqrt{r}} A\tilde{s}[t - (r + S)/c_0]. \tag{1.26}$$

According to this formula, a sound pulse is of the same shape at any range, up to the range-dependent factor $A/\sqrt{r}$. The pulse arrives at an observation point at time

$$t = (r + S)\,/c_0. \tag{1.27}$$

Thus, the eikonal determines the travel times of the acoustic sound pulse.

[1]Strictly speaking, a dimensional quantity cannot be considered as a large or small parameter. A real large parameter in the method of geometrical optics is a dimensionless product $k_0 L_{\text{inh}}$ [Kravtsov and Orlov (1990)].

1.3.2 *Momentum-position variables*

The method of characteristics allows one to reduce partial differential equations (1.24) and (1.25) to a system of ordinary differential equations describing the ray trajectories (paths) [Kravtsov and Orlov (1990); Abdullaev and Zaslavsky (1991); Jensen *et al.* (1994)]. In the scope of the Hamiltonian formalism, the ray trajectory at each range point r is characterized by not only its coordinate (depth) z, but also the so-called generalized momentum (or, simply, momentum)

$$p = \frac{\partial S}{\partial z}. \tag{1.28}$$

The momentum p is connected to the ray grazing angle χ via

$$p = \tan \chi. \tag{1.29}$$

The ray dynamics is determined by the Hamiltonian H related to the eikonal by

$$H = -\frac{\partial S}{\partial r}. \tag{1.30}$$

Substituting Eqs. (1.28) and (1.30) into eikonal equation (1.24) yields an explicit expression for the Hamiltonian

$$H(r,p,z) = H_0(p,z) + V(r,z), \tag{1.31}$$

where

$$H_0(p,z) = \frac{p^2}{2} + U(z) \tag{1.32}$$

and V is defined by Eq. (1.19). The ray paths are governed by the Hamilton equations

$$\frac{dp}{dr} = -\frac{\partial H}{\partial z}, \quad \frac{dz}{dr} = \frac{\partial H}{\partial p}. \tag{1.33}$$

A derivation of these equations is presented in Appendix B.

In what follows we will use functions $p(r, p_0, z_0, r_0)$ and $z(r, p_0, z_0, r_0)$ to denote the momentum and vertical coordinate, respectively, of a ray path at range r. Arguments p_0 and z_0 are the initial momentum and depth, respectively, at $r = r_0$. The symbols p and z (without arguments) will be used to denote the momentum and coordinate of a current ray point. Thus, the ray path will be described by the equations

$$p = p(r, p_0, z_0, r_0), \quad z = z(r, p_0, z_0, r_0). \tag{1.34}$$

If $r_0 = 0$, then the last argument of the functions in the right-hand sides will be omitted. Where it does not cause confusion, we will also omit the arguments p_0 and z_0 and describe a ray path by the functions $p(r)$ and $z(r)$.

In an unperturbed (range-independent) waveguide with $V = 0$, the Hamiltonian $H = H_0$ remains constant along the ray path. This property expresses the energy conseravtion law in mechanics and the Snell law in the geometrical optics. From Eqs. (1.32) and (1.33) we find

$$\frac{dz}{dr} = p = \pm\sqrt{2\left[H_0 - U(z)\right]}. \tag{1.35}$$

The function $z(r, p_0, z_0, r_0)$ is implicitly defined by the relation

$$r - r_0 = \text{const} \pm \int_{z_0}^{z} \frac{dz}{\sqrt{2\left[H_0 - U(z)\right]}}, \tag{1.36}$$

which is valid within the range intervals where a ray path monotonically increases or decreases.

The eikonal, or principal Hamilton's function, is also expressed through solutions of the Hamilton (ray) equations (1.33). It is given by the integral

$$S = \int_{\Gamma} (p dz - H dr) \tag{1.37}$$

running over a ray path Γ. The eikonal S, which is considered a function of range r, initial coordinate z_0, and final coordinate z, satisfies the relations [Landau and Lifshitz (1976)]

$$\frac{\partial S}{\partial z_0} = -p_0, \qquad \frac{\partial S}{\partial z} = p, \tag{1.38}$$

where p_0 and p are the momenta of a ray path at the starting and final points, respectively.

It is well known that the ray trajectories are stationary paths of the functional in the right-hand side of Eq. (1.37). The first variation of this functional vanishes at solutions of the Hamilton equations (1.33). In classical mechanics, this statement is called Hamilton's principle of stationary action, while in wave theory it is called Fermat's principle. The Hamilton (ray) equations (1.33) can be derived directly from this principle [Landau and Lifshitz (1976)].

Ray amplitude. The solution to transport equation (1.25) is also expressed through parameters of the ray paths. In the two-dimensional problem considered in this book, the wave field is described by a one-parameter family of rays. Let α be a parameter that labels the ray paths. In Appendix C, it is shown that the ray amplitude is given by the formula

$$A = \frac{C}{\sqrt{|\partial z/\partial \alpha|}}, \tag{1.39}$$

where C is a factor that does not depend on range. The value of C (generally, it is a function of α) is determined by the starting wave field. In the case of a point source corresponding to starting field (1.20), it is convenient to label the ray paths by a starting momentum, that is, to take $\alpha = p_0$. Substituting Eq. (1.39) into Eq. (1.23) and comparing with Eq. (1.21) yields

$$A = \sqrt{\frac{k_0}{2\pi i\,|\partial z/\partial p_0|}}. \tag{1.40}$$

The geometrical optics approximation fails at *caustics*, where the derivatives $\partial z/\partial \alpha$ and $\partial z/\partial p_0$ vanish.

Any waveguide represents a multipath environment. In order to evaluate the wave field at a given observation point, one should sum up the contributions from all rays arriving at this point. These rays are called the *eigenrays*. The total wave field at the observation point is

$$u(r,z) = \sum_j A_j e^{ik_0 S_j - i\mu_j \pi/2}, \tag{1.41}$$

where the index j numbers the eigenrays, μ_j is the Maslov index, or the integral number of times that the j-th ray passes through caustics. According to this formula, each time the ray passes through caustic its amplitude is multiplied by the phase factor $\exp(-i\pi/2)$ [Gutzwiller (1967)].

Transient wave field. In underwater acoustics, both ray eikonal and amplitude do not depend on the frequency Ω, and the same is true for the Maslov index. Therefore, insertion of Eq. (1.41) into Eq. (1.22) yields

$$\tilde{v}(r,z,t) = r^{-1/2} \sum_j A_j e^{-i\mu_j \pi/2} \tilde{s}(t - (r + S_j)/c_0). \tag{1.42}$$

According to this formula, the signal at the observation point is the superposition of pulses — we will call them the *ray pulses* — coming through different eigenrays. The arrival time of a ray pulse is given by Eq. (1.27) and is called the *ray travel time*.

1.3.3 *Action-angle variables in a range-independent waveguide*

Ray paths in a waveguide represent oscillating curves. For their description (as for description of a mechanical particle oscillating in a potential well), besides the momentum-position variables (p, z), it is convenient to use the action-angle variables (I, ϑ) [Landau and Lifshitz (1976); Abdullaev and Zaslavsky (1991)]. Loosely, the action variable I defines the shape of a cycle of the ray path, while the angle variable ϑ indicates the position of the current ray point inside a cycle. If we compare an oscillating path with the sinusoid, then ϑ plays the role of its phase. In this subsection we introduce the action-angle variables in a range-independent waveguide.

1.3.3.1 *Canonical transformation*

The action variable I is determined as an integral of momentum p over a period (cycle) of the ray path

$$I = \frac{1}{2\pi} \oint p\, dz = \frac{1}{\pi} \int_{z_{\min}}^{z_{\max}} dz \sqrt{2\,[H_0 - U(z)]}, \tag{1.43}$$

where $z_{\min}$ and $z_{\max}$ are ray turning depths (solutions to equation $H_0 = U(z)$). Equation (1.43) implicitly defines the Hamiltonian H_0 as a function of the action variable. We will denote this function by $H_0(I)$.

The canonical transformation from the momentum-position (p, z) to the action-angle variables (I, ϑ) is determined by the equation [Landau and Lifshitz (1976)]

$$dS = pdz - Hdr = dG - \vartheta dI - \bar{H}dr, \tag{1.44}$$

where $\bar{H}(I, \vartheta, r)$ is a new Hamiltonian and $G(I, z)$ is the so-called generating function. In the so-defined new variables, the Hamilton equations preserve their canonical form, and we have

$$\frac{dI}{dr} = -\frac{\partial \bar{H}}{\partial \vartheta}, \quad \frac{d\vartheta}{dr} = \frac{\partial \bar{H}}{\partial I}. \tag{1.45}$$

Taking the generating function

$$G(I, z) = \begin{cases} \displaystyle\int_{z_{\min}}^{z} dz \sqrt{2\,[H_0(I) - U(z)]}, & p > 0, \\ \displaystyle 2\pi I - \int_{z_{\min}}^{z} dz \sqrt{2\,[H_0(I) - U(z)]}, & p < 0, \end{cases} \tag{1.46}$$

from Eq. (1.44), we obtain

$$p = \frac{\partial G(I, z)}{\partial z}, \quad \vartheta = \frac{\partial G(I, z)}{\partial I}. \tag{1.47}$$

Equations (1.43)–(1.46) define the direct

$$I = I(p, z), \quad \vartheta = \vartheta(p, z) \tag{1.48}$$

and inverse

$$p = p(I, \vartheta), \quad z = z(I, \vartheta) \tag{1.49}$$

canonical transformations.

The choice of the integration limits in Eq. (1.46) implies that the ray cycle begins at a minimum of the ray path and ends at the next one. At the beginning of a cycle, ϑ takes the value $\vartheta = 0$, passes through π at the maximum, and reaches 2π at the end of the cycle. Then it jumps again to $\vartheta = 0$ and all repeats. The so-defined ϑ will be called the *wrapped* angle variable. To make ϑ continuous, its value should be increased by 2π at the beginning of each new cycle. This is a standard procedure [Landau and Lifshitz (1976)]. This makes the angle variable ϑ *unwrapped.* Then both functions on the right of Eqs. (1.49) are periodic in ϑ with period 2π.

The connection between unwrapped and wrapped angle variables is given by the formula

$$\vartheta = \vartheta_0 - \vartheta_1 + 2\pi N, \tag{1.50}$$

where ϑ is an unwrapped variable at the end point of a ray path with N minima, and ϑ_0 and ϑ_1 are wrapped variables at the starting and final points of this path, respectively.

According to the Liouville theorem the Jacobian of any canonical transformation is equal to unity [Landau and Lifshitz (1976)]. Thus, we have

$$\frac{\partial(I(p, z), \vartheta(p, z))}{\partial(p, z)} = \frac{\partial(p(I, \vartheta), z(I, \vartheta))}{\partial(I, \vartheta)} = 1. \tag{1.51}$$

1.3.3.2 *Ray path in the unperturbed waveguide*

Since the generating function (1.46) does not depend on r, it follows from Eq. (1.44) that in the unperturbed waveguide $H = \bar{H} = H_0$. Then the Hamilton equations take the simpest form

$$\frac{dI}{dr} = 0, \quad \frac{d\vartheta}{dr} = \omega(I), \tag{1.52}$$

where

$$\omega(I) = \frac{dH_0(I)}{dI} \tag{1.53}$$

is the spatial frequency of the ray path oscillations. The cycle length of the ray path is

$$D(I) = \frac{2\pi}{\omega(I)}. \tag{1.54}$$

Equations (1.52) have elementary solutions

$$I = I_0, \ \vartheta = \vartheta_0 + \omega(I)r, \tag{1.55}$$

where I_0 and ϑ_0 are the starting values of the action and angle at $r = 0$. For a better understanding of the interrelation between the (p, z) and (I, ϑ) variables, let us introduce a function

$$L(I, z) = \int_{z_{\min}}^{z} \frac{dz}{\sqrt{2\left[H_0(I) - U(z)\right]}}. \tag{1.56}$$

According to Eq. (1.36), $L(I, z)$ determines the range interval between the minimum of the ray path and the closest ray point at depth z (see Fig. 1.3). Using Eqs. (1.46) and (1.53), the second of Eqs. (1.47) can be represented in the form

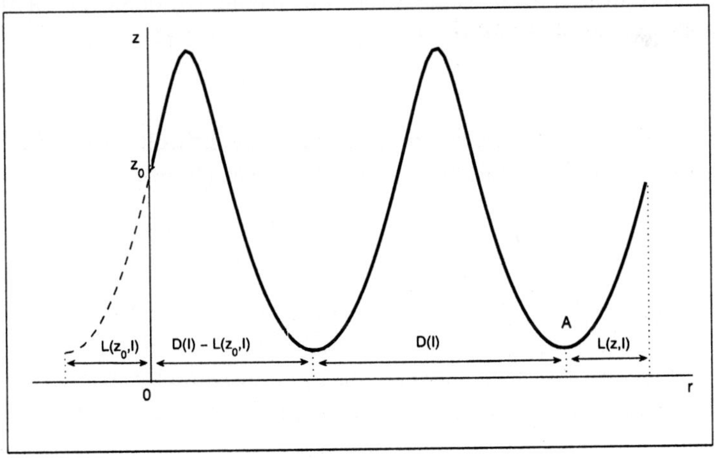

Fig. 1.3 Plot illustrating the meaning of the function $L(I, z)$. The part of the ray path on the right of point A is described by the equation $r = -L(I, z_0) + 2D(I) + L(I, z)$.

$$\vartheta = 2\pi \begin{cases} \dfrac{L(I,z)}{D(I)}, & p > 0, \\ 1 - \dfrac{L(I,z)}{D(I)}, & p < 0. \end{cases} \tag{1.57}$$

This formula implicitly defines the function $z(I, \vartheta)$ on the right-hand side of the second of Eqs. (1.49). It follows from Eq. (1.57) that

$$\frac{\partial z(I,\vartheta)}{\partial \vartheta} = \gamma \frac{D(I)\sqrt{2\left[H_0(I) - U(z)\right]}}{2\pi}, \tag{1.58}$$

$$\gamma = \begin{cases} 1, & 0 < \vartheta < \pi, \\ -1, & \pi < \vartheta < 2\pi. \end{cases} \tag{1.59}$$

The ray trajectory with action I in a range-independent waveguide can be presented in the form

$$r = R(I, z) \tag{1.60}$$

with

$$R(I,z) = \pm L(I, z_0) + ND(I) \pm L(I, z), \tag{1.61}$$

where N is the number of minima of the ray path, and z_0 is the initial depth. Hereafter, for brevity sake, we omit z_0 and N in the list of arguments of R. Figure 1.3 illustrates the meaning of the function $L(I, z)$ and Eq. (1.61).

We now turn to evaluation of the ray eikonal. In one cycle of the ray path, the eikonal increases by

$$\Delta S = 2\pi I - H(I)D(I). \tag{1.62}$$

Integrating Eq. (1.44) over a ray path, it is not difficult to show that the eikonal of a path with N minima connecting points $(0, z_0)$ and (r, z) is given by the formula

$$S(r, z_0, z) = 2\pi NI - H(I)r - G(I, z_0) + G(I, z). \tag{1.63}$$

Let us emphasize that the action I on the right-hand side of Eq. (1.63) is determined by Eq. (1.60) and is, therefore, a function of r, z, z_0, and N.

At a long range r, the eikonal can be approximated by $\left[I\omega(I) - H(I)\right] r$. Then, according to Eq. (1.27), an approximate estimate for the ray travel time is

$$\bar{t}(I, r) = \left[1 + I\omega(I) - H(I)\right] r/c_0. \tag{1.64}$$

1.3.3.3 *Canonical transformation in the form of Fourier series*

Since the right-hand sides of Eqs. (1.49) are periodic functions of ϑ, they can be expanded in Fourier series. According to our definition of generating function (1.46), $z(I,\vartheta) = z(I, 2\pi - \vartheta)$ and $p(I,\vartheta) = -p(I, 2\pi - \vartheta)$. Therefore we have

$$z(I,\vartheta) = \sum_{j=0}^{\infty} Z_m(I) \cos(j\vartheta) \tag{1.65}$$

and

$$p(I,\vartheta) = \sum_{j=0}^{\infty} P_m(I) \sin(j\vartheta). \tag{1.66}$$

Similarly, the arbitrary function $F(z)$ expressed through the action-angle variables (that is, with z replaced by $z(I,\vartheta)$) becomes a periodic function of ϑ and can also be represented as a Fourier series

$$F(z(I,\vartheta)) = \sum_{j=0}^{\infty} F_j(I) \cos(j\vartheta). \tag{1.67}$$

The coefficients $Z_j(I)$, $P_j(I)$, and $F_j(I)$ are expressed analytically only for a few special examples of $U(z)$ and $F(z)$. However, they can easily be calculated numerically. For the ray trajectory $z(r)$ with action variable I we have

$$\begin{aligned} F_j(I) &= \frac{\epsilon_j}{\pi} \int_0^{\pi} d\vartheta\, F(z(I,\vartheta)) \cos(j\vartheta) \\ &= \frac{\epsilon_j}{D(I)} \int_0^{D(I)} dr\; F(z(r_{\min} + r)) \cos(j\omega(I)(r_{\min} + r)), \end{aligned} \tag{1.68}$$

where $\epsilon_0 = 1$, $\epsilon_j = 2$ for $j > 0$, and $r_{\min}$ is the range corresponding to the minimum of the ray path. The last integral is readily computed using a standard ray code. The function $P_j(I)$ can be calculated in the same manner. Equations (1.67) and (1.68) are readily generalized for the case of function F dependent on both p and z.

1.3.4 *Action-angle variables in a range-dependent waveguide*

We will consider two methods of introducing the action-angle variables in a range-dependent waveguide. Both methods are rigorous, but the first of them is efficient when the range variation is strong and slow, while the second method is more convenient in the case of a weak range dependence.

1.3.4.1 *Slow range variation of the sound speed field*

In order to define the action and angle variables in a range-dependent environment, each range point is associated with an imagined range-independent waveguide which is called the *reference waveguide.* Generally, different range points r are associated with different reference waveguides. The connection between pairs (p, z) and (I, ϑ) at range r is determined by the above formulas for the range-independent environment applied in the corresponding reference waveguide. The functions $p(I, \vartheta)$, $z(I, \vartheta)$, $G(I, z)$, $H(I)$, $\omega(I)$, and $F_j(I)$ are generally different at different ranges and therefore acquire an additional argument r.

If the parameters of a medium slowly vary with the distance at any range point r, then it is convenient to take a reference waveguide whose cross section coincides with that of a real waveguide at this range. According to Eq. (1.44) which remains valid in a range-dependent environment, we have

$$\bar{H}(I, \vartheta, r) = H(I, r) + \Lambda(I, \vartheta, r), \tag{1.69}$$

where

$$\Lambda(I, \vartheta, r) = \left. \frac{\partial G(I, z, r)}{\partial r} \right|_{z=z(I,\vartheta,r)}. \tag{1.70}$$

It should be emphasized that the argument r of the functions $G(I, z, r)$, $H(I, r)$ and $z(I, \vartheta, r)$ (as well as in other functions determining the canonical transformation) just labels the reference waveguide. It only indicates the potential U to be inserted into Eq. (1.46).

The Hamilton equations take the form

$$\frac{dI}{dr} = -\frac{\partial \bar{H}}{\partial \vartheta} = -\frac{\partial \Lambda}{\partial \vartheta}, \qquad \frac{d\vartheta}{dr} = \frac{\partial \bar{H}}{\partial I} = \omega + \frac{\partial \Lambda}{\partial I}, \tag{1.71}$$

where

$$\omega(I, r) = \frac{\partial H(I, r)}{\partial I} \tag{1.72}$$

is the generalization of Eq. (1.53) and $\omega(I, r)$ is the spatial frequency of ray path oscillations in a reference waveguide. As in the case of a range-independent waveguide, the ray eikonal can be found by integrating Eq. (1.44). After some algebra, we obtain

$$S = \int_0^r (I d\vartheta - (H + \Lambda)\, dr) \\ - G(z(0), I(0), 0) + G(z(r), I(r), r) + \vartheta(0)I(0) - \vartheta(r)I(r), \quad (1.73)$$

where $\vartheta(0)$ and $\vartheta(r)$ are the wrapped angle variables at the end points of a ray path.

The action variable I, defined using the above choice of the reference waveguide, has a remarkable property. If the spatial scales of a medium range variation are large compared to the cycle length D, then I remains approximately constant along the ray path. In other words, the so-defined action is an adiabatic invariant [Landau and Lifshitz (1976)]. However, in this book, we will restrict our attention to environmental models with weak and relatively rapid range variations. In this case, it is more convenient to apply another method for introducing the action-angle variables discussed in the next subsection.

1.3.4.2 *Weak variation of the sound speed field*

The canonical tranformation in a range-independent waveguide described in Sec. 1.3.3 formally can be used in the presence of a range-dependent perturbation $\delta c(r, z)$, as well. This means that at any range point the unperturbed waveguide is used as the reference one. This method of introducing the action-angle variables will be used throughout the book. Such a technique is especially convenient when the range-dependent perturbation is weak. Then the ray dynamics can be studied in terms of a perturbation theory based on the smallness of the sound speed fluctuations.

In the case a range-independent generating function G, the new Hamiltonian $\bar{H}$, according to Eq. (1.44), is equal to $H + V$. The Hamilton (ray) equations take the form

$$\frac{dI}{dr} = -\frac{\partial V}{\partial \vartheta}, \quad (1.74)$$

and

$$\frac{d\vartheta}{dr} = \omega + \frac{\partial V}{\partial I}. \quad (1.75)$$

The eikonal S is

$$S = S_{\mathrm{a}} + S_{\mathrm{b}}, \quad (1.76)$$

where

$$S_{\mathrm{a}} = -G(I_0, z_0,) + G(I_1, z_1) + \vartheta_0 I_0 - \vartheta_1 I_1, \tag{1.77}$$

$$S_{\mathrm{b}} = \int_0^r (I d\vartheta - (H_0 + V)\, dr)\,, \tag{1.78}$$

and I_0, ϑ_0, and z_0 are the starting values of the action, wrapped angle variable, and depth, respectively, while similar symbols with subscript 1 denote the same quanitities at the final point of the ray path.

In order to describe the ray path, along with the functions $p(r, p_0, z_0, r_0)$ and $z(r, p_0, z_0, r_0)$ we will use similar functions $p(r, I_0, \vartheta_0, r_0)$ and $z(r, I_0, \vartheta_0, r_0)$, where the ray starting parameters are determined by the initial values I_0 and ϑ_0 of the action and the angle variables, respectively.

1.3.5 *Geometrical optics for the Helmholtz equation*

The geometrical optics solution for the Helmholtz equation is obtained along the same lines as has been done for the parabolic equation. Many formulas, especially those expressing the Hamiltonian formalism, even retain their form.

1.3.5.1 *Eikonal and transport equations*

In the geometrical optics approximation, the solution to the Helmholtz equation is sought in the form (cf. Eq. (1.23))

$$v = r^{-1/2} A e^{i k_0 S}, \tag{1.79}$$

where A is the ray amplitude and S is the eikonal. Inserting this into Eq. (1.10) and collecting terms of like order in k, we obtain the eikonal equation

$$\left(\frac{\partial S}{\partial r}\right)^2 + \left(\frac{\partial S}{\partial z}\right)^2 = n^2 \tag{1.80}$$

and the transport equation

$$2\left(\frac{\partial A}{\partial r}\frac{\partial S}{\partial r} + \frac{\partial A}{\partial z}\frac{\partial S}{\partial z}\right) + A\left(\frac{\partial^2 S}{\partial r^2} + \frac{\partial^2 S}{\partial z^2}\right) = 0. \tag{1.81}$$

Here again, the eikonal equation has the form of the Hamilton-Jacobi equation and reduces (by the method of characteristics) to a system of ordinary differential (Hamilton) equations which determine the ray trajectories (paths). As in the case of the parabolic equation, both the eikonal S and the amplitude A are expressed through parameters of the ray paths.

The transient wave field can be found by substituting Eq. (1.79) into Eq. (1.14). Since the eikonal S is frequency-independent, using Eq. (1.15) we find (cf. Eq. (1.26))

$$\tilde{v}(r, z, t) = \frac{1}{\sqrt{r}} A\tilde{s}(t - S/c_0). \tag{1.82}$$

According to this relation, the arrival time t of a sound pulse is given by

$$t = S/c_0 \tag{1.83}$$

(cf. Eq. (1.27)).

1.3.5.2 *Momentum-position variables*

As in the case of the parabolic equation, each ray point is characterized by its vertical coordinate z and momentum

$$p = \frac{\partial S}{\partial z}. \tag{1.84}$$

The connection between the eikonal and the Hamiltonian

$$H = -\frac{\partial S}{\partial r} \tag{1.85}$$

also retains its form. Insertion of Eqs. (1.84) and (1.85) into eikonal equation (1.80) gives an explicit expression for the Hamiltonian

$$H(r, p, z) = -\sqrt{n^2(r, z) - p^2}. \tag{1.86}$$

Equation (1.29) relating the momentum p to the ray grazing angle χ changes to

$$p = n \sin \chi. \tag{1.87}$$

The relationship between the Hamiltonian H and the ray grazing angle is

$$H = -n \cos \chi. \tag{1.88}$$

The Hamilton equations retain their canonical form:

$$\frac{dp}{dr} = -\frac{\partial H}{\partial z}, \quad \frac{dz}{dr} = -\frac{\partial H}{\partial p}. \tag{1.89}$$

The derivation of these equation is given in Appendix B. Here again we represent the Hamiltonian as a sum of the range-independent constituent

$$H_0(p, z) = -\sqrt{\bar{n}^2(z) - p^2} \tag{1.90}$$

and the range-dependent perturbation

$$V(r, z) = H - H_0 \simeq \frac{\delta c(r, z)}{c_0}. \tag{1.91}$$

Thus, for small δc, perturbation is expressed by the same function as in the parabolic approximation. Again, eikonal S is the analogue of Hamilton's principal function (or mechanical action integral) and is expressed by Eq. (1.37), but with a different Hamiltonian.

In a range-independent environment the Hamiltonian remains constant along the ray path. As in the parabolic approximation, this analogue of the energy conservation law in mechanics yields the Snell law [Brekhovskikh and Lysanov (1991)]. According to Eq. (1.88), the latter can be written as

$$n \cos\chi = \text{const.} \tag{1.92}$$

This equation determines the unperturbed ray path.

As earlier, the Hamilton equations can be derived from Fermat's (Hamilton's) principle, that is from the fact that the ray paths are stationary trajectories of the functional on the right-hand side of Eq. (1.37) [Born and Wolf (1968)]. Equation (1.38) also remains valid. In Appendix B, it is shown that the ray path is orthogonal to the wave front (line of constant phase of the wave field (1.79)). In the geometrical optics obtained via the parabolic approximation, the rays are not exactly perpendicular to the wave front.

The applicability of the geometrical optics approach based on the standard parabolic equation requires that [Tappert (1977)]

$$|\chi| \ll 1, \qquad |1 - n| \ll 1. \tag{1.93}$$

If these conditions are met, then $|p| \ll 1$ and Eqs. (1.87) and (1.29) give the same result. According to Eqs. (1.18) and (1.19), $n^2 = 1 - 2(U + V)$. Substituting this into Eq. (1.86) yields

$$H(r, p, z) \simeq -1 + \frac{1}{2}p^2 + U(z) + V(r, z). \tag{1.94}$$

An expression for the Hamiltonian, following from Eqs. (1.31) and (1.32), differs from Eq. (1.94) by -1. This constant does not affect the ray equations. However, the expressions for eikonal S given by Eq. (1.37) with

Hamiltonians (1.31) and (1.94) differ by r. This explains the difference between Eqs. (1.27) and (1.83) for the ray travel time.

Ray amplitude. It is shown in Appendix C, that in the case of the Helmholtz equation, general formula (1.39) for the ray amplitude translates to

$$A = \frac{C}{\sqrt{|H\partial z/\partial\alpha|}}, \tag{1.95}$$

where again α is a parameter labeling the ray paths and C is a factor (independent of range r) determined by the starting field. For the wave field excited by a point source and determined by the forced Helmholtz equation (1.12), the above expression takes the form

$$A = \frac{1}{\sqrt{|H\partial z/\partial p_0|}} = \frac{1}{\sqrt{|\partial z/\partial \chi_0|}}, \tag{1.96}$$

where p_0 and χ_0, are the starting momentum and grazing angle, respectively.

The wave field at a given observation point is represented as the sum of contributions from all the eigenrays arriving at this point

$$v(r,z) = r^{-1/2}\sum_j A_j e^{ik_0 S_j - i\mu_j\pi/2}, \tag{1.97}$$

where the index j numbers the contributing eigenrays and μ_j is the Maslov index.

Transient wave field. Consider a point source exciting a pulse signal $\tilde{s}(t)$. Substituting Eq. (1.97) into Eq. (1.14) yields

$$\tilde{v}(r,z,t) = r^{-1/2}\sum_j A_j e^{-i\mu_j\pi/2}\tilde{s}(t - S_j/c_0). \tag{1.98}$$

As earlier, the transient wave field is the superposition of ray pulses coming to the observation point through the eigenrays.

1.3.5.3 *Action-angle variables*

Action-angle variables are introduced in exactly the same manner as in the case of the parabolic equation. All relations presented in Secs. 1.3.3 and 1.3.4 remain valid except for Eqs. (1.43) and (1.46) which depend on the explicit form of the Hamiltonian. Equation (1.43) changes to

$$I = \frac{1}{2\pi} \oint p \, dz = \frac{1}{\pi} \int\limits_{z_{\min}}^{z_{\max}} dz \, \sqrt{\bar{n}^2(z) - H_0^2}, \tag{1.99}$$

where $z_{\min}$ and $z_{\max}$ are the upper and lower ray turning depths, respectively (solutions to the equation $\bar{n}(z) = -H_0$). Equation (1.46) determining the generating function transforms to

$$G(I, z) = \begin{cases} \int\limits_{z_{\min}}^{z_{\max}} dz \, \sqrt{\bar{n}^2(z) - H_0^2(I)}, & p > 0, \\ 2\pi I - \int\limits_{z_{\min}}^{z_{\max}} dz \, \sqrt{\bar{n}^2(z) - H_0^2(I)}, & p < 0. \end{cases} \tag{1.100}$$

We should emphasize a somewhat unexpected fact that the canonical transformations corresponding to the Helmholtz and standard parabolic equations for the given profile $\bar{n}(z)$ **coincide** [Virovlyansky (2006)]. Indeed, note that

$$\sqrt{2\,[H_0 - U(z)]} = \sqrt{\bar{n}^2(z) - 1 + 2H_0} \tag{1.101}$$

and substitute this relation into Eqs. (1.43) and (1.46). Then it is easy to see that for a given refractive index profile $\bar{n}(z)$ Eqs. (1.43), (1.46), and (1.47) define exactly the same connection between (p, z) and (I, ϑ) as Eqs. (1.99), (1.100), and (1.47).

1.4 Ray Travel Times

According to Eqs. (1.42) and (1.98), a signal arriving at an observation point is the superposition of what we have called the ray pulses, that is, sound pulses coming through different eigenrays. Arrival times of these pulses, i.e., ray travel times, are extensively studied in the ocean acoustics for the following two reasons. First, these quantities to a significant extent determine the temporal structure of the received signal. Second, they are considered as main observables in different schemes of acoustic monitoring of the ocean temperature fields [Munk and Wunsch (1979); Dushaw (1999)].

In this section we analyze general properties of the ray travel times in a range-independent waveguide. In the next chapters, it will be shown that some of these properties remain valid in a range-dependent environment even under conditions of ray chaos.

1.4.1 *Timefront*

In the ocean acoustics, a distribution of ray travel times at a given observation range most frequently is characterized by the so-called *timefront* representing ray arrivals in the time–depth, $(t - z)$, plane. Figure 1.4 presents two typical examples of the timefront. They depict ray arrivals at ranges of 750 km (upper panel) and 1500 km (lower panel) in a range-independent

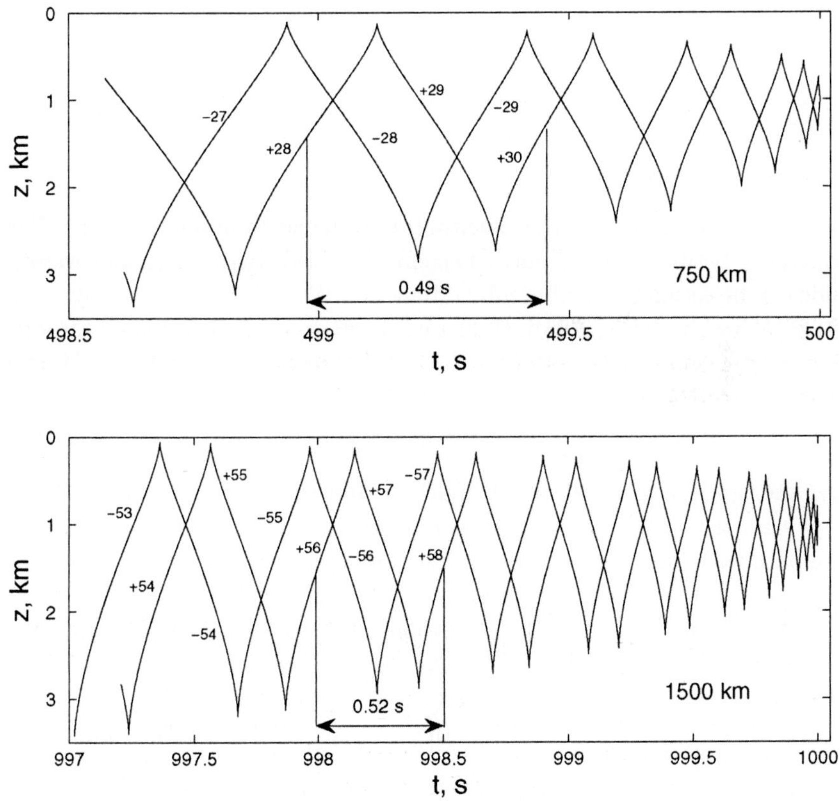

Fig. 1.4 Timefronts at 750 km (upper panel) and 1500 km (lower panel) in the unperturbed waveguide. Identifiers of rays forming some particular segments are indicated next to the corresponding segments.

waveguide with the Munk canonical sound-speed profile

$$\bar{c}(z) = c_0 \left[1 + \varepsilon \left(e^{2(z_a - z)/B} + 2(z - z_a/B - 1\right)\right], \tag{1.102}$$

where $c_0 = 1.5$ km/s, $z_a = 1$ km, $\varepsilon = 0.0057$, and $B = 1$ km. This profile is described in Appendix A. The timefronts have been constructed by tracing numerically 50000 ray paths escaping depth $z_0 = 1$ km with starting momenta p_0 that uniformly fill an interval corresponding to launch angles $\pm 12°$. The timefront in, practically, any deep water waveguide is qualitatively of the same shape. It is represented by two broken lines and sometimes is mentioned as double accordion-shaped. Each broken line is formed by rays with the same sign of launch angles and consists of smooth segments. Each segment is formed by rays with the same identifier $\pm M$, where M is the number of ray turning points and the sign + (−) indicates that the ray starts toward the surface (bottom). Since the z axis is directed downward rays with positive (negative) identifiers have negative (positive) launch angles. In Fig. 1.4, some identifiers are indicated next to the corresponding segments.

The travel times of eigenrays arriving at a fixed depth z_r are determined by intersections of the timefront with the horizontal line $z = z_r$. The difference between travel times of eigenrays in a range-independent waveguide can be estimated analytically [Munk and Wunsch (1983); Virovlyanskii (1985); Virovlyansky (1995)]. Take two eigenrays with identifiers $\pm(M+2)$ and $\pm M$. Denote their actions $I + \delta I$ and I, respectively. We will assume that

$$M \gg 1, \tag{1.103}$$

i.e., the range of observation r is much greater than the cycle length. Both rays start from depth z_0 and arrive at depth z. The number of minima for the path with identifier $\pm M$ will be denoted by N. The path with identifier $\pm(M + 2)$ has an extra minimum.

We now make use of Eqs. (1.60) and (1.61) which determine ray trajectories in a range-independent waveguide. Functions $R(I, z)$ describing the paths with $N + 1$ and N minima will be denoted by $R_{N+1}(I, z)$ and $R_N(I, z)$, respectively. Similarly, the eikonals of these paths, which are determined by Eq. (1.63), will be denoted by $S_{N+1}(I, z)$ and $S_N(I, z)$.

The difference δI between actions of the two eigenrays under consideration can be found from the equations

$$R_{N+1}(I + \delta I, z) = R_N(I, z) = r. \tag{1.104}$$

This yields an approximate relation given by

$$D + \frac{\partial R_N}{\partial I}\delta I = 0. \tag{1.105}$$

Equation (1.105) fails at caustics where $\partial R_N/\partial I$ vanishes. It follows from Eq. (1.105) that far from caustics $\delta I = O(1/M)$, i.e., for fixed I the value of δI decreases with increasing range r.

Take the difference between eikonals of the eigenrays under consideration $\Delta S = S_{N+1} - S_N$. Using Eq. (1.63) and assuming that $|\delta I|$ is sufficiently small, we find

$$\Delta S = 2\pi(I + \delta I) + \frac{\partial S_N}{\partial I}\delta I + \frac{1}{2}\frac{\partial^2 S_N}{\partial I^2}\delta I^2. \tag{1.106}$$

Equations (1.53), (1.61), (1.56), and (1.63) yield

$$\frac{\partial S_N}{\partial I} = \omega(R_N - r), \; \frac{\partial^2 S_N}{\partial I^2} = \frac{d\omega}{dI}(R_N - r) + \omega\frac{\partial R_N}{\partial I}. \tag{1.107}$$

From Eqs. (1.104) and (1.105) it follows that

$$\frac{\partial S_N}{\partial I} = 0, \; \frac{\partial^2 S_N}{\partial I^2}\delta I^2 = 2\pi\delta I. \tag{1.108}$$

Substituting Eqs. (1.108) into Eq. (1.107) yields a simple expression for the difference between travel times of our eigenrays

$$\Delta t = \Delta S/c_0 = 2\pi(I + \delta I/2)/c_0. \tag{1.109}$$

Thus, the difference between travel times of eigenrays with identifiers $\pm(M+2)$ and $\pm M$ is proportional to the half sum of their action variables. At long ranges we get for most rays $|\delta I| \ll I$ and Eq. (1.109) simplifies to

$$\Delta t = 2\pi I/c_0. \tag{1.110}$$

According to Eqs. (1.109) and (1.110), a simple evaluation of the function

$$\tau(\chi_s) = 2\pi I(\chi_s)/c_0, \tag{1.111}$$

where $I(\chi_s)$ expresses the action as a function of the launch angle, gives a significant quantitative information on the temporal structure of a pulse signal at long ranges [Virovlyanskii (1985); Virovlyansky (1995)]. Note that $I(\chi) = I(-\chi)$ and $\tau(\chi) = \tau(-\chi)$. Figure 1.5 shows the function $\tau(\chi_s)$ for a source set at a depth of 1 km in a waveguide with the canonical sound speed profile. Each circle or triangle shows the difference in travel times of two eigenrays with identifiers $+(M+2)$ and $+M$ arriving at $z = z_a$

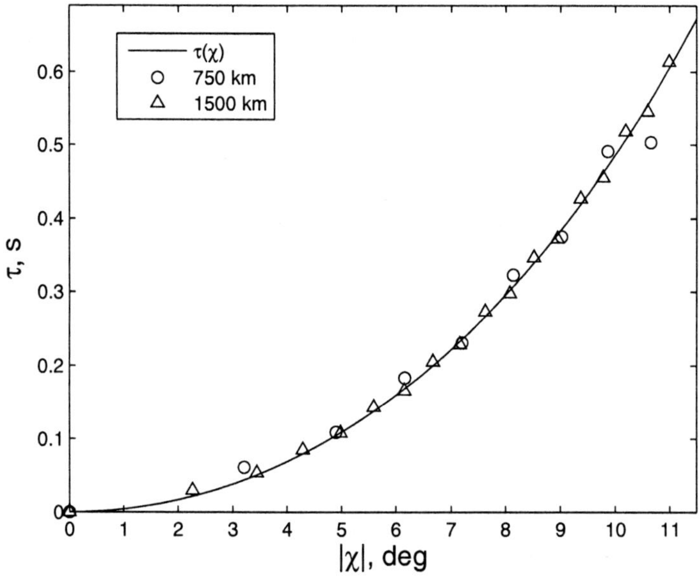

Fig. 1.5 Solid line corresponds to the function $\tau(\chi) = 2\pi I(\chi)/c_0$. Circles and triangles show delays between the segments of the timefront formed by the rays with identifiers $+(M+2)$ and $+M$ at a depth of 1 km at ranges 750 km and 1500 km, respectively.

at $r = 750$ km (circles) and 1500 km (triangles) as a function of the half sum of their launch angles. In agreement with Eq. (1.109), all circles and triangles lie in the close vicinity of the $\tau(\chi)$ curve. Looking at the $\tau(\chi)$ curve, we conclude, e.g., that the difference in travel times of two eigenrays with identifiers $+(M+2)$ and $+M$ escaping the source at grazing angles close to, say, $-10°$ will be about $\tau = 0.5$ s at **any** (long enough) distance and at **any** depth (provided the rays with launch angles close to $-10°$ reach this depth). Therefore, at long ranges, the $\tau(\chi_s)$ curve estimates the delay between the segments associated with identifiers $\pm(M+2)$ and $\pm M$ formed by the rays with launch angles close to χ_s. This statement is illustrated in Fig. 1.4. The rays with $\chi_s \simeq 10°$ at 750 km and 1500 km form segments of the timefront with identifiers close to ± 28 and ± 56, respectively. Consistent with our expectation, delays between the segments with identifiers $+30$ and $+28$ at 750 km and those with identifiers $+60$ and $+58$ at 1500 km are close to 0.5 s.

This result, in particular, means that an average delay between the neighboring segments remains approximately the same at any (long enough) range r. On the other hand, the total length of the timefront along the t-axis grows linearly with r. This occurs due to an increase of the number of segments.

Since the action I is positive (by definition), it follows from Eq. (1.109) that $\Delta t > 0$. This means that the travel time of a ray is the greater the more turning points it has [Virovlyansky (1995)]. This general property manifests itself in Fig. 1.4. The segments corresponding to early arrivals with large cycle lengths correspond to comparatively small M. The late portion of the timefront is formed by axial rays with small cycle lengths and, correspondingly, large M. Note that a different case, in which the flat rays arrive earlier than the steep ones, is possible. It occurs, for example, in a waveguide with reflecting boundaries and a uniform sound speed field. In this case, the travel time is proportional to the length of the ray path. Since the steep rays have longer paths than the flat ones, the former have greater travel times. However, the steep rays have more reflections from the waveguide boundaries and, hence, have more turning points M than the flat ones. Thus, in accord with the above general rule, the travel time once again grows with M.

A simple analytical description can be obtained not only for segment delays. In the same manner, we can derive an approximate expression for the shape of a segment. Take a particular segment and denote by ΔT the time interval between its lowest point located at depth $z_{\max}$ and an arbitrary point at depth $z < z_{\max}$. Actions of the rays belonging to the selected segment and arriving at the depths $z_{\max}$ and z will be denoted by I and $I + \delta I$, respectively. Both rays have the same number of lower turning points N. Using Eq. (1.63), we express ΔT as

$$c_0 \Delta T = S_N(I + \delta I, z_0, z) - S_N(I, z_0, z_{\max}). \tag{1.112}$$

Since $z_{\max}$ is a turning depth of the ray with action I, we have $G(I, z_{\max}) = 0$ and $L(I, z_{\max}) = 0$. An analogue of Eq. (1.105) has the form

$$c_0 \Delta T = G(I, z) + \left. \frac{\partial S_N(I, z_0, z)}{\partial I} \right|_{z=z_{\max}} \delta I. \tag{1.113}$$

It can be shown (as it was done when deriving Eqs. (1.109) and (1.110)) that the last term on the right-hand side vanishes. Since all rays forming the segment have close actions, we find from Eq. (1.113) [Smirnov *et al.* (2002)]

$$\Delta T = G(\bar{I}, z)/c_0, \tag{1.114}$$

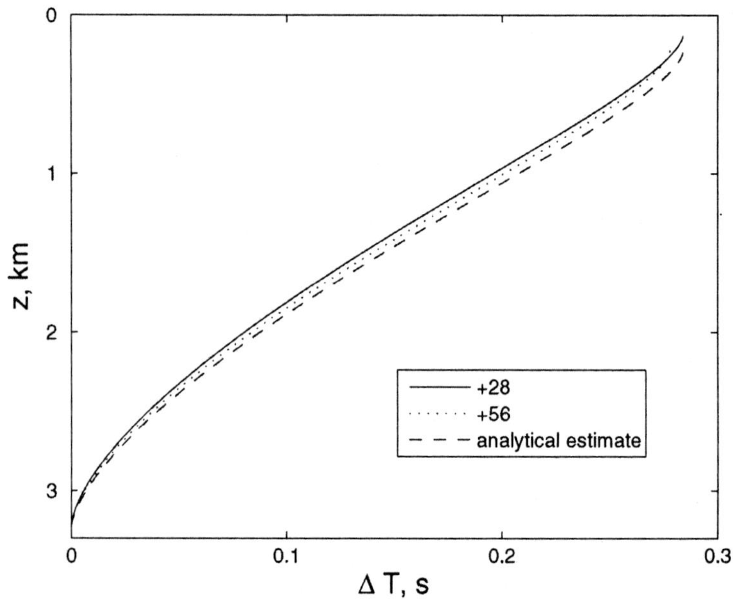

Fig. 1.6 Segments of the timefront formed by rays with launch angles χ close to 10°. Prediction of Eq. (1.114) (dashed curve) is compared with the segments associated with identifier +28 at 750 km (solid curve) and identifier +56 at 1500 km (dotted curve).

where $\bar{I}$ is the mean action of a segment. Set $z = z_{\rm min}$, where $z_{\rm min}$ is the depth of an upper point of the segment that is close to the upper turning depth of the ray with action $\bar{I}$. Therefore, Eq. (1.114) reduces to

$$\Delta T = \pi \bar{I}/c_0. \tag{1.115}$$

This quantity estimates the spread of a segment along the t axis. The total spread of two segments with identifiers $+M$ and $+(M+1)$ (or $-M$ and $-(M+1)$) is approximately equal to the time delay between segments $\pm M$ and $\pm(M+2)$. Accordingly, Eq. (1.115) yields Eq. (1.110) (but not a more accurate formula given by Eq. (1.109)).

Since the right-hand sides of Eqs. (1.114) and (1.115) do not depend on range we have a **conservation law**: if the condition (1.103) is met (i.e., at long distances), then the shape of a segment formed by rays with launch angles close to some given value does not depend on range. This statement is confirmed in Fig. 1.6 showing the dependence of ΔT on coordinate z for a segment with identifier +28 at a range of 750 km (solid curve) and for a segment with identifier +56 at 1500 km (dotted curve). Both segments are formed by rays with close launch angles, and, hence, they have almost the same mean action $\bar{I}$. Both solid and dotted curves are close to the dashed curve representing the prediction of Eq. (1.114).

1.4.2 *Travel time as a function of starting momentum*

An alternative representation of travel times at an observation range is given by the distribution of ray arrivals on the "time — starting momentum" ($t-p_0$) plane. In the case of a range-independent waveguide the arrivals are concentrated in a smooth regular curve which we will call the $t(p_0)$ curve. An example of this curve is shown by a solid line in Fig. 1.7. The curve was computed in a waveguide with Munk canonical sound speed profile at a range of 1500 km. As above, the rays escape a point source set at a depth of 1 km. Therefore, the $t(p_0)$ curve in Fig. 1.7 and the timefront in the lower panel of Fig. 1.4 present in different forms the same ray arrivals. The rays with equal identifiers forming a segment of the timefront, form a portion of the $t(p_0)$ curve which we will also call the segment. Two segments of the $t(p_0)$ curve, AB and BC, which are associated with identifiers +31 and +30, respectively, are shown in the inset.

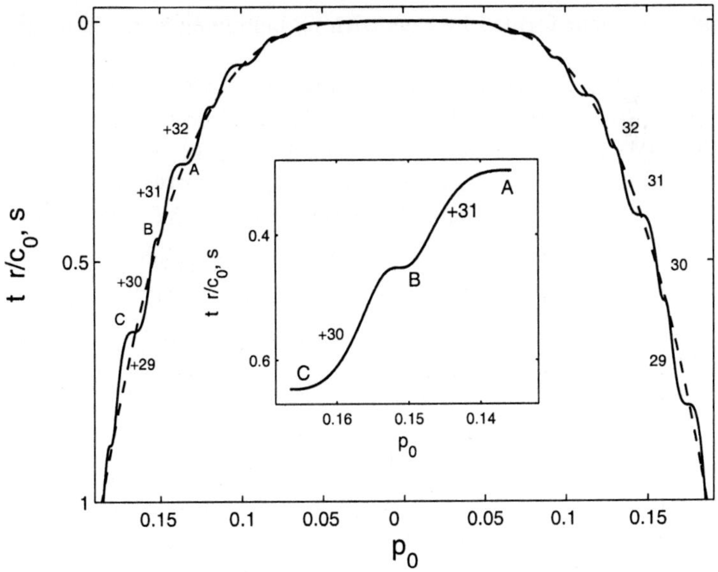

Fig. 1.7 Ray travel time (reckoned from r/c_0) versus starting momentum in the unperturbed waveguide at 1500 km (solid line) and its approximation given by Eq. (1.64). Inset: Magnified view of segments AB and BC associated with identifiers +31 and +32.

The $t(p_0)$ curve has "shelves" in which its inclinations with respect to the horizontal are small. The "shelves" are located at the ends of the segments. For example, there are "shelves" centered at points A, B, and C. At intervals of the starting momenta corresponding to the "shelves" the ray travel time t is most stable (least sensitive) with respect to small variations in p_0. The dashed line in Fig. 1.7 shows an approximate dependence of the ray travel time on starting momentum p_0 predicted by Eq. (1.64). In this approximation, the "shelves" are smoothed out.

Note that the eikonal S as a function of range r, initial depth of ray path z_0, and its final depth z satisfies Eqs. (1.38). Exploiting the second of these equations combined with Eq. (1.27) yields

$$\frac{dt(p_0)}{dp_0} = \frac{1}{c_0}\frac{\partial S}{\partial z}\frac{\partial z}{\partial p_0} = \frac{p}{c_0}\frac{\partial z}{\partial p_0}. \tag{1.116}$$

The last expression vanishes for rays that have turning points at the given range r ($p = 0$) or pass at this range through caustics ($\partial z/\partial p_0 = 0$). It is known that at long ranges the ray passes through a caustic only in the vicinity of its turning point [Brekhovskikh (1980)]. Thus, zeros of dt/dp_0 should come in closely spaced pairs, determining the positions of the centers of the "shelves" [Smirnov *et al.* (2002)].

Take a segment formed by rays with p_0 belonging to an interval

$$p_0^* \le p_0 \le p_0^{**}. \tag{1.117}$$

The trajectories with starting momenta p_0^* and p_0^{**} at the observation range r have turning points. Assume that the ray path corresponding to p_0^* has a minimum $z_{\min}$ at this range. Then the quantity ΔT for the ray corresponding to arbitrary p_0 from the interval (1.117) is equal to $t(p_0) - t(p_0^*)$. Let us consider ΔT, a function of three arguments $\Delta p = p_0 - p_0^*$, r, and $\bar{p}_0 = (p_0^* + p_0^{**})/2$. Denote by $I(p_0)$ the dependence of the ray action on the starting momentum. An approximate expression for the function $\Delta T(\Delta p, r, \bar{p}_0)$ is given by Eq. (1.114) with z being the depth of the ray corresponding to p_0 at range r, and $\bar{I} = I(\bar{p})$. To find a connection between Δp and z, we use Eqs. (1.60) and (1.61). Using

$$R(I(p_0^*), z_0, z_{\min}) = R(I(p_0^* + \Delta p), z_0 + z) = r \tag{1.118}$$

and

$$L(I(p_0^*), z_{\min}) = 0 \tag{1.119}$$

one gets

$$L(I(p_0^*), z) + \Delta p \left. \frac{\partial I}{\partial p_0} \frac{\partial R}{\partial I} \right|_{p_0 = p_0^*} = 0. \tag{1.120}$$

If the condition (1.103) is met, then $I(p_0^*) \simeq \bar{I}$ and the derivative $\partial R_N / \partial I$ far from caustics can be approximated by $N \partial D(I)/\partial I$. Replacing N by r/D, we obtain from Eq. (1.120)

$$\Delta p = r^{-1} L(\bar{I}, z) D(\bar{I}) \left(\frac{\partial I}{\partial p_0} \right)^{-1}_{p_0 = \bar{p}} \left(\frac{\partial D}{\partial I} \right)^{-1}_{I = \bar{I}}. \tag{1.121}$$

According to this expression, the function $\Delta T(\Delta p, r, \bar{p}_0)$ satisfies a scaling law [Smirnov *et al.* (2002)]

$$\Delta T(\Delta p, r, \bar{p}_0) = \Delta T(\mu^{-1} \Delta p, \mu r, \bar{p}_0), \tag{1.122}$$

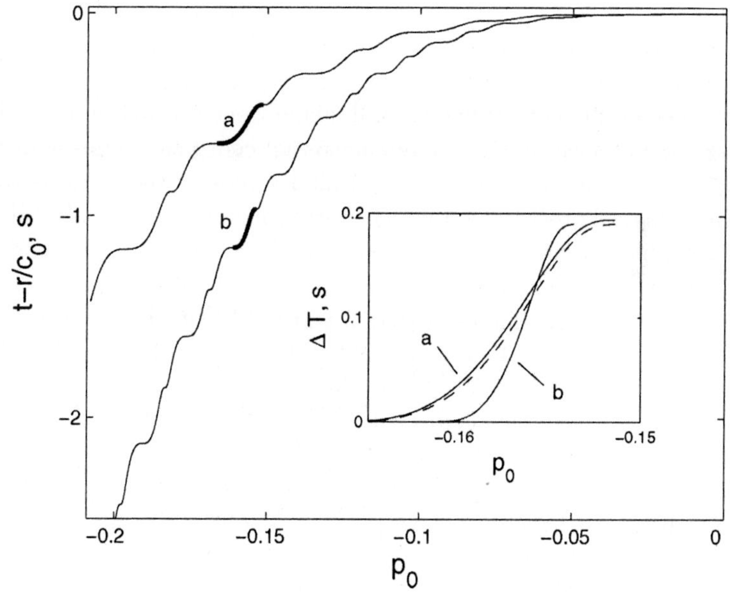

Fig. 1.8 Travel times against starting momenta for an unperturbed waveguide at 750 km (upper curve) and 1500 km (lower curve). Segments a and b marked by heavy lines are formed by the rays with identifiers +30 and +60 (respectively). Inset: Segments a and b (solid lines) shown on the same plot with travel times reckoned from their earliest points. The dashed line presents segment b extended along the p_0 axis by a factor of 2 and shifted along this axis.

where μ is a positive number. This law describes the range variation of a segment of the $t(p_0)$ curve formed by the rays with starting momenta close to some fixed value $\bar{p}_0$.

Figure 1.8 shows two $t(p_0)$ curves at ranges 750 km (the same solid curve as in Fig. 1.7) and 1500 km. At each curve we have selected a segment marked by a heavy line. They are the segments a and b formed by the rays with starting momenta close to the same value, specifically, $\bar{p}_0 = -0.155$. We use these segments to check the scaling law (1.122). This is done in the inset. The segments a and b are shown on the same time scale with travel times ΔT reckoned from their minima. In accordance with Eq. (1.122), both segments span almost the same temporal interval, while the interval of starting momentum p_0 for the segment b is two times smaller than that for the segment a. Furthermore, the segment b extended by a factor of 2 along the p_0 axis (dashed curve) after an appropriate shift along this axis almost coincides with the segment a.

1.5 Range-Dependent Environments

Ocean is essentially non-stationary and inhomogeneous environment, therefore, the sound speed varies along the spatial coordinates and with time. There is a lot of factors of the ocean variability, most of them are associated with hydrological processes within the water column. In acoustics, the temporal variability is usually eliminated, and one adopts the approximation of a "frozen" environment. The reason is that significant ocean changes occur on much larger timescales than a travel time of sound pulses.

Spatial variability makes remarkable influence on the sound propagation, especially at long ranges. All the ocean inhomogeneities caused by hydrological process can be conditionally divided into three classes according to their length scale. The first class involves microstructures, such as bubbles and fluctuations of temperature and salinity due to microturbulence or some small-scale convective processes. They don't affect refraction of low-frequency sound waves strongly and reveal themselves mainly as a factor of volume attenuation. This class will not be considered here. The second class are the inhomogeneities with the scale comparable with acoustic wavelength. They relate to the ocean fine structure associated with thermohaline processes or short-wavelength internal waves, and are usually treated as a stochastic environment impact. Their influence on sound propagation is still a subject of extensive research. In this case we encounter a mixture of diffractive and refractive effects which frequently cannot be separated from each other. The third class are large-scale inhomogeneities: mesoscale (synoptic) eddies, currents, long-wavelength internal waves, Rossby waves, and the ocean variability due to geographic variations, for instance, due to global changes of temperature and salinity. Most of them influence mainly on wave refraction and therefore can be described within the one-way ray approach.[2]

1.5.1 *Internal waves*

Internal waves and internal tides are known as important factors of ocean variability that affect sound propagation and thought of as the main source of fluctuations in acoustic fields. Despite the long history, the problem of influence of internal waves onto sound propagation stands on the foreground of ocean acoustics yet. As it will be shown in the subsequent chapters, sound-speed variations induced by internal waves lead to ray and wave chaos, being the main object of the present book.

Physical origin of internal waves can be described in the following way. Consider a fluid with a stable density stratification, i.e., with the growing of the density with increasing depth. Displacement of a fluid parcel from its equilibrium state due to some external perturbation leads to the density difference between the parcel and the background media. This difference induces a restoring Archimedeanian force. Acceleration of the displaced parcel in depth from the equilibrium position for distance ζ is given by the formula

$$\frac{d^2\zeta}{dt^2} = -g\frac{\Delta\rho}{\rho}, \tag{1.123}$$

where ρ is the density, $g = 9.8$ ms^{-2} is the gravitational acceleration, t is time, $\Delta\rho$ is the density difference of the displaced parcel and the background media

$$\Delta\rho = \zeta\left(\frac{d\rho}{dz} - \frac{\rho g}{c^2}\right). \tag{1.124}$$

Substitution of (1.124) into (1.123) yields

$$\frac{d^2\zeta}{dt^2} + N^2\zeta = 0, \tag{1.125}$$

[2]The exception is the so-called fronts discussed below.

where N is the buoyancy frequency

$$N = \sqrt{\frac{g}{\rho}\frac{d\rho}{dz} - \frac{g^2}{c^2}}. \tag{1.126}$$

In the ocean, the buoyancy frequency is a function of depth with the maximum at the thermocline. Solution of (1.125) is harmonic oscillations in the vertical direction. The stable density stratification is the necessary condition for generation of internal waves. This condition is fulfilled almost everywhere in the ocean, therefore, internal waves are ubiquitous phenomena.

In general case trajectories of fluid parcels have horizontal components and are expressed in the following form:

$$\zeta(z, r, y, t) = \mathrm{Re}\{\bar{\zeta}(z)\exp[i(k_x x + k_y y - \Omega t)]\}, \tag{1.127}$$

where $\bar{\zeta}$ is a solution of the Sturm-Liouville problem

$$\frac{d^2\bar{\zeta}}{dr^2} + k_z^2\bar{\zeta} = 0 \tag{1.128}$$

with the boundary conditions

$$\bar{\zeta} = 0 \quad \text{at} \quad z = 0, \tag{1.129}$$

$$\bar{\zeta} = 0 \quad \text{at} \quad z = h. \tag{1.130}$$

The frequency obeys the expression

$$\Omega^2 = N^2\cos\theta + f_i^2\sin\theta, \tag{1.131}$$

where θ is the angle of inclination, $f_i = 2\Omega_\mathrm{e}\sin\phi$, Ω_e is the angular velocity of the Earth, and ϕ is the latitude. Since $f_i \ll N$, the Earth rotation is necessary to be taken into account only for almost horizontal trajectories. The link between the frequency Ω and the wave numbers k_x, k_y, and k_z is described by the dispersion relation

$$k^2 = k_x^2 + k_y^2 = k_z^2\frac{\Omega^2 - f^2}{N^2 - \Omega^2}, \tag{1.132}$$

where k is the horizontal component of the wave number vector.

Colosi and Brown (1998) proposed a simple numerical scheme for generation of a sound-speed perturbation induced by internal waves. That scheme is based on the Garrett-Munk model of internal-wave field [Garrett and Munk (1972)]. The basic equation relating sound-speed fluctuations and internal waves is

$$\frac{\delta c}{c_0} = \frac{\mu}{g} N^2(x, y, z)\zeta(x, y, z), \tag{1.133}$$

where $\mu = 24.5$ is a dimensionless constant. The squared buoyancy frequency N^2 is a slowly varying quantity playing the role of an envelope function of the sound-speed perturbation. In numerical simulations concerned with the deep ocean, one often uses the idealized expression (A.2) with $B = 1$ km. Vertical displacement ζ is a random quantity, though theoretical studies of various acoustical problems are often being carried out using deterministic models for the function $\zeta(z, r)$ [Palmer *et al.* (1988); Smith *et al.* (1992); Simmen *et al.* (1997)]. Those oversimplified deterministic models are used for a qualitative description of acoustical effects caused by the internal waves and provide a cornerstone for studying sound propagation in more realistic environments.

Function ζ can be represented as the sum of the modes (1.127)

$$\begin{aligned}\zeta(x, y, z, t) = \mathrm{Re}\biggl\{\int\int dk_x\, dk_y \sum_j A(j, k_x, k_y)\bar{\zeta}(z) \\ \times \exp\Bigl[i(k_x x + k_y y - \Omega(j, k)t)\Bigr]\biggr\}.\end{aligned} \tag{1.134}$$

We use the buoyancy-frequency profile (A.2) and solve the problem (1.128) in the Ventzel-Kramers-Brillouin approximation. The respective solution is

$$\bar{\zeta}(z) = \sqrt{\frac{2}{N_0 N(z) B}}\sin[j\pi\xi(z)], \tag{1.135}$$

where

$$\xi(z) = \frac{1}{N_0 B}\int_z^h N(z')\, dz'. \tag{1.136}$$

The mode intensities $A(j, k_x, k_y)$ are complex-valued Gaussian variables with zero mean and the correlation function

$$\langle A(j, k_x, k_y)A^*(j', k_x', k_y')\rangle = G_\zeta(j, k_x, k_y)\delta_{jj'}\delta(k_x - k_x')\delta(k_y - k_y'), \tag{1.137}$$

determined by the well-known empiric Garrett-Munk spectrum

$$G_\zeta(j,\,k_x,\,k_y) = \frac{2B^3N_0^2E}{\pi^2M}\frac{1}{(j^2+j_*^2)}\frac{k_jk}{(k^2+k_j^2)}, \tag{1.138}$$

where $j_* = 3$, $E = 6.3\cdot 10^{-5}$,

$$k_j = \frac{f_i}{N_0}\frac{\pi j}{B}, \tag{1.139}$$

$$M = \sum_{j=1}^{\infty}(j^2+j_*^2)^{-1} \simeq j_*^{-2}\frac{\pi j_* - 1}{2}. \tag{1.140}$$

In the real ocean internal waves cause sound-speed variations with the amplitude of approximately 1 m/s. The strongest variations take place at low depths. Internal-wave field consists of a large number of modes. Horizontal periods of internal waves vary from hundreds meters to hundreds of kilometers. For the sake of universality, in acoustic modeling one frequently invokes the Garrett-Munk model of an internal wave field. In spite of the empiric nature of this model, its relevance is confirmed by numerous natural experiments. High-number modes of internal waves lead to a fine structure of a sound-speed profile, though their spectral weights are relatively small.

1.5.2 *Rossby waves*

Different types of waves may be generated in the ocean. *Rossby* or planetary waves following the parallels propagate from east to west. They can exist only in the presence of a potential-vorticity gradient which, in turn, is caused by variation of the horizontal component of the Coriolis force with latitude. The need to conserve potential vorticity forces water columns to oscillate about a line of latitude and generate a westward propagating wave [Pedlosky (1987)].

Their horizontal scale is of the order of hundreds of kilometers and the amplitude of the vertical oscillations at the sea surface is about 10 cm. The speed of Rossby waves varies with latitude and is of the order of a few tens of cm/s for the barotropic waves and a few cm/s for the baroclinic ones. Typical time scales are of the order of days for the barotropic waves and a few months or even years for the baroclinic waves.

In contrast to internal waves, it is reasonable to treat Rossby waves as a deterministic periodic or quasiperiodic process than a stochastic one: they can be fairly modeled as a superposition of a few modes. The lowest of the modes, the barotropic one, doesn't influence on the curvature of a sound-speed profile, though it leads to significant horizontal currents. The first baroclinic mode induces sound speed variations of the order of 10 m/s, mainly near the channel axis. The second baroclinic mode weakly distorts the upper part of the sound-speed profile.

1.5.3 *Currents*

Large-scale oceanic currents, like Antarctic Circumpolar Current, the Gulf Stream, the Kuroshio, and many others, when crossing the distance between the source and the receiver, appreciably affect the sound signals. They cause strong variations of the sound-speed profile, such as large shifts of the channel axis, transformation to the double underwater sound channel or complete disappearance of the waveguide. Currents generate eddies, stationary and non-stationary ones, which act as strong scatterers of sound waves. In addition, boundaries of the large-scale currents, the so-called fronts, where large step-like horizontal gradients of hydrological parameters take place, cause horizontal refraction and backscattering of sound. Since fronts separate water masses with different temperature and salinity, there occur extensive convective mixing producing a specific "spicy" fine structure which leads to diffractive scattering of sound waves.

Another kind of currents to be noted is the so-called stream-like currents, i.e. jets with strong vertical gradients of the horizontal speed. Their vertical length scale is of the order of tens and hundreds meters, and the horizontal length ranges from a few to tens of kilometers. Sound-speed variations associated with the streamers may be of the order of 10 m/s. They can cause the microchannels capturing sound rays.

1.5.4 *Eddies*

Eddies or rings occur everywhere in the ocean. Permanent quasi-stationary or topographic eddies are anticyclonic structures (with the horizontal scale of the order of a few tens of kilometers) that appear in a background flow over seamounts. Moving *mesoscale eddies* have a diameter of the order of 100-200 km, the vertical size of the order of a few hundreds of meters, and their lifetimes may vary from a month to a year. Energetic mesoscale eddies are commonly found in the vicinity of the strong large-scale currents like the Gulf Stream and the Kuroshio. The core of a cyclonic (anticyclonic) eddy is colder (warmer) than outer waters. Drifters, floats, and satellite images of sea surface are used to identify and track topographic and mesoscale eddies. Temperature differences between the core and outer waters cause strong inhomogeneities in the sound-speed profile which, in turn, influence on the propagation of sound across the eddies.

1.6 Acoustic Ocean Tomography

Various hydrological processes render remarkable changes into the spatial distribution of the sound speed. Obviously, acoustic wavefields and their measurable characteristics are influenced by hydrological variability of the environment in the propagation direction as well. There arises a possibility to extract information about intervening environment from measured acoustic data. In other terms, sound propagation can be used for the *tomography* of the ocean, or for solving *the inverse problem.*

Munk and Wunsch in their seminal work [Munk and Wunsch (1979)] proposed a scheme of ocean reconstruction (hereafter MW scheme), based upon the analysis of acoustic ray travel times along different paths. Nowadays the MW scheme is the main technique used in acoustic tomography.

Before a description of the MW scheme, some words should be said about the inverse problems like the acoustic tomography. Generally speaking, the problem of ocean reconstruction is incorrect in the strict mathematical sense. It means that finding of the solution requires, firstly, introducing some *a priori* information about environment and, secondly, some procedures of regularization [Tikhonov and Arsenin (1977)]. As *a priori* information, one often uses a reference sound-speed profile $c_{\mathrm{ref}}(y, z, r)$, where y is the transversal horizontal coordinate, which is close to the real one $c(y, z, r)$. Reference profile is usually constructed using climatological databases.

The link between sound-speed and travel time variations can be described as follows [Spiesberger (1985)]. In the reference waveguide, travel time along a sampling ray connecting a fixed acoustic source and a receiver is given by

$$t_0 = \int\limits_{\Gamma_0} \frac{ds}{c_{\mathrm{ref}}(y, z, r)}, \tag{1.141}$$

where the integral is evaluated along the initial ray path Γ_0. The exact (measured) travel time is given by

$$t_1 = \int\limits_{\Gamma_1} \frac{ds}{c_{\mathrm{ref}}(y, z, r) + \delta c(y, z, r)}, \tag{1.142}$$

where $\delta c = c - c_{\text{ref}}$, and the integral is evaluated along the real ray path Γ_1. The travel time difference is

$$\delta t = t_1 - t_0 = \int_{\Gamma_1} \frac{ds}{c_{\text{ref}}(y,\, z,\, r) + \delta c(y,\, z,\, r)} - \int_{\Gamma_0} \frac{ds}{c_{\text{ref}}(y,\, z,\, r)}. \tag{1.143}$$

Since $\delta c \ll c_{\text{ref}}$, the exact travel time t_1 can be expanded as the converging series

$$t_1 = \int_{\Gamma_1} \frac{ds}{c_{\text{ref}}} \left[1 - \frac{\delta c}{c_{\text{ref}}} + \left(\frac{\delta c}{c_{\text{ref}}} \right)^2 - \ldots \right]. \tag{1.144}$$

The travel time difference δt now can be written as

$$\delta t = \delta t' - \int_{\Gamma_1} \frac{ds}{c_{\text{ref}}^2} \delta c \left(1 - \frac{\delta c}{c_{\text{ref}}} + \ldots \right), \tag{1.145}$$

where

$$\delta t' = \int_{\Gamma_1} \frac{ds}{c_{\text{ref}}} - \int_{\Gamma_0} \frac{ds}{c_{\text{ref}}}. \tag{1.146}$$

The last term in (1.145) can be recast as

$$- \int_{\Gamma_1} \frac{ds}{c_{\text{ref}}^2} \delta c \left(1 - \frac{\delta c}{c_{\text{ref}}} + \ldots \right) = \delta t_{\text{lin}} + \delta t'', \tag{1.147}$$

i.e., as the sum of the linear term with respect to δc

$$\delta t_{\text{lin}} = - \int_{\Gamma_0} ds \frac{\delta c}{c_{\text{ref}}^2} \tag{1.148}$$

and the higher-order correction given by

$$\delta t'' = \int_{\Gamma_0} ds \frac{\delta c}{c_{\text{ref}}^2} - \int_{\Gamma_1} ds \frac{\delta c}{c_{\text{ref}}^2} + \int_{\Gamma_1} ds \frac{\delta c^2}{c_{\text{ref}}^2 (c_{\text{ref}} + \delta c)}. \tag{1.149}$$

Thus we arrive to the formula

$$\delta t = t_{\text{lin}} + \delta t' + \delta t''. \tag{1.150}$$

On practice, one often considers only the linear term (1.148) that simplifies remarkably solving the inverse problem, although a respective solution becomes less accurate. A reduction of the parameter search space can be achieved by representing a sound-speed perturbation as a sum of empirical orthogonal functions [Tolstoy *et al.* (1991)] which are introduced by means of the covariance matrix

$$K_{ij} = \frac{1}{N} \sum_{n=1}^{J} [c_n(z_i) - c_{\text{ref}}(z_i)] * [c_n(z_j) - c_{\text{ref}}(z_j)], \tag{1.151}$$

where $\{c_n(z)\}$ are sample profiles which are obtained using historical data or from oceanographical measurements. Here discrete depths $z_1, ..., z_K$ are the points at which sound speed is sampled. By definition, the orthogonal functions are the eigenvectors $\mathbf{f}(z_i) = (f_1(z_i), f_2(z_i), ..., f_K(z_i))$ of the covariance matrix $\mathbf{K} = \{K_{ij}\}$

$$\mathbf{Kf} = \mathbf{\Lambda f}, \tag{1.152}$$

where $\mathbf{\Lambda}$ is a diagonal matrix of eigenvalues. Then the sound speed can be written as

$$\delta c(y, r, z_i) = c_{\text{ref}}(z_i) + \sum_k \alpha_k(y, r) f_k(z_i). \tag{1.153}$$

The expansion in terms of the orthogonal functions is also known as the Karhunen–Loeve expansion [Van Trees (1980)]. It has remarkable advantages for solving tomographic problems. Firstly, we just need to find the coefficients α_k to reconstruct an environment. Secondly, only a few orthogonal functions may be sufficient to describe accurately a sound-speed profile. The problem of specifying a suitable basis for the representation of environment is discussed by Burov *et al.* (2007).

The first experimental test of the MW scheme has been performed in 1981, when the acoustic tomography was used for mapping of the sound-speed field in a 300 km × 300 km square in the Western North Atlantic [Cornuelle *et al.* (1985)]. The reference ocean was modeled using the climatological data. The maps, constructed using the MW scheme, were found to be in a good agreement with the maps calculated using independent measurements of the CTD-characteristics (CTD — acronym for a conductivity, temperature, and depth sensor).

After the work of Munk and Wunsch, the issue of ocean acoustic tomography became a subject of extensive research, both theoretical and experimental. On relatively short ranges, up to 200–300 km, acoustic tomography can be performed using numerous spatially distributed moored or moving sources and receivers that enable effective imaging of eddies within an insonified area. With greater distances, one typically uses one source and a few number of receivers.

At low frequencies, the standard ray approximation is not applicable, and it is reasonable to use modal tomography based upon the analysis of phase velocities of different modes. In the case of weak inhomogeneities phase variations of detected modes with respect to an unperturbed waveguide are restricted in the interval $[-\pi : \pi]$, therefore the problem is non-degenerate, and the modal tomography can be performed using the monochromatic sources. With stronger inhomogeneities, the inverse problem becomes degenerate, and there occurs an uncertainty in reconstruction of an environment. This degeneracy is removed if we exploit pulse sources. Then the role of measurable characteristics can be played by time delay of a mode. The linear approximation to this quantity is given by

$$\delta t_i = -\int \frac{\delta \nu_i}{\nu_i^2}\, dl, \tag{1.154}$$

where ν_i and $\delta\nu_i$ are the group velocity of the i-th mode and its perturbation, respectively. Notably, time delay (1.154) looks similar as that in the ray-based scheme (1.148). A different, non-perturbative scheme for the modal tomography was proposed by Munk and Wunsch (1983) and then developed by Jones *et al.* (1993) and is not discussed here.

Actually, the acoustic tomography should be considered as a complementary technique to other methods. It has important advantages which are not accessible for other methods of distant probing the ocean. For instance, abilities of satellite remote sensing are restricted by near-surface layer, while the acoustic methods are capable to give insight into the ocean interior. One hopes that cooperative studies using different methods can lead to a remarkable success, as that is demonstrated in subsection 1.7.5.

1.7 Experiments on Long-Range Sound Propagation

1.7.1 *The Heard Island Feasibility Test*

The Heard Island Feasibility Test (hereafter HIFT) conducted in January 1991 is famous as the transmission of coherent acoustic signals over recordingly long ranges: the distance from the source to the furthermost receiver was of about 18 megameters.

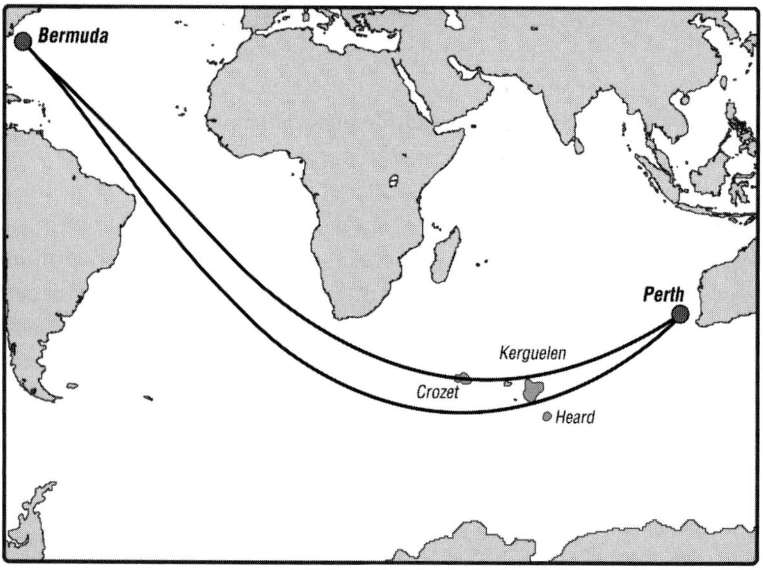

Fig. 1.9 Paths connecting Perth to Bermuda [Munk *et al.* (1988)].

Before the HIFT, the propagation over tens megameters was achieved only by use of explosive charges, in the seminal experiment on sound transmission from Perth to Bermuda, on Spring 1960 (Fig. 1.9). That experiment consisted of two sessions. The first session was conducted on 1 March 1960, when nine 200- and 300-lb Amatol charges were deployed from the research vessel Vema, belonging to the Lamont Geological Observatory of Columbia University, Palisades, New York. The explosions occurred at different depths between 732 and 1800 m, because the pressure detonators were not exact. The sounds were detected by hydrophones off Bermuda, about 19820 km away. The signals were weak, therefore an additional session was requested. It was conducted twenty days later, when three 300-lb charges were deployed from the Australian Navy's HMAS Diamantina. The locations of explosions in both the cases were about the same, just west of Perth. Each reception on the hydrophone Juliet suspended at the sound channel axis consisted of two peaks with a 20-sec difference. Seemingly, the peaks correspond to different propagation paths. Due to the incoherent nature of the signals transmitted, each of the peaks had a duration of ten seconds, and single arrivals, ray or modal, couldn't be resolved. When results of the experiment were analyzed, there occurred a difficulty in finding exact the propagation path. It was not clear to what extent horizontal temperature gradients refracted the propagation path [Munk *et al.* (1988)]. This uncertainty had limited the accuracy of calculations of travel times using the ray tracing. Additional difficulties were concerned with bathymetric interactions, which seemed to appear near the Kerguelen Plateau, near Cape Agulhas, South Africa, and near Cape Saõ Roque, Brasil. The bathymetric features seemed to filter out all the acoustic modes except for the lowest ones. Unfortunately, an analysis of modal spectrum at the receivers was not allowable, mainly due to excessively long duration of the signals detected and the few number of hydrophones.

In contrast to the Perth-Bermuda experiment, in the HIFT only carefully crafted acoustic signals were used. The signal allowed for the resolution of travel times down to the millisecond level. That was conditioned by the specific challenges of the HIFT, concerned with hydroacoustic tomography. In fact, HIFT was a preliminary stage before the series of experiments on long-range acoustic thermometry. It was necessarily to be convinced that sound signals can preserve sufficient energy to be detected along very long paths, and the respective arrivals can be stable, well-resolved, and identifiable with specific ray paths.

The HIFT was actually an example of effective international collaboration: scientists from nine countries had joined their efforts to obtain the result. The key role was played by USA scientists, which provided signal generation and prepared four receiving sites. Four receiving sites were Australian, two were Canadian. South Africa, India, France, USSR, New Zealand, and Japan participated each by one receiving site.

The signals were generated at the Heard Island, a subantarctic Australian island in the Indian Ocean. The main advantage of this site is the possibility to ensonify the Atlantic and Pacific Oceans simultaneously. The desirable signal level was obtained by using a vertical array consisting of ten sources. Each source was capable of transmitting level of 206 dB. The array was carried by the research vessel Cory Chouest and suspended to the depth of 175 m, where the local sound channel axis was placed. To avoid overheating, only five sources were active at a moment. Thus, the maximal source level was of approximately 220 dB. The source frequency was chosen to be 57 Hz. Three types of signals were used: the 57-Hz cw tone (a signal comprising a single frequency), a m-sequence yielding a pentaline spectrum (five principal spectral lines), and a narrowband signal consisted of four different m-sequences. The cw signal had the highest signal-to-noise ratio among all of the HIFT signals.

Since the source level was high, U.S. and Australian Marine Fishery Services demanded for procedures to avoid any harmful effect on marine mammals. A transmission would be aborted if any marine mammals were within 1 km of the source ship at the transmission start time. There was no such instance. In addition, the second research vessel, Amy Chouest, was chartered for biological observations. No indication of any harmful effect on local marine life was detected, albeit changes in behavior were observed [Bowles *et al.* (1994)].

Receivers were located at 16 sites (see Fig. 1.10). Unfortunately one of the receivers, the one at Bermuda, sank before the start of the transmissions. The Japanese receiver off Samoa had not detected any signals due to the unexpected acoustic blockage in the Tasman Sea. Other 14 receivers had successively detected the signals from the HIFT source. The furthermost of them were the Canadian receiver near the Cape Cod and the American receiver at Whidbey Island, near Seattle. Both were at ranges of about 18 Mm. The scheduled start of transmissions was 26 January. Then the transmissions were carried out until, on 31 January, the Cory Chouest encountered a gale with 10-m seas, and one of the suspended sources was

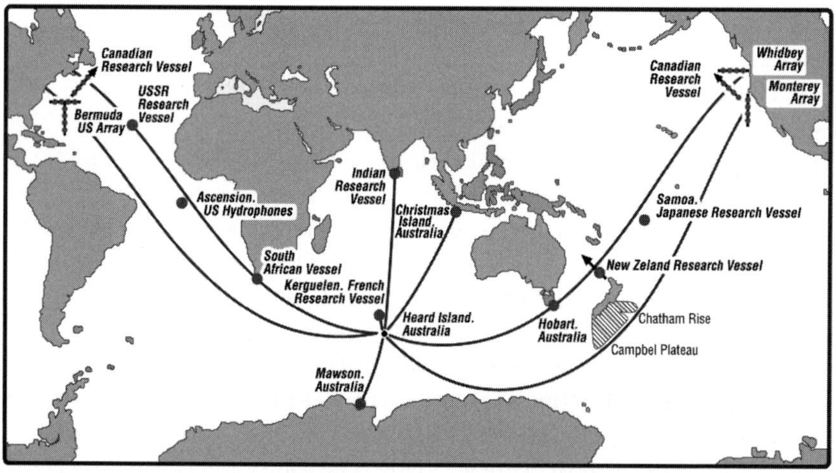

Fig. 1.10 Propagation paths from source to receiver sites in the HIFT. Single dots indicate sites with single receivers. Dots connected by horizontal lines designate horizontal bottom-mounted arrays, vertical lines designate vertical arrays, and slanted lines designate arrays towed in the direction of the arrow. Signals were received at all sites except for the vertical array at Bermuda (which sank) and the Japanese station off Samoa (Munk *et al.*, 1994).

failed. The others were severely damaged, and the experiment had been ultimately terminated.

Since the propagation length was excessively large, modeling of ray paths from the source to the receivers requires to take into account the Earth curvature. It should be expected that rays must follow the so-called geodesics, i.e., great circles which correspond to the shortest distance between two points on a spheroid. In fact, the Earth is not perfectly spherical, therefore, the actual ray paths trace the refracted geodesics, which remarkably differ from the great circles.

The major contribution into the sound attenuation was expected from geometric cylindrical spreading, producing r^{-1}-decay for the field intensity. Geometric spreading was accompanied by spreading in the time domain due to geometric dispersion between different ray paths. Combination of space and time spreadings was estimated to yield attenuation of roughly 135 dB at 18 Mm range. The next factor of attenuation to be considered was surface scattering, causing transmission losses up to 60 dB. Volume attenuation was sufficiently weak and estimated to be 5 dB in the Atlantic ocean and 3 dB in the Pacific Ocean. Under those estimates, the source transmitting level of 220 dB seemed to be enough for successful signal receptions. The actual transmission loss, the difference of the transmitting and received levels, had been found 10-45 dB above the expectations taking into account only cylindrical spreading. The main contribution to the additional transmission loss was given from the partial or complete acoustic blockages due to the interaction with underwater ridges [Birdsall *et al.* (1994)].

Although the measured arrival patterns were at whole repeatable, the problem of resolution and identification of individual arrivals has not been resolved in the HIFT. That was partially due to the source motion because of bad weather.

The complete acoustic blockage in the Tasman Sea is one of the most intriguing results of the HIFT. The Doppler-derived launch azimuth technique yielded that signals arriving at Whidbey are emitted from Heard Island at an azimuth of 130°, not 115° as derived for the Tasman geodesics. It was realized that the acoustic path went south and east of New Zealand through the "Polynesian window". It is consistent with measurements off the coast of southern California, which indicated that arrivals detected corresponded to rays passing through the Polynesian window [Munk *et al.* (1994)]. It is still not well recognized by which reason the direct path through the Tasman Sea appeared closed. It was suggested that the Tasman Sea path was blocked by the ridges lying in the northeastern sector of the

Tasman Sea, between New Zealand and Fiji. The New Zealandian receiver in the center of the Tasman Sea did successively receive signals, but the Japanese receiver near Samoa didn't received any signals. A partial acoustic blockage occurred also in the Atlantic, perhaps by the mid-Atlantic ridge.

Another issue to be addressed is the interaction with the Antarctic front, named as the Antarctic Circumpolar Convergence. At small distances from the source, the sound channel is the near-surface one, as it is typical for the polar ocean. It means that almost all the rays emitted from the source are surface-reflected. After crossing the Antarctic front, the upper ocean layer becomes much warmer, and the channel axis comes down from the surface to a deeper layer. Just as it appears, the near-surface USC transforms into an interior USC, and surface-reflected rays become purely refracted. This situation is demonstrated in Fig. 1.11. In the HIFT data, the transformation of the channel was followed by enhancing of mode coupling and occurrence of higher modes, those were detected by the vertical array off Monterey. It contradicts to the presupposition made before the HIFT that only lowest modes are able to propagate over very large distances.

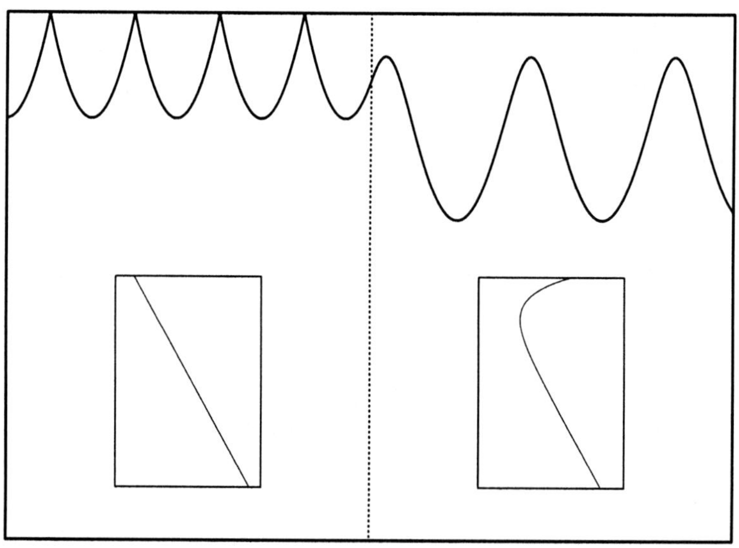

Fig. 1.11 Cartoon for a transition from surface-reflected to purely refracted propagation across the Antarctic Circumpolar Convergence. The high-latitude sound-speed profile (left impact) is associated with an axis along the surface; the temperate sound-speed profile (right impact) is associated with an interior sound channel.

1.7.2 *Experiments with bottom-mounted sources*

1.7.2.1 *Downsloped bathymetry near a source*

We proceed with a series of experiments carried out using sources mounted into the bottom in a shallow-water region. Bottom-mounted sources have important advantages in exploitation as compared with suspended ones, such as the absence of source motion and longevity. However, in this case sound propagation suffers from the constraints posed by the non-negligible interaction with the ocean bottom.

Typically, the sound speed in shallow water environments takes on the lowest value at the bottom, so there occurs a near-bottom waveguide and refraction tends to deflect ray paths downwards. Under a reflection from the bottom, some amount of the acoustic energy penetrates into the bottom and departs from the waveguide. The energy loss depends on the incidence angle and is of the lowest value for small grazing-angle sound waves. There are two reasons for that. The first one is that high grazing-angle waves more frequently interact with the lossy bottom. The second one is that the modulo of the reflection coefficient for small incidence angles is close to unity, i.e., almost all the energy remains within the water column.

A downslope diminishes the energy loss. Each ray reflection from a downsloped searfloor reduces the grazing angle, therefore, steeper ray paths transform into less lossy flatter ones, as that is illustrated in Fig. 1.12. There also can appear a non-trivial regime when sound rays slide along the bottom downward until they reach the USC axis in deep water. Then rays detach from the sloping bottom and continue to propagate near the sound

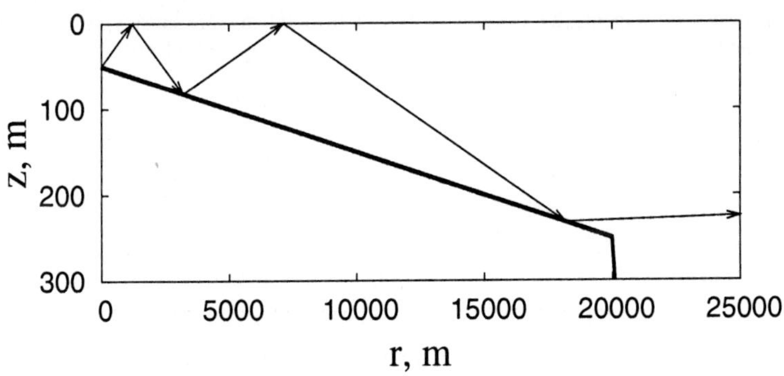

Fig. 1.12 Flattening of a ray path due to reflections off the downsloped bottom. The bold line represents the bottom profile, arrows indicate the ray trajectory.

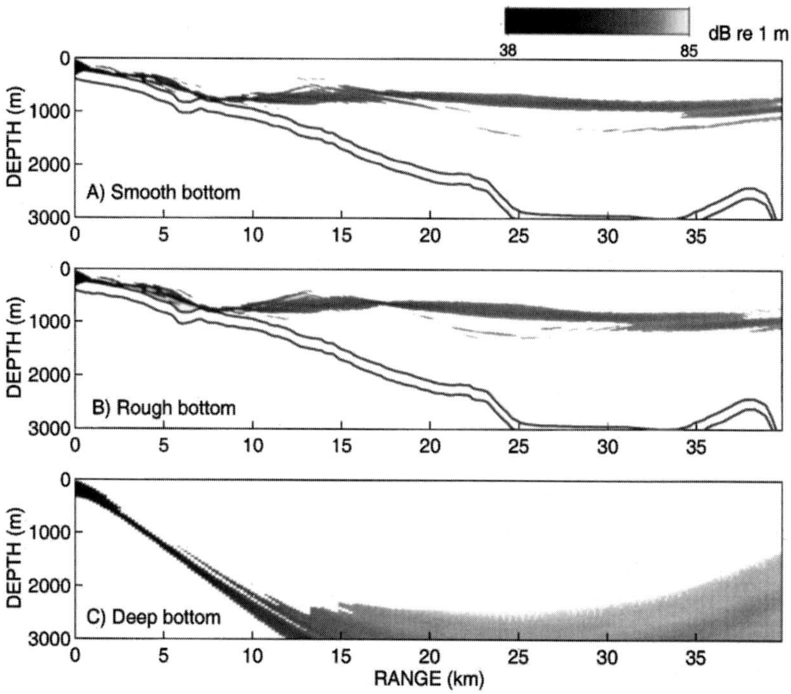

Fig. 1.13 The mudslide effect. (A) Downward propagating sound inclined near 5 degrees in about a one degree-beam width is modeled from the depth of the Kaneohe source into the ocean along smooth water/sediment and sediment/basement interfaces. Only the top 6 dB of sound pressure levels are shown at each range step to emphasize the location of the loudest sound. The top and bottom lines show the water/sediment and sediment/basement interfaces. (B) Same as (A) except the standard deviation of the water/sediment and sediment/basement interfaces are 2 and 8 m, respectively. (C) Same as (A) except the bottom is moved to 5000 m. No energy is trapped near 800 m depth without the mudslide effect arising from the sound speed gradients near the water/sediment interface [Tappert *et al.* (2002)].

channel axis. This effect is referred to as the mudslide effect [Tappert *et al.* (2002)] and demonstrated in Fig. 1.13. Such a behavior is a consequence of a down-refracting form of the sound-speed profile in shallow seas. Mudslide effect is found to be robust with respect to the bottom roughness (compare Fig. 1.13(A) and (B)). There is no acoustic energy trapped near the depth of minimum sound speed when the source is placed at the same depth in deep water, i.e., in the absence of the downslope (see Fig. 1.13(C)).

These peculiarities suggest that the sound emitted from a downslope should propagate predominantly horizontally in the vicinity of the axis of the deep-water waveguide. However, this presupposition was criticized after the experiment with the source mounted in Kaneohe Bay. This experiment was conducted in the years 1983–1989. A broadband 133-Hz source was mounted on the bottom at 183-m depth in Kaneohe Bay, Oahu island. The omnidirectional source transmitted sound over 3709 km to a bottom-mounted horizontal array at 1433-m depth offshore of California. All the signal paths were reflecting from the slope near Oahu before becoming trapped in the sound channel.

The signal reception comprised five distinct early arrivals and the transmission finale consisted of non-resolved arrivals (Fig. 1.14). The finale was followed by a cut-off. The transmission finale appeared to be strongly spread in depth, that contradicts to the expectations of axially trapped energy. The main contribution to the depth spread is given from scattering on internal waves. A similar result concerning vertical spreading was obtained by Colosi *et al.* (1994) for the experiment on 1000-km sound propagation in the North-Eastern Pacific. That experiment is known as SLICE89.

The distinct arrivals in the early part were formed by ray clusters consisted of steep rays inclined at about 15 degrees to the horizontal at the channel axis. No individual ray arrivals were detected. Spiesberger and Tappert [Spiesberger and Tappert (1996)] pointed out that the clusters were formed by *chaotic* rays. To explain remarkably small travel time spreads of each cluster, they conjectured that these rays belong to the so-called bounded chaotic seas in the phase space like those considered in Chap. 2.

The near-axial trapping of sound is much more relevant with smaller distances between the source and the receiver. For an illustration of that, let us consider the experiment in the Sea of Japan (or the East Sea), which was conducted in August 2006 [Akulichev *et al.* (2007); Bezotvetnykh *et al.* (2009)]. The sources with the center frequencies of 366 and 600 Hz were mounted at the depth of 40 m, on the shallow-water shelf near Cape Shultz, to the south of the Vladivostok harbor (see Fig. 1.15). The receiver was a single hydrophone suspended from a yacht to the depth near the channel axis. Signal receptions were carried out at six moorings of the yacht, the ranges of moorings are from 55 km to 368 km. The furthermost mooring was nearby Kita-Yamato Bank.

There were observed two dominant arrivals with the travel time difference of 15–30 ms (Fig. 1.16). The dominant arrivals spread with increasing

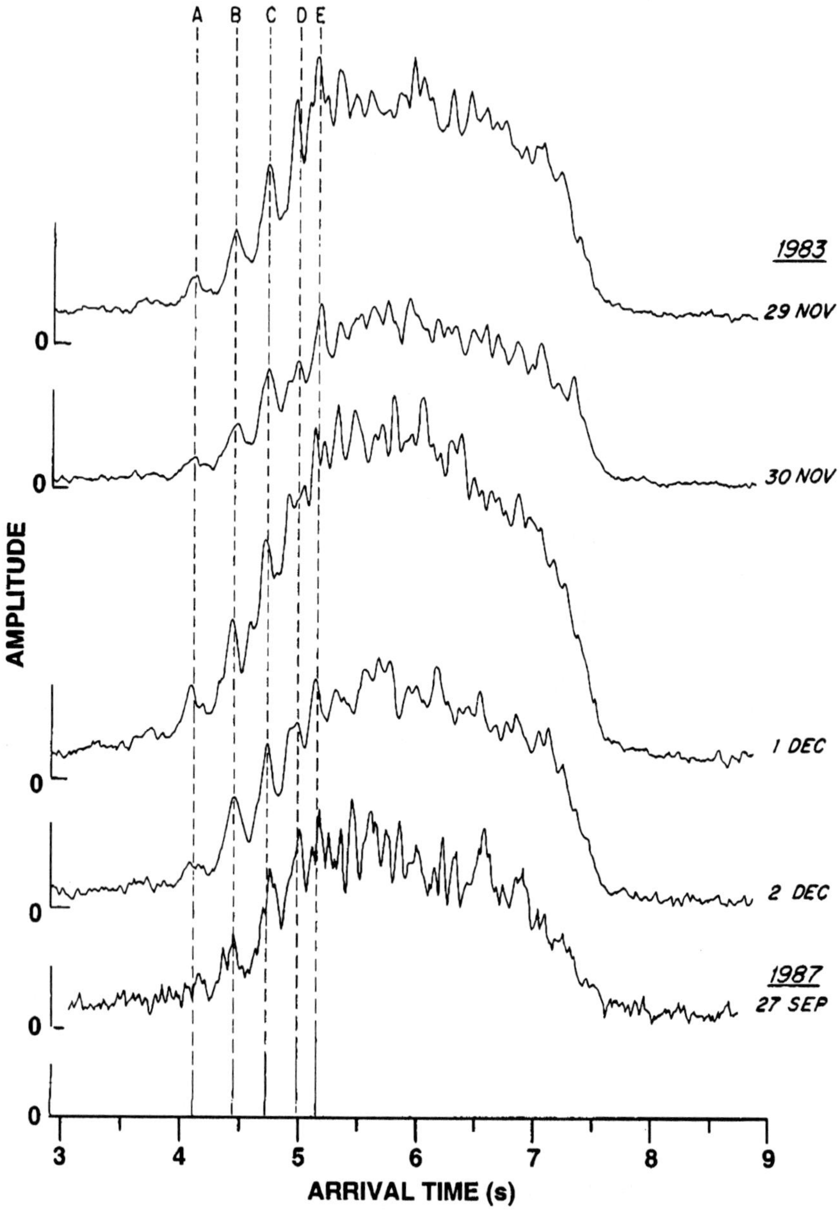

Fig. 1.14 The experiment with a bottom-mounted source in Kaneohe Bay. The daily incoherent average of acoustic multipaths for four days in 1983 and one day in 1987. The vertical dashed lines join together acoustic arrivals which are similar [Spiesberger and Tappert (1996)].

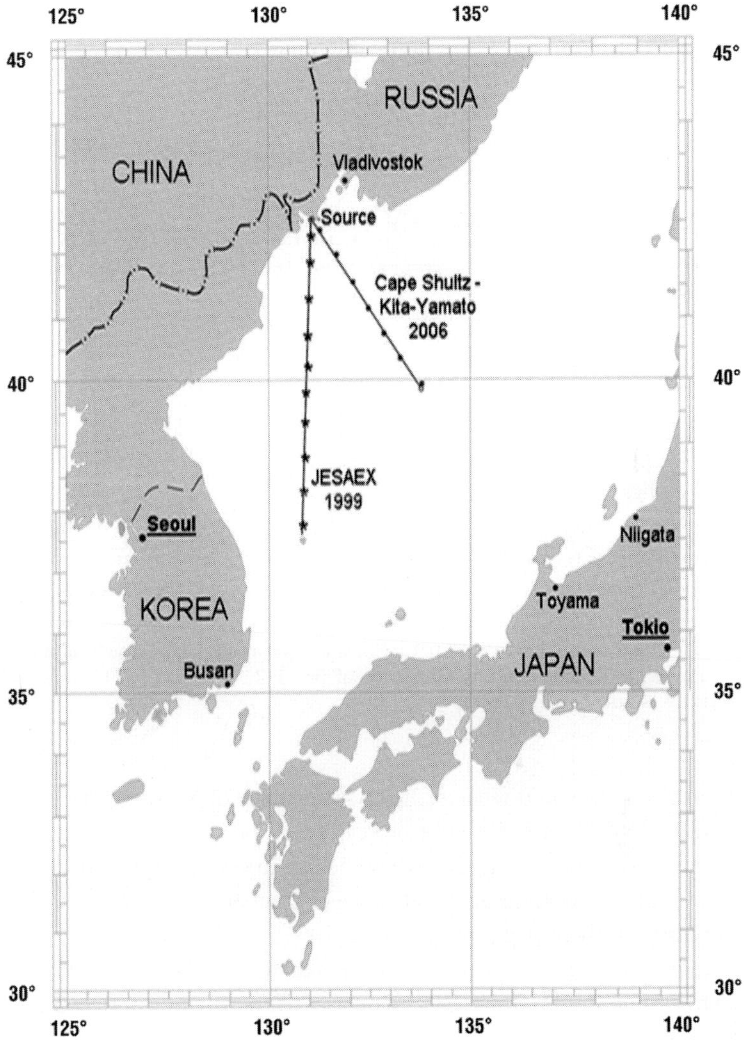

Fig. 1.15 Schematic diagram of the experiments with the sources near Cape Shultz.

the distance between a source and a receiver, and chaotic microarrivals appear in their vicinity, probably, because of scattering at internal waves. The surprising result was that simple multiplying of the travel time by the range-averaged sound speed at the channel axis, after minor corrections concerned with the shelf, yielded an excellent estimate for the distance

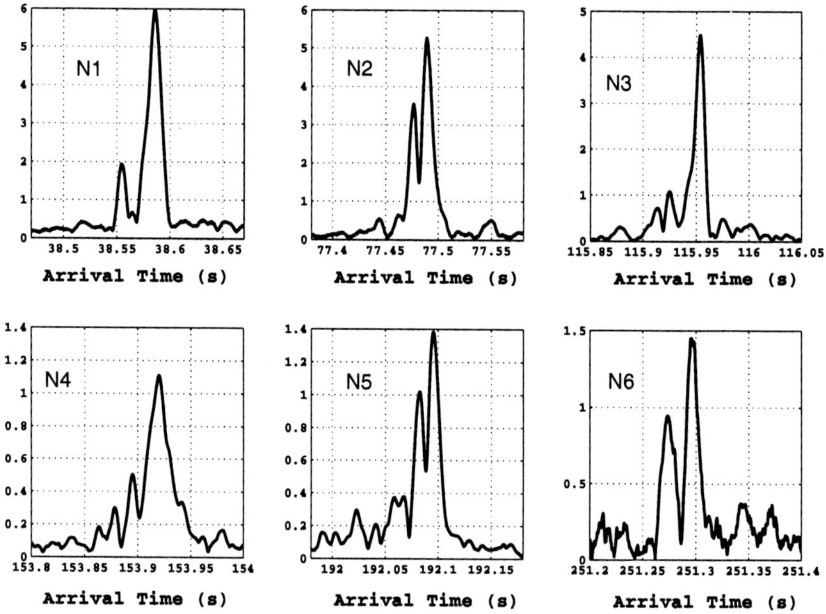

Fig. 1.16 The measured arrival structure in the experiment near Cape Shults in August 2006. Points 1–6 correspond to distances of 56, 113, 169, 224, 279.5, and 368 km, respectively [Akulichev *et al.* (2007)].

between the source and receiver, with inaccuracy of 0.1 percents [Akulichev *et al.* (2007)]. Perhaps this phenomenon may be regarded as the sequence of the mudslide effect providing the efficient injection of sound energy into the axial paths. Seemingly, the absence of steeper paths was caused by the strong bottom absorption on the shelf.

Almost the same results were obtained for the experiment named as JESAEX (The Japan/East Sea Acoustics Experiment) and conducted in 1999 [Spindel *et al.* (2003)]. Broadband sources with center frequencies between 250 and 634 Hz were installed on the shelf at almost the same area as in the experiment described previously. Two vertical receiving arrays were deployed near Ulleung-do island, 558 km to the south. The observed arrival pattern was very similar to that shown in Fig. 1.16 and consisted of two dominant well-resolved arrivals with a time difference of approximately 30 ms. The data from the JESAEX experiment were also found to allow for such an accurate estimate of the distance despite the presence of a front along the propagation path.

1.7.3 *Acoustic Engineering Test*

The Acoustic Engineering Test (AET) was conducted for seven days on November 1994. The acoustic source was suspended at 652-m depth, near the sound channel axis, from the floating instrument platform which was moored in the North Pacific Ocean. The source transmitted a broadband signal with a center frequency of 75 Hz and 37.5-Hz bandwidth. The transmissions were received on two vertical line arrays, on the numerous U.S. Navy receivers in the North Pacific and on a deep sonobuoy near New Zealand. The transmission path to Hawaii is depicted on the map presented in Fig. 1.17.

Here we shall review the results obtained by one of the vertical arrays moored east off Hawaii. That one consisted of 20 hydrophones at 35-m

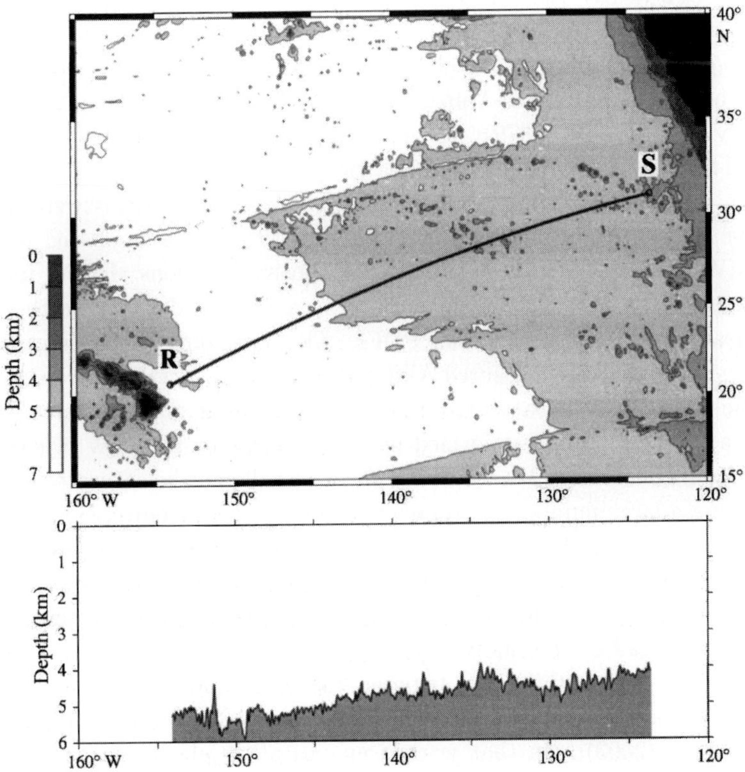

Fig. 1.17 The upper plot is an acoustic path from the acoustic source **S** to the vertical receiving array **R** moored east of Hawaii in the AET experiment. The lower plot is the bathymetry along the path [Colosi *et al.* (2005)].

spacing (1.75 wavelengths at 75 Hz) between the depths of 900 and 1600 m. There also was an upper 20-hydrophone subarray between the depths of 200 and 900 m, but it had been failed before the experiment. The receiver allowed for measurements of the vertical structure of the signals and was designed to be able to resolve low order modes of a wavefield. The transmission path to Hawaii does not cross any significant hydrological structures such as fronts or currents. All the path is deep, without significant underwater ridges along (see the lower plot of Fig. 1.17). The length of the path was equal to 3252 km. The path was a refracted geodesics slightly deviated from the unrefracted one due to horizontal gradients in sound speed.

The analysis of the AET was carried out by [Worcester *et al.* (1999); Colosi *et al.* (1999, 2001); Beron-Vera *et al.* (2003)]. It had been found that the early and late AET arrivals possess qualitatively different properties. The early AET arrivals were found to be temporally resolved, were stable over the duration of the experiment and could be identified with specific ray paths. Travel time statistics was compared with predictions of the Flatté-Dashen-Munk-Watson-Zachariasen (or shortly FDMWZ) theory [Flatté *et al.* (1979)] which considers internal-wave-induced sound-speed fluctuations along deterministic ray paths in the background medium. Quite unexpectedly, time spreads of single observed arrivals were found to be 40–1000 times smaller than the statistical estimates based on the FDMWZ theory. Measured probability density functions (PDFs) for the peak intensity were close to lognormal, in a marked contrast to the theoretically predicted exponential PDFs. The surprising conclusion of this analysis was that the measured AET pulse statistics suggested propagation in the statistically unsaturated regime. Physically it implies that a wavefield at a given point was formed by only one ray or by a few rays which interfered coherently. The FDMWZ theory predicted the saturated regime implying a large number of rays with incoherent interference.

In contrast to the early arrivals, the AET arrivals in the last 1 to 2 sec could not be temporally resolved or identified. The intensity PDF is close to the fully saturated exponential distribution indicating strong modal coupling caused by internal waves.

The qualitative difference between the early and late portions of an arrival pattern was elucidated by invoking the theory of ray chaos [Beron-Vera *et al.* (2003)]. In that work a numerical simulation of ray dynamics in the AET environment internal waves with the Garrett-Munk spectrum imposed was carried out. Although some amount of instability was found to appear due to the mesoscale variability, it was found that the ray

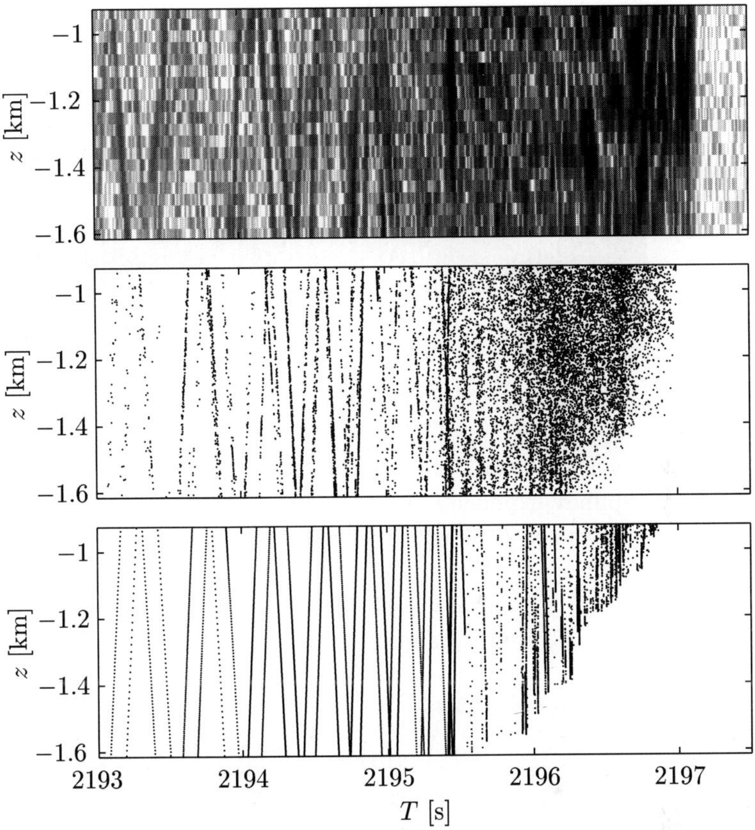

Fig. 1.18 Measured and simulated AET wavefields in the time-depth plane. Upper panel: a typical measurement of the wavefield intensity shown on a gray scale plot with a dynamic range of 30 dB. Middle panel: ray simulation with internal waves. Lower panel: ray simulation without internal waves. The wavefield intensity in the simulations is approximately proportional to the local density of dots, each corresponding to a ray [Beron-Vera *et al.* (2003)].

modeling involving internal waves captures the main qualitative features of the measured data (see Fig. 1.18). Under chaotic conditions, the separation between neighboring rays grows exponentially, on average, with range. The infinite-range asymptotics of the rate of the growth is the so-called Lyapunov exponent. Calculations of ray divergence over finite ranges yield finite-range Lyapunov exponents, which depend on ray launch angle χ_0 and more relevant for describing the underlying physics. Numerical simulations revealed that the flat rays ($|\chi_0| \leq 5°$) have finite-range Lyapunov exponents

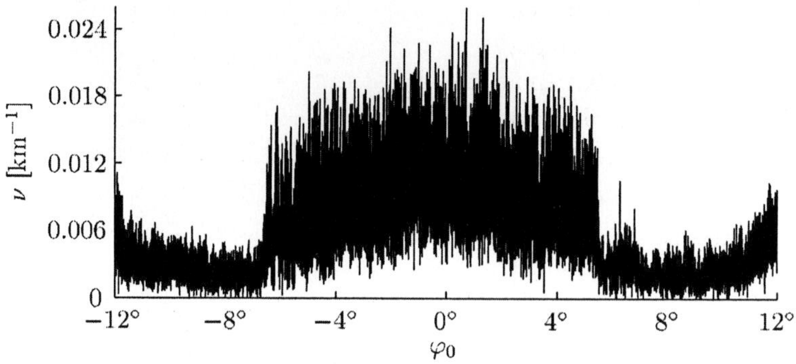

Fig. 1.19 Finite-range Lyapunov exponents as a function of ray launch angle in the AET environment with internal waves [Beron-Vera *et al.* (2003)].

of approximately 1/(100 km), while the steeper rays, $6^\circ \leq \chi_0 \leq 11^\circ$, have finite-range Lyapunov exponents of approximately 1/(300 km) (Fig. 1.19). Thus, the steep rays are significantly less chaotic than the flat ones. This partially explains the stability of the early arrivals coming through steep rays in the presence of internal waves. More detailed analysis of this issue will be presented in Chapters 3 and 5.

1.7.4 *Alternate Source Test*

The Alternate Source Test (AST) provided a good opportunity to compare wavefield properties at different signal frequencies. It was conducted in summer 1996. That was a period of the ATOC (Acoustic Thermometry of Ocean Climate) experiment with the 75-Hz source mounted on the Pioneer Seamount, 88 km west of Pillar Point, California. The source was wideband with 3-dB bandwidth of 37.5 Hz. Receivers were two vertical line arrays, one at Hawaii and another one at Kiritimati island. Each of them consisted of 40 hydrophones with 35-m spacing. Transmissions from the 75-Hz source were temporarily discontinued during the AST experiment which was conducted over 9 days from 30 June to 8 July 1996. The AST source was suspended from the military vessel Independence to a depth of 652 m, at a position about 100 km west of the coast of California. The source transmitted signals with two carrier frequencies, 28 Hz and 84 Hz, simultaneously. The 3-dB bandwidth of the AST source was 10 Hz. The ranges from the AST source to Hawaii and Kiritimati receivers were 3502 and 5157 km, respectively. The respective ranges from the 75-Hz ATOC source were 3515 and 5171 km. Unfortunately, the lower half of the Hawaii array was failed prior to the AST.

The observations indicated high sensitivity to the geographic separations of the transmission paths. Surprisingly, the signal transmissions along the longer paths to Kiritimati appeared to be more stable than the transmissions to Hawaii. The 28-Hz transmissions exhibited much weaker fluctuations than the transmissions at 75 and 84 Hz. The spectra of phase and log intensity fluctuations were found to be broad, without apparent cutoffs at high and low frequencies (Fig. 1.20). At 75 and 84 Hz, the spectra agree with the Dyson model of random multipath propagation [Dyson *et al.* (1976)]

$$S_\iota \propto \frac{1}{(\beta + \omega^2/\nu_t^2)^{3/2}}, \tag{1.155}$$

where β is an empirical constant chosen to be 2.43, ν_t is the root mean squared phase rate. As it follows from Fig. 1.20, the data from the AET experiment also fits good the Dyson model. At 28 Hz, the spectra of fluctuations remarkably depart from the Dyson model indicating different physics of fluctuations [Colosi *et al.* (2005)].

Timefronts at 75 and 84 Hz reveal the pattern similar to those observed in the long-range experiments considered above, i.e., it had the accordion-like early part followed by the crescendo of unresolved arrivals in the final. As opposed to the higher center frequencies, 28-Hz AST receptions had regular form and were dominated by strong individual arrivals, even in the arrival final.

Long vertical receiving arrays allowed for the filtering of the lowest modes of a wavefield. At 75 and 84 Hz, the first ten modes were successfully extracted. The failure of the half of hydrophones impeded appreciably modal analysis of the Hawaii receptions at 28 Hz, and only the first mode was extracted robustly. In the Kiritimati 28-Hz receptions, the first four modes were successfully filtered. After the filtering, the modal signals were obtained and analyzed.

Arrival patterns of the first mode at different center frequencies obviated different physics of mode scattering at different frequencies. In the cases with 75 and 84 Hz, the arrival patterns of the first modes revealed the multipath regime of propagation, when each modal signal consists of multiple temporally unresolved arrivals. In contrast to that, at the 28-Hz Hawaii receptions one arrival appeared to dominate in the receptions. At the Kiritimati, the arrival pattern of the first mode at 28 Hz consisted of four distinct arrivals [Wage *et al.* (2005)].

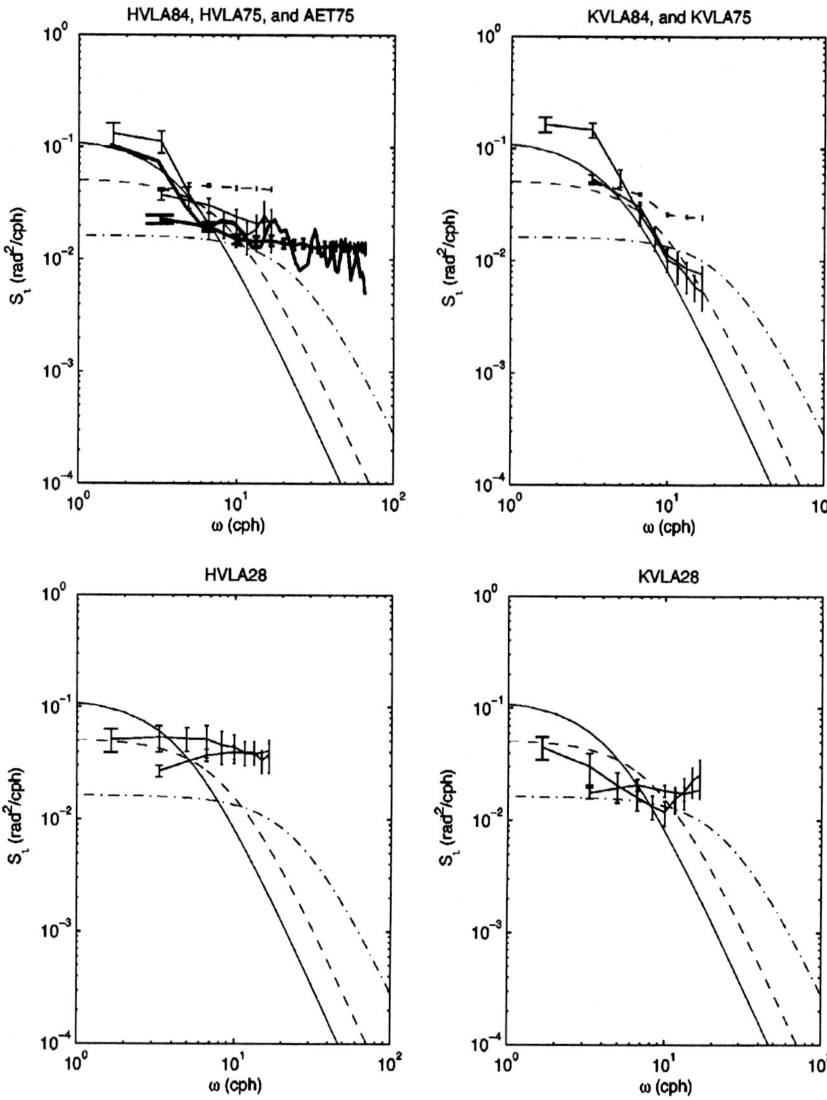

Fig. 1.20 Frequency spectra of log intensity. Here the Hawaii and Kiritimati data are denoted as HVLA and KVLA, respectively. In the upper left panel the AET (denoted as AET75) spectra are plotted with heavy lines, the HVLA spectra at 84 Hz are plotted with light lines, and the HVLA spectra at 75 Hz are plotted with dashed lines. In the upper right panel the KVLA data at 84 Hz are plotted with light lines while the KVLA data at 75 Hz are plotted with dashes. In all panels, the log intensity spectra from the Dyson model (1.155) are displayed (smooth curves) with $\nu_t = 5$ (solid), $\nu_t = 10$ (dashed), and $\nu_t = 30$ (dash-dotted) rad/s [Colosi *et al.* (2005)].

1.7.5 *Acoustic Thermometry of Ocean Climate*

We complete the review of experiments by briefly discussing a complex study of changes in oceanic heat storage involving acoustic tomography, satellite altimetry and direct temperature measurements. Fifteen months of acoustic data from the Acoustic Thermometry of Ocean Climate (ATOC) project in combination with altimeter data were compared with an oceanic general circulation model (GCM) and then used to constrain the model [The ATOC Consortium (1998)]. Sea level change was used as a convenient measure for comparing the different ATOC observations. It is directly measured by the satellite altimetry and can be also estimated from acoustic data under certain assumptions.

The acoustic part of the experiment was conducted using the acoustic 75-Hz source deployed at a depth of 939 m on Pioneer Seamount, 100 km west of San Francisco. Transmissions began in December 1995, and the transmitted signals have been received by U.S. Navy Sound Surveillance System and other arrays. The early parts of the observed data consisted of resolvable, identifiable and stable ray-like arrivals. The arrivals were tracked and then used to infer range-averaged profiles of sound speed and temperature along each section. Despite the presence of mesoscale eddies and internal waves, the arrival times vary smoothly through the months.

The altimetric component of the experiment was conducted using the the TOPEX/POSEIDON satellite altimeter in a 10-day repeating orbit. Direct temperature measurements were made by means of expendable bathythermographs (XBT) which provide direct measure of the heat content in the upper ocean (< 800 m).

For all sections except one marked as v1, acoustic measurements roughly track the climatological annual cycle, as well as the altimetric and GCM records, but they don't exhibit the short-period fluctuations (see Fig. 1.21). These fluctuations are primarily caused by barotropic processes such as wind-forced barotropic Rossby waves which don't affect sound propagation. The amplitude differences are suggested from the deficiency of information about salinity.

Combining acoustic, altimetric and XBT results, one can deduce corrections to the GCM. The uncorrected GCM underestimates the magnitudes of changes in heat content and the strength of the seasonal cycle. This results from insufficient information about mixing processes inside the water volume and inaccuracies in the surface boundary conditions. Thus, acoustic thermometry proved to be a useful complementary method to examine gyre-scale changes in the state of the ocean.

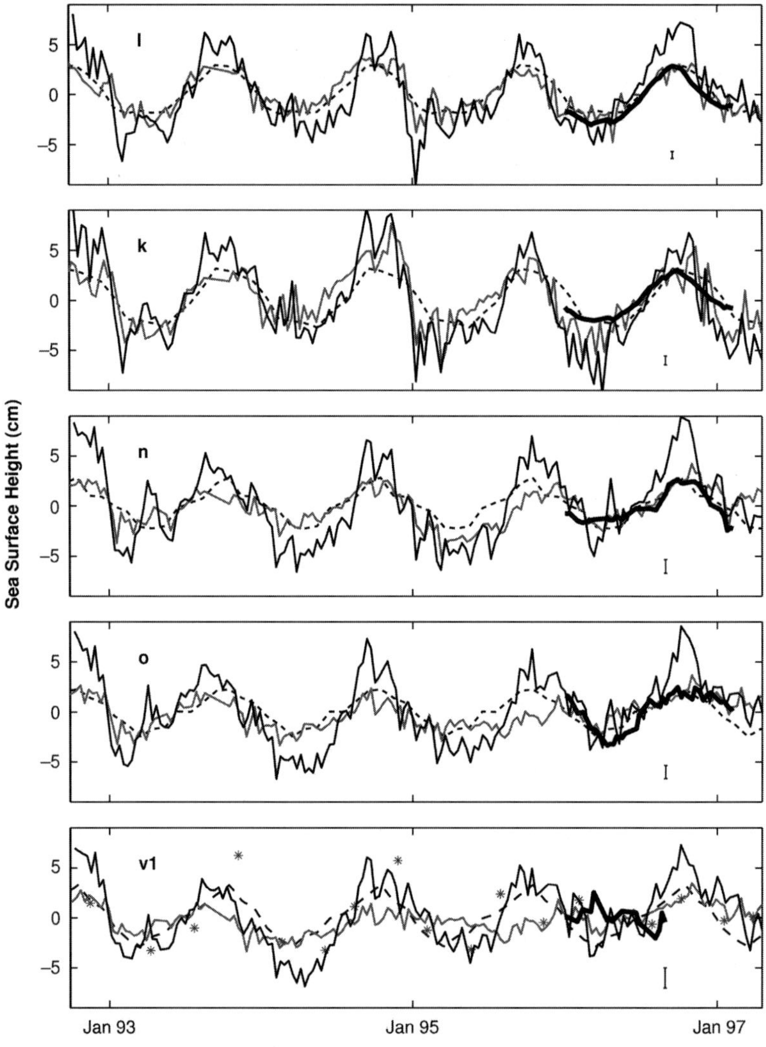

Fig. 1.21 The range-averaged sea level anomaly along the acoustic sections (the liters in the upper left corners specify the section) inferred by several independent methods: (i) thick black lines indicate the ATOC acoustic measurements converted to equivalent sea surface height for comparison with the altimeter data, (ii) thin black lines are from the TOPEX/POSEIDON altimeter data, (iii) dashed lines represent the climatological thermal anomaly converted to sea surface height, (iv) grey lines are GCM estimates, and (v) the asterisks along v1 are the XBT data. Uncertainties are indicated by vertical bars (in the lower left corners) for the acoustic estimates: the possible errors are largest along section v1 because the upper ocean variability is unresolved due to a lack of surface-reflecting rays near the receiver [The ATOC Consortium (1998)].

1.8 Summary

This chapter contains a background material. We began with a discussion of an idealized environmental model used throughout the book. In this model the ocean is represented as a 2-D frozen waveguide medium. At distances of order of hundreds and thousands kilometers, the sound field is formed by waves propagating at small grazing angles. Only such waves are captured in the underwater sound channel and survive at long ranges. Therefore, the Helmholtz equation describing the sound field can be approximately replaced by the standard parabolic equation. The latter formally coincides with the Schrödinger equation from quantum mechanics. In the limit of small wavelength the wave field can be examined using the geometrical optics approximation or ray approach. We introduced the Hamiltonian formalism for description of the ray dynamics. The Hamilton (ray) equations and main characteristics of the ray paths were expressed in terms of both momentum-position and action-angle variables. A special attention was paid to the ray travel times, that is, arrival times of sound pulses coming via individual ray paths. These characteristics of the sound field play an important role in the ocean acoustics. In some way they involve information about the waveguide condition and are used as observables in acoustic monitoring of the ocean temperature fields. In this chapter we considered general properties of the ray travel times in an unperturbed (range-independent) waveguide. Below we will see that some of them remain valid in the presence of perturbation giving rise to ray chaos.

We briefly considered the main factors affecting the propagation of the low frequency sound waves in a deep ocean. These are internal and Rossby waves, fronts, currents, streamers, and eddies. We also described a method of the ocean acoustic tomography which allows one to reconstruct some of these inhomogeneities on the basis of acoustic measurements.

The final part of the chapter was devoted to the review of experiments on the long-range sound propagation. These experiments demonstrate that the low-frequency acoustic signals can propagate over multimegameter ranges with comparatively small attenuation. It turns out that even at so long distances important features of the sound fields can be predicted in the scope of the geometrical optics approximation. In particular, sound pulses coming to the observation point through steep ray paths and forming the early portion of the arrival pattern in many cases may be resolved and identified. On the other hand, the experiments show that the sound speed fluctuations induced by random internal waves and other medium inhomogeneites strongly affect the long range wave propagation. These results suggest that the chaotic ray dynamics caused by the medium inhomogeneities should be taken into account for obtaining a comprehensive description of the sound wave fields at megameter ranges.

Chapter 2

Ray Chaos

Numerous experiments on long range sound propagation in the ocean, including those described in the preceding chapter, demonstrate that the sound speed fluctuations strongly influence the wave field. In the ocean acoustics it is believed that the dominant factor which determines the sound field variability at megameter ranges is the random internal waves [Flatté *et al.* (1979)]. In the 1990s it was realized that the internal-wave-induced fluctuations of the sound speed cause Lyapunov instability of the ray paths which is one of the main attributes of the ray chaos.

In this chapter we discuss mechanisms giving rise to ray chaos and consider manifestations of this phenomenon in a waveguide with periodic range dependence. The choice of this highly-idealized environmental model has two important advantages. First, the perturbation is expressed by a simple deterministic function. It allows for an analytical description in terms of theory of periodically-driven dynamical systems. Second, similar systems are studied in the theory of dynamical and quantum chaos and, therefore, we can apply methods derived in this theory. In other words, we can use a close analogy between ray/wave and classical/quantum phenomena. We have an opportunity to study wavefield counterparts of physical effects which are observed in quantum-mechanical systems and associated with quantum chaos. This issue will be addressed in the next chapter. This chapter is devoted entirely to the study of chaotic ray motion.

2.1 Hamiltonian Chaos

With the aim of self-consistency of the text, we begin this chapter with a brief description of Hamiltonian dynamics and chaos theory. Some of the concepts and notations of that theory are described in more detail in the Glossary in the end of the book. Rigorous results and their proofs can be found in a number of textbooks and monographs on dynamical systems theory and Hamiltonian chaos [Arnold *et al.* (1988); Lichtenberg and Liberman (1983); Zaslavsky (2005)].

2.1.1 *Dynamics of Hamiltonian systems*

Each state of a Hamiltonian system with N degrees of freedom in the phase space is described by N generalized positions $(q_1, \ldots, q_N)$ and momenta $(p_1, \ldots, p_N)$ which are pairwise canonically conjugated variables. The equations of motion of a Hamiltonian system are specified with the help of a Hamiltonian function of the generalized positions and momenta

$$\dot{\mathbf{q}}_i = \frac{\partial H}{\partial \mathbf{p}_i}, \quad \dot{\mathbf{p}}_i = -\frac{\partial H}{\partial \mathbf{q}_i}, \tag{2.1}$$

where the dots denote differentiation over time. If the Hamiltonian $H(p, q, t)$ depends on time, then the corresponding system can be studied in the enlarged $(2N+1)$-dimensional phase space $(q_1, \ldots, q_N; p_1, \ldots, p_N; t)$ where it has $N + 1/2$ degrees of freedom. The equations (2.1) satisfy to the incompressibility condition

$$\sum_{i=1}^{N} \left(\frac{\partial \dot{q}_i}{\partial q_i} + \frac{\partial \dot{p}_i}{\partial p_i} \right) = 0. \tag{2.2}$$

If one specifies a volume (a phase drop) of initial conditions, then this expression means that the phase fluid conserves its volume in course of time (the Liouville theorem). There are no attractors in the phase space of a Hamiltonian system. In spite of a conservation of a phase volume, it can be transformed in the process of evolution in a very complicated way to occupy all the energetically accessible region in the phase space.

The Hamilton equations (2.1) are not, in general, integrable. The Liouville-Arnold theorem (see the article "Invariant tori" in the Glossary) states that a Hamiltonian system with N degrees of freedom is *fully integrable* if there exist N linearly independent first integrals of motion C_i in involution, i.e., with zero Poisson brackets $\{C_i, C_j\} \equiv 0, i, j = 1, 2, \ldots, N$. Equations of motion (2.1) of a fully integrable system can be always transformed to the following form:

$$\dot{I}_i = -\frac{\partial H}{\partial \theta_i} = 0, \quad \dot{\theta}_i = \frac{\partial H}{\partial I_i} \equiv \omega_i \quad (I_1, \ldots, I_N), \tag{2.3}$$

where I_i and θ_i are pairwise canonically conjugated variables known as *action* and *angle*, respectively.

Trajectories of a Hamiltonian system with N integrals of motion lie on N dimensional invariant *manifolds* in the $2N$ dimensional phase space. The manifolds have torus topology and are called *invariant tori*. Any trajectory, starting on a given torus, stays on it all the time. If a Hamiltonian system is fully integrable, then the representation in terms of I_i and θ_i is global, i.e., the phase space is partitioned to invariant tori, and any trajectory is located on some torus. If a system is not integrable, then some trajectories do not lie on invariant tori. Up to now, there is no complete theory of behavior of nonintegrable Hamiltonian systems. However, there exists very important theorem about the behavior of Hamiltonian systems under weak perturbations.

The Kolmogorov-Arnold-Moser theorem (KAM theorem) states that under a sufficiently small conservative Hamiltonian perturbation a majority of *nonresonant invariant tori* does not disappear but they are slightly deformed in such a way that there appear invariant tori (called *KAM tori*) in the phase space of a nonintegrable Hamiltonian system filled up everywhere densely with (quasi)periodic trajectories [Arnold and Avez (1968); Arnold *et al.* (1988)]. The KAM theorem says nothing about the fate of resonant tori. It has been shown in numerous studies that they may break down and produce seeds of *Hamiltonian chaos*. Since resonant tori is a set of zero measure (the probability to find such a torus under a random choice of initial conditions is equal to zero), the KAM theorem can be formulated as follows: under a sufficiently small perturbation, almost all invariant tori of the integrable system under consideration are conserved. When proving the theorem, it is stated what is it "sufficiently small" and "almost all" [Arnold and Avez (1968); Arnold *et al.* (1988)].

A set of trajectories in the phase space with a number of initial conditions is called "a phase portrait" of a dynamical system. As an example, let us consider a nonlinear oscillator, namely, a physical pendulum, with the unit mass

$$H = \frac{1}{2}\dot{x}^2 - \omega_0^2 \cos x \tag{2.4}$$

and the equation of motion

$$\ddot{x} + \omega_0^2 \sin x = 0, \tag{2.5}$$

where ω_0 is the frequency of small oscillations. From Eq. (2.4) one can easily find the dependence of the velocity $\dot{x}$ on the position and construct the

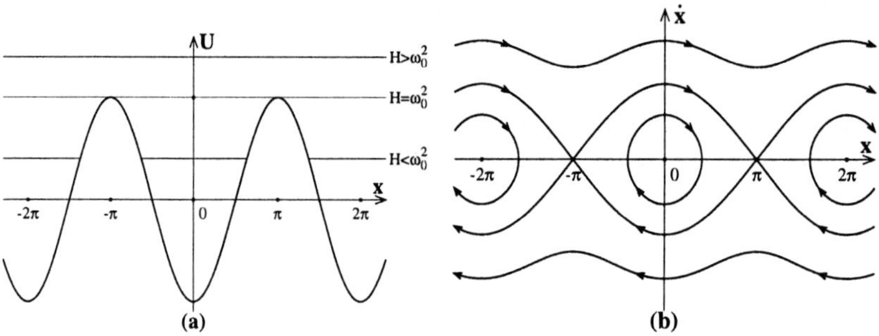

Fig. 2.1 (a) Potential energy of the physical pendulum U and (b) its phase portrait.

phase portrait shown in Fig. 2.1(b). The points of equilibrium (or stationary points) are found from the equation of motion (2.5): the elliptic points have coordinates ($x = 2\pi n$, $\dot{x} = 0$) and the hyperbolic ones (saddles) have coordinates ($x = 2\pi(n+1)$, $\dot{x} = 0$), where $n = 0, 1, \ldots$. If $H < \omega_0^2$, then the motion is oscillatory-like. If $H > \omega_0^2$, then the pendulum rotates around the fixed point. At the value $H = \omega_0^2$, the phase point moves along a separatrix which is a special trajectory connecting hyperbolic points and separating bounded and unbounded regimes of motion. The period of motion on the separatrix is infinite since the velocity in equilibrium points is zero by definition. The phase portrait of the physical pendulum is of a heteroclinic type with the separatrix leaving one saddle point and coming in the other one. After computing the phase portrait, one studies stability of motion in the neighbourhood of stationary points using linearized equations of motion (see the analysis in the article "Stationary points" in the Glossary). If small deviations from a stationary point rotate around it and the corresponding phase trajectories are ellipses, the stationary point is called *a stable* or *an elliptic* one. If the motion nearby such a stationary point is unstable, it is called *a hyperbolic* or *a saddle* point. Chaotic motion is characterized by an exponential sensitivity to small variations in initial conditions/parameters. It means that initially closed trajectories may diverge exponentially fast in time. *Lyapunov exponents* serve a measure of mean velocity of exponential divergence (convergence) of close trajectories. In course of time a small initial phase volume stretches mostly in the direction corresponding to a maximal Lyapunov exponent which is to some extent a quantitative measure of chaos.

Even a small perturbation of an integrable system may cause strong changes in the system's motion. As we know from the KAM theorem, non-resonant invariant tori does not disappear under a small perturbation but they are slightly deformed. The fate of resonant tori is more dramatic: some of them break down producing at their places chains of resonances or *islands of stability* (see the corresponding article in the Glossary). So called *Poincaré map* helps to visualize such a complicated motion. Instead of tracking individual trajectories, which may be very entangled, let us fix coordinates of a phase point at specified time moments or when it crosses a given surface in the phase space. The corresponding trajectory is represented by a set of points in the phase space or on one of its projections. If the system is perturbed by an external force with the period T, it is convenient to fix coordinates of phase points at the moments of time multiplied by the period $t_n = nT$ $(n = 1, 2, \ldots)$ and project them to a plane constructing in such a way *a Poincaré section*. The Poincaré sections are usually computed with a large number of initial conditions and a few values of the control parameters.

Hamiltonian chaos is dynamical chaos in Hamiltonian systems. A deterministic dynamical system is called chaotic if it is locally unstable, i.e., it has at least one positive Lyapunov exponent and globally mixed (see the definition of mixing below). Local instability produces an exponential sensitivity of trajectories to small variations of initial conditions and/or control parameters. Manifestations of dynamical chaos become evident when computing Poincaré sections, Melnikov integrals, intersections of stable and unstable manifolds, and maximal Lyapunov exponents. The phase space in a typical Hamiltonian system is not homogeneous, i.e., the regions with regular motion coexist with chaotic ones. KAM tori are stable invariant manifolds with boundaries that are impenetrable to particle's transport in the phase space. The special topological objects, *cantori* (see the corresponding article in the Glossary), are Cantor-like unstable invariant sets with gaps transport through which is possible but difficult. Hamiltonian chaos is a special type of motion with the properties both of regular motion (due to determinism of equations of motion) and stochastic motion (due to local instability of trajectories).

Chaotic Hamiltonian systems are typical whereas integrable ones are exceptional. A Hamiltonian system with one degree of freedom, for example, a nonlinear oscillator described by Eq. (2.4), may be chaotic under a periodic perturbation. Typically, a nonstationary perturbation breaks an unperturbed separatrix, and a narrow stochastic layer appears at its place even at a small perturbation amplitude. In an integrable Hamiltonian system separatrices connect either two hyperbolic stationary points in such a way that the stable separatrix of one point $W_{\rm s}^{(0)}(h_1)$ coincides with the unstable separatrix of the other point $W_{\rm u}^{(0)}(h_2)$ and vice versa (see Fig. 5.2(a)) or $W_{\rm s}^{(0)}$ and $W_{\rm u}^{(0)}$ of the same hyperbolic point coincide (see Fig. 5.2(b)). In the former case one gets *heteroclinic connection*, whereas in the latter one — *homoclinic connection.*

Under a periodic perturbation, hyperbolic (saddle) points h of the unperturbed system become unstable periodic trajectories $\gamma(t)$ with stable and unstable separatrix branches which are called stable $W_{\rm s}(\gamma)$ and unstable $W_{\rm u}(\gamma)$ *manifolds* of the corresponding hyperbolic trajectory (see the corresponding article in the Glossary). In the enlarged phase space the manifolds $W_{\rm s}(\gamma)$ and $W_{\rm u}(\gamma)$ are surfaces which do not coincide but intersect each other (see Fig. 5.3(c)). The corresponding curves $W_{\rm s}(\gamma)$ and $W_{\rm u}(\gamma)$ intersect each other on a Poincaré section in *homoclinic points.* H. Poincaré proved that given one homoclinic point there is an infinite number of homoclinic points of intersections of stable and unstable manifolds of a hyperbolic trajectory. Any point of an invariant manifold maps, by definition, on the Poincaré section to another point of the same manifold. When moving away from the hyperbolic point, the amplitude of oscillations of the curve $W_{\rm u}$ increases. When approaching the same or another hyperbolic point, "the period" of oscillations decreases (the successive distances between the points of intersections of $W_{\rm s}$ and $W_{\rm u}$ decrease when approaching to h) because of slowing down of the motion nearly h. It results in a complicated *homoclinic* or *heteroclinic* structure and forming a stochastic layer.

The Melnikov's method (see the corresponding article in the Glossary) provides an analytical study of the motion nearby the unperturbed separatrix of a dynamical system close to an integrable one. This method allows to prove analytically intersections of stable and unstable manifolds and calculate a distance between them which is a measure for the width of a stochastic layer. The idea and the results obtained by Melnikov (1963) are based on using global solutions of the corresponding unperturbed system when calculating perturbation solutions.

2.1.2 *Statistical description of Hamiltonian chaos*

Given initial positions $\vec{q}_0$ and momenta $\vec{p}_0$ of a dynamical system, one can solve the Hamilton equations and compute the corresponding trajectory. Instead of that we may study a set of identical dynamical systems with a variety of initial conditions. The Gibbs ensemble is a cloud of points in the phase space with different initial conditions that is described by a probability density function (PDF) $\rho(\vec{q}, \vec{p}, t)$ with the normalization

$$\int_{\Gamma} \rho(\vec{q}, \vec{p}, t) d\Gamma = 1 \tag{2.6}$$

and integration over all the accessible phase space Γ. The quantity $\rho(\vec{q}, \vec{p}, t) d\Gamma$ means the probability to find a phase point at the moment of time t in a small volume of the phase space $d\Gamma$ in a neighbourhood of the point $(\vec{q}, \vec{p})$. In particularly, one gets for one particle

$$\rho(\vec{q}, \vec{p}, t) = \delta(\vec{q} - \vec{q}(t)) \delta(\vec{p} - \vec{p}(t)), \tag{2.7}$$

where $\vec{q}(t) = q(t, \vec{q}_0, \vec{p}_0)$ and $\vec{p}(t) = p(t, \vec{q}_0, \vec{p}_0)$ defines a trajectory of that particle. The PDF $\rho(\vec{q}, \vec{p}, t)$ is a density of the phase fluid whose streamlines are trajectories of the ensemble elements. The Liouville equation follows from the incompressibility condition (2.2)

$$\frac{\partial \rho}{\partial t} + \dot{\vec{q}} \frac{\partial \rho}{\partial \vec{q}} + \dot{\vec{p}} \frac{\partial \rho}{\partial \vec{p}} = 0, \tag{2.8}$$

and does not contain any new information as compared with the Hamiltonian equations. In other words, description on the language of individual trajectories is equivalent to the statistical description.

The fruitfulness of statistical approach is well known for describing systems with astronomically large degrees of freedom. Description in terms of individual trajectories is practically impossible because of the lack of knowledge of exact values of initial positions and moments of a large number of particles. Introducing a randomness in the Hamiltonian equations (random-phase approximation for colliding particles), we get a kinetic equation for the PDF providing a coarse-grained description of a system with a large number of degrees of freedom.

When studying chaotic dynamical systems even with a few degrees of freedom, we are forced to leave attempts to describe chaotic motion in terms of individual trajectories. In spite of the fact that equations of motion are fully deterministic, their solutions may be so unstable that it is practically impossible to predict the behavior of chaotic trajectories, and one is forced to use statistical methods. In a chaotic system a distance between trajectories $\|\delta \vec{r}(t)\|$ grows exponentially in time for almost all initial conditions $\delta \vec{r}(0)$

$$\|\delta \vec{r}(t)\| = \|\delta \vec{r}(0)\| e^{\nu t}, \tag{2.9}$$

where ν is a positive Lyapunov exponent and $\|\cdot\|$ is a norm of the vector. Exact direction of the vector $\delta\vec{r}(0)$ does not matter because majority of such vectors (in the sense of nonzero measure) evolves in the direction of a maximal stretching of the corresponding phase volume. The quantity ν characterizes asymptotically (at $t \to \infty$) the average rate of that stretching and is called sometimes *a global Lyapunov exponent.* In a Hamiltonian system that volume shrinks along another direction with the same rate but of the opposite sign. The sum of all the Lyapunov exponents in a Hamiltonian system is zero. It follows from Eq. (2.9) that chaos restricts our ability to forecast the behavior of the system to the time interval of the order

$$\tau_p \approx \frac{1}{\nu} \ln \frac{\Delta a}{\Delta a(0)}, \tag{2.10}$$

where Δa is a confidence interval of a variable a and $\Delta a(0)$ is a practically inevitable inaccuracy in measuring of that quantity. The time interval τ_p is called *a predictability horizon.* It should be stressed that the dependence of the predictability horizon on the lack of knowledge is logarithmic, i.e., it is much weaker than on the measure of instability. Simply speaking, with any degree of accuracy in initial conditions there is a time interval beyond which the predictability is impossible, and that time interval may be rather small for chaotic dynamical systems.

Instead of tracking an individual trajectory, it is reasonably, for example, to ask how frequently "a typical" trajectory visits a specified domain in the phase space. If a dynamical system is a discrete map, for example, *the Bernoulli shift* (see the corresponding article in the Glossary), the computational algorithm may be the following. Choose in a random way "a typical" initial point, equipartition the unity interval, and compute how many times iterations of the initial point visit a specified segment of the unity interval. Dividing the number obtained to the total number of iterations (it should be large enough), one gets a number meaning the frequency of visiting the segment by "a typical" trajectory of the Bernoulli shift.

The phase space of a continuous dynamical system is partitioned by cells with an equal volume $\Delta\Gamma$. One computes an amount of time "a typical" trajectory spends in the i-th cell $\Delta\Gamma_i$, in other words, one computes the ratio of an amount of time t_i the trajectory spends in the i-th cell to the total amount of time t: $\rho_i(t) = \lim_{t\to\infty}(t_i/t)$. It is *a probability density,* and its integral $\mu(\Delta\Gamma_i) = \int_{\Delta\Gamma} \rho\, d\Gamma$ is a probability for trajectories to come to a given region in the phase space $\Delta\Gamma_i$. A real nonnegative number μ is *a measure* of that region. A dynamical system can be prescribed a measure in the phase space just like in probability theory one introduces a measure in the space of random events that is equal to a probability of an event in the case of the discrete space and to a probability density in the continuous space.

A method of prescribing a real nonnegative number μ to every subset A in the phase space, may be called "a measurement", and the very number $\mu(A)$ is called *a measure of the set A*. It is possible to introduce a large number of different measures. A measure is "good" if almost each initial condition generates a trajectory which "measures" in the limit $t \to \infty$ every subset in the phase space in an equal way. There are, however, initial conditions generating periodic trajectories which have no such a property. If such and other similar initial conditions (of zero measure) are ignored, then a limit value of the amount of time, a trajectory spends in a given region on the phase space, does not depend on the initial conditions. Measures with such a property are called *natural* or physical ones. If the measure of all the phase space is equal to 1, it is called *a probabilistic* one. If it is not changed under time shifts, the corresponding measure is *invariant*. Assuming the existence of a natural probabilistic invariant measure, we can use statistical methods to specify dynamical systems, namely, to introduce and compute different distribution functions, mean values, variations and high moments, correlation function, entropies, etc. The existence of an invariant measure allows to compute Lyapunov exponents because time averages along a given direction are the same with all the initial conditions excepting a zero-measure set.

The motion is called *ergodic* if the time average of a given quantity, for example, a smooth function of positions and momenta $\vec{f}(\vec{q},\vec{p},t)$, is equal to its average over the phase space

$$\lim_{t\to\infty} \frac{1}{T} \int\limits_{t}^{t+T} \vec{f}(\vec{q}(\tau),\vec{p}(\tau))\, d\tau = \int\limits_{\Gamma} \vec{f}(\vec{q},\vec{p})\, d\Gamma. \tag{2.11}$$

The limit in this expression does not depend on the initial point of the interval of averaging. The simple example of ergodic motion is a motion of two coupled harmonic oscillators with irrational ratio of their natural frequencies. A trajectory in the phase space covers the surface of a two-dimensional torus uniformly and densely everywhere. Ergodicity means that each invariant subspace of the phase space has either a zero or full measure. In autonomous Hamiltonian systems there is, at least, one first integral of motion, energy, which reduce dynamics to the motion on an energetic hypersurface. In an ergodic dynamical system a drop of the phase fluid, conserving its volume and form, moves over all the accessible phase space. Energetic spectra of ergodic systems are discrete.

Mixing is a key notion in ergodic theory. Let B is a region with dye in the phase space A. The volume of B at $t = 0$ is $V(B_0)$. Let C is another region in A. The amount of dye in C is $V(B_t \cap C)$ at the moment of time t and its concentration in C is $V(B_t \cap C)/V(C)$. The definition of global mixing is: $V(B_t \cap C)/V(C) - V(B_t)/V(A) \to 0$ at $t \to \infty$, i.e., the concentration of dye in any region C is the same as in the whole phase space A. Recall that in Hamiltonian systems the phase fluid is incompressible, i.e., $V(B_t) = V(B_0)$. In a system with mixing a drop of phase fluid conserves its volume in course of time but is deformed strongly becoming a long thin filament penetrating in all the energetically accessible regions in the phase space. Energetic spectra of systems with mixing are continuous. Mixing means ergodicity but not vice versa.

Another key notion in ergodic theory is entropy. The entropy $S \sim \ln \Delta\Gamma(t)$ of a drop of phase fluid with the volume $\Delta\Gamma_0$ is a constant in time because of its conserving volume. However, the form of the drop changes strongly due to mixing becoming more "porous" with increasing effective volume. To characterize such an inflation of the effective phase volume a coarse graining of the phase space is introduced. All the points of that phase volume are replaced with spheres of a small radius ε. Joining all the spheres, we obtain a coarse-grained volume $\Delta\bar{\Gamma}(t)$, whose magnitude changes in time at the expense of inflation in difference from the true volume. In accordance with the formula for a local instability (2.9), we get $\Delta\bar{\Gamma}(t) = \Delta\Gamma_0 \exp(h_0 t)$ $(h_0 > 0)$. Taking the initial volume to be $\Delta\Gamma_0 = \varepsilon$, we define the quantity

$$h = \lim_{\varepsilon \to 0} \lim_{t \to \infty} \frac{1}{t} \ln \Delta\bar{\Gamma}(t), \tag{2.12}$$

which is called *Krylov-Kolmogorov-Sinai entropy* or simply *KS entropy*. It does not depend on ways of coarse graining and partition of the phase space, i.e., it is a metric invariant of the system under consideration. The KS entropy gives a volume-averaged increment of local instability $h = \langle h_0 \rangle$. Pesin proved that the KS entropy h is a sum of all positive Lyapunov exponents.

Dynamical systems with positive values of the KS entropy are called *K-systems*. In fact, K-systems are the systems with chaos (or mixing). Almost all dynamical systems are K-systems, i.e., systems, which have in a finite range of control parameters regions in the phase space with $h > 0$, constitute a set of the full measure in the space of dynamical systems. Systems with global mixing ($h > 0$ everywhere in the phase space) are called *hyperbolic*. Majority of Hamiltonian dynamical systems are nonhyperbolic systems, i.e., they have a mixed phase space with coexisting regions of chaotic and regular motion.

Movement of phase points from one region of the phase space to another one is called *transport*. In integrable systems transport is trivial in the sense that the motion is regular and fully predictable. In chaotic deterministic systems due to local instability, the motion becomes practically unpredictable on the time scale longer than the predictability horizon (2.10) defined, mainly, by the maximal Lyapunov exponent. Transport in chaotic systems should be described in statistical terms.

The common statistical measure of transport is *a dispersion* which is simply $\sigma^2 = \langle x^2 \rangle$ for a one-dimensional process. The averaging is over an ensemble of initial conditions (trajectories). In hyperbolic dynamical systems the dispersion is normal, it increases in time linearly, $\sigma^2 = 2Bt$, with a well-defined diffusion coefficient $B = \lim_{t\to\infty} \sigma^2/2t$. Statistics of transport in hyperbolic systems is equivalent to the statistics of Brownian motion. The PDF $P(x,t)$ for a Brownian particle is governed by the standard diffusion equation

$$\frac{\partial P}{\partial t} = \frac{B}{2} \frac{\partial^2 P(x,t)}{\partial^2 x}, \tag{2.13}$$

with the Gaussian as its solution with the initial condition $x_0 = 0$

$$P(x,t) = \frac{1}{\sqrt{2\pi B t}} \exp(-x^2/2Bt). \tag{2.14}$$

All the moments of this function are finite. The normal distribution is a narrow distribution with the tails decaying fast. The area under a Gaussian (2.14), limited by straight lines $x = \pm 3$, is equal to 99.7 percents of the total area. In other words, probability to find a Brownian particle at the distance $|x| > 3$ from its average distance $\langle x \rangle = 0$ is exponentially small. Thus, transport in hyperbolic Hamiltonian systems has the properties of normal diffusion.

Transport in typical Hamiltonian systems with mixed phase space is much more complicated, mainly, due to dynamical traps [Zaslavsky (2005)]. *A dynamical trap* is a domain in the phase space where a particle (and its trajectory) may spend an arbitrary long but finite time in spite of the fact that the trajectory is chaotic in any relevant sense. Strictly speaking, it is the definition of a quasitrap. Absolute traps, where particles could spend an infinite time, are not possible in Hamiltonian systems which have no attractors. The dynamical traps are caused by *stickiness* of trajectories, mainly, to the boundaries of islands of stability where *cantori* (see the corresponding article in the Glossary) are situated. Dynamical traps influence significantly transport in Hamiltonian systems specifying its anomalous statistical properties. In particular, tails of the corresponding PDFs decay slower than a Gaussian, i.e., algebraically but not exponentially.

Probability to find very large deviations in the process with algebraic PDF is not negligibly small as for the normal distribution. Let us describe shortly general features of the algebraic distribution

$$P(x,t) = Ax^{-(1+\alpha)}, \tag{2.15}$$

where α is a positive number. It is easily to check that at $\alpha \leq 1$ the average value of x is infinite. At $\alpha \leq 2$, one gets the diverging second moment, $\langle x^2 \rangle = \infty$, and at $\alpha \leq k$ the diverging k-th moment $\langle x^k \rangle = \infty$. The power-law function (2.15) looks like a straight line in a doubly logarithmic scale, i.e., it scales uniformly in all ranges of the argument. Such an invariant scaling means that a trajectory has no dominant scale in difference from Brownian particles for which σ plays the role of the dominant scale. A power-law function is self-similar, i.e., changing the scale $x \to lx$ (a similarity transformation) one changes only the coefficient in (2.15). A random walking process with the PDF having an infinite second moment is called *a Levy flight*. Frequently, quasiregular fragments of a single chaotic trajectory are called Levy flights. Sticking of trajectories to the boundaries of stability islands is one of the reasons of Levy flights in chaotic Hamiltonian systems.

2.2 Lyapunov Instability

Lyapunov instability plays an especial role in the occurrence of ray chaos and is worthy of a detailed consideration. Let's explore the evolution of a narrow ray bundle in an arbitrary USC. For the sake of simplicity, we restrict ourselves by the two-dimensional case. Assume that the bundle is centered at the ray with initial depth z_0 and the ray momentum p_0. The phase space distance from some arbitrarily chosen probe ray in the bundle to the center ray can be expressed as the vector $\vec{d} = (\delta z,\, \delta p)$. Since the rays obey

the Hamiltonian equations, the range evolution of $\vec{d}$ can be expressed as

$$\vec{d}(r) = \hat{A}(r)\vec{d}(0), \tag{2.16}$$

where $\hat{A}$ is the Jacobian matrix of the Hamiltonian flow

$$\hat{A}(r) = \begin{pmatrix} \left.\dfrac{\partial z(r)}{\partial z(0)}\right|_{p_0} & \left.\dfrac{\partial z(r)}{\partial p(0)}\right|_{z_0} \\ \left.\dfrac{\partial p(r)}{\partial z(0)}\right|_{p_0} & \left.\dfrac{\partial p(t)}{\partial p(0)}\right|_{z_0} \end{pmatrix}. \tag{2.17}$$

$\hat{A}$ is also known as a *stability matrix.* Its determinant is equal to 1. This reflects the conservation of the phase-space volume along the Hamiltonian flow. One of the consequences of this property is the conservation of the acoustic energy flux along a ray path. Evolution of the stability matrix is governed by the equation

$$\frac{d\hat{A}}{dr} = \hat{J}\hat{A}, \tag{2.18}$$

with the initial condition

$$\hat{A}(r = 0) = \hat{I} = \begin{pmatrix} 1 & 0 \\ 0 & 1 \end{pmatrix}. \tag{2.19}$$

The matrix $\hat{J}$ is given by the equation

$$\hat{J} = \begin{pmatrix} -\dfrac{\partial^2 H}{\partial p \partial z} & -\dfrac{\partial^2 H}{\partial z^2} \\ \dfrac{\partial^2 H}{\partial p^2} & \dfrac{\partial^2 H}{\partial p \partial z} \end{pmatrix}. \tag{2.20}$$

In the small-angle approximation, the ray Hamiltonian looks as

$$H = \frac{p^2}{2} + U(z, r), \tag{2.21}$$

where $U(z, r) = -1 + c(z, r)/c_0$, and the matrix (2.20) reduces to

$$\hat{J} = \begin{pmatrix} 0 & -\dfrac{\partial^2 U}{\partial z^2} \\ 1 & 0 \end{pmatrix}. \tag{2.22}$$

Eigenvalues of the stability matrix can be found from the characteristic equation

$$\det[\hat{A} - \lambda\hat{I}] = 0. \tag{2.23}$$

In a two-dimensional waveguide, equation (2.23) has two roots, λ_1 and λ_2, linked by the relation $\lambda_1\lambda_2 = 1$. There are the following three possibilities:

(1) λ_1 and λ_2 is a complex-conjugated pair $e^{i\alpha}$ and $e^{-i\alpha}$. Then the probe ray winds nearby the center ray. It implies that the respective bundle is *stable.*

(2) λ_1 and λ_2 are real and positive with the sum $\lambda_1 + \lambda_2 > 2$. Then the vector $\vec{d}$ evolves as

$$\vec{d}(r) = [\lambda_1(r) + \lambda_2(r)]\vec{d}(0) = (e^{\nu r} + e^{-\nu r})\vec{d}(0), \quad \nu > 0, \tag{2.24}$$

so that the probe and center rays diverge rapidly. Thus, the motion nearby the center ray is *unstable*: any compact bundle of rays will spread with an exponential rate. The quantity ν is called the *stability exponent* or *finite-range Lyapunov exponent.* It depends on initial conditions for a ray trajectory, as well as on the range r.

(3) $\lambda_1 = \lambda_2 = 1$. This is a boundary case between the stable and unstable regimes.

The long-range asymptotics of the finite-range exponent is the so-called global Lyapunov exponent

$$\nu_0 = \lim_{\substack{r\to\infty \\ \|\vec{d}(0)\|\to 0}} \frac{1}{r} \ln \frac{\|\vec{d}(r)\|}{\|\vec{d}(0)\|}. \tag{2.25}$$

In some cases, it is suitable to express the global Lyapunov exponent by means of the trace of the stability matrix

$$\nu_0 = \lim_{r\to\infty} \frac{\ln\left\|\mathrm{Tr}(\hat{M})\right\|}{r}. \tag{2.26}$$

In the case of ergodic ray motion, the global Lyapunov exponent doesn't depend on ray initial conditions and has an unique value characterizing the whole system. Ergodicity implies that the system's phase space doesn't contain regions with impenetrable boundaries, and in the long-range limit ray trajectories cover the phase space uniformly. If the ergodicity is violated, and the phase space is divided into different areas with impenetrable boundaries, then each such area is characterized by its own global Lyapunov

exponent. Obviously, the long-range limit is unphysical in ocean acoustics, because we always deal with finite ranges. Therefore, finite-range Lyapunov exponents seem to be more relevant characteristics of ray stability.

However, when we deal with finite-range Lyapunov exponents, we must take care about the severe constraints posed by ray chaoticity. Any finite-range Lyapunov exponent is determined by a particular ray trajectory. If a ray considered is unstable, and, moreover, chaotic, then its trajectory is extremely sensitive to infinitesimal variations of initial conditions. All analytical schemes of solving the ray equations fail under chaotic conditions. Generally speaking, the only hope to reproduce such a ray is to integrate numerically the ray equations. However, now one encounters the following problem. Any computer has a finite accuracy in setting the initial conditions, therefore, an undesirable small error inevitably occurs. Under integration, this error will grow exponentially with range, deflecting the computed ray from the "exact" one farther and farther. In the case of strong instability the difference may appear significant even at small ranges.

However, this difficulty can be partially overcome in the way ensuing from the *shadowing theorem*. In a simplified version, it can be expressed in the following way: although a numerically computed chaotic trajectory diverges exponentially from the true trajectory with the same initial coordinates, there exists an errorless trajectory with a slightly different initial condition that stays near ("shadows") the numerically computed one [Ott (1993)]. Owing to that, an efficient modeling of chaotic rays infers a consideration of an ensemble of rays with close initial conditions instead of a computation of a single ray. In other words, computation of an unstable ray should be thought of as a numerical experiment which returns only some approximation to the exact trajectory, and any physical conclusions can be made only after sampling over an ensemble of rays. A finite-range Lyapunov exponent calculated along a particular unstable ray doesn't carry significant information about ray dynamics (at least, if the range is not very large!), because it is an unstable quantity as well.

To illustrate that, consider the model of an USC with the sound-speed profile given by the expression

$$c(z,\,r) = c_0\left[1 + \frac{b^2}{2}\left(e^{-az} - \eta\right)^2\right] + \varepsilon c_0 \frac{z}{B} e^{2z/B} \sin k_z z \sin k_r r. \qquad (2.27)$$

This is the deep-water biexponential model, introduced in Sec. A.4, with a range-periodic perturbation imposed. The parameters of the perturbation are: $\varepsilon = 0.002$, $B = 1$ km, $k_z = 2\pi/200\,\mathrm{m}^{-1}$, $k_r = 2\pi/5000\,\mathrm{m}^{-1}$.

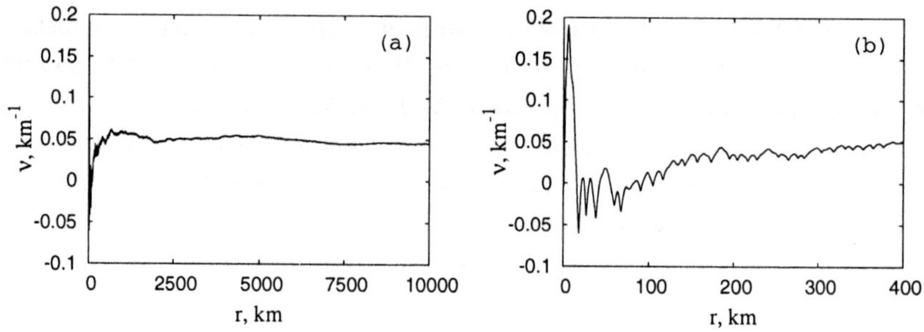

Fig. 2.2 (a) Finite-range Lyapunov exponent as a function of range for the range-periodic deep-water BEP model described by Eq. (2.27). (b) The enlarged view of the transient regime at low ranges.

This model is characterized by strong instability of rays propagating with small angles with respect to the channel axis (this issue will be discussed in Sec. 2.5). Figure 2.2(a) shows how the finite-range Lyapunov exponent varies with range for one of near-axial rays. We can see that, after a transient regime with strong fluctuations, the finite-range Lyapunov exponent achieves a plateau with $\nu_0 \simeq 0.046$. It should be noted that the transient regime lasts over several hundreds of kilometers (see Fig. 2.2(b)). It is very long even compared with typical oceanographical length scales. Rather surprisingly, the plateau doesn't correspond to the global Lyapunov exponent. Extending the range of intergration to 10^6 km, we have found that $\nu_0 \simeq 0.03$. The difference indicates on the intermittent ray motion: a more unstable regime is alternating with a less unstable, may be even almost

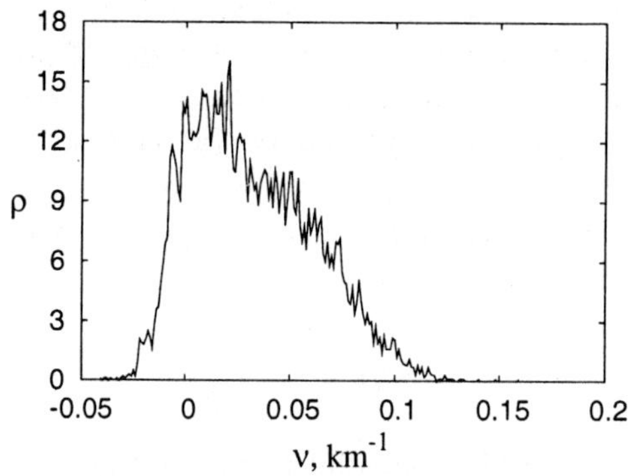

Fig. 2.3 Probability density function of finite-range Lyapunov exponents calculated over the distance of 100 km.

stable, and each of these regimes persists for very long ranges. Intermittency is well revealed in the distribution of finite-range Lyapunov exponents. For example, the distribution of finite-range Lyapunov exponents calculated over 100 km is concentrated in the range of low or even negative values, as it is demonstrated in Fig. 2.3, although the distribution is broad. With increasing range that distribution becomes more and more narrow, and in the limit $r \to \infty$ it collapses to the delta-function with the peak at $\nu = \nu_0$.

2.3 Ray-Medium Resonance

In this and the next sections we shall consider the simplest scenario of the onset of global ray chaos in an USC. Here the term "global" means that a chaotic ray can move inside a wide domain in the phase space, that domain is called as a "chaotic sea". During this motion, the topology of the ray can undergo significant changes.

In the paraxial approximation, the ray Hamiltonian looks as

$$H(p,z,r) = \frac{p^2}{2} + U(z) + V(z,r), \tag{2.28}$$

where

$$U(z) = \frac{\Delta c(z,r)}{c_0}, \quad V(z,r) = \frac{\delta c(z)}{c_0}. \tag{2.29}$$

Assume that a sound-speed perturbation has a small amplitude, and is a periodic function of range, that is

$$V(z, r+\lambda_r) = V(z,r) = \varepsilon F(z,r), \quad \varepsilon \ll 1. \tag{2.30}$$

In addition, let the sound-speed perturbation be a smoothly-varying function of depth, i.e.

$$\frac{dV}{dz} = \mathrm{O}(\varepsilon). \tag{2.31}$$

Such a sound-speed perturbation simulates the effect of the first-order mode of the internal-wave field. Transforming the depth-momentum ray variables $z-p$ to the action and angle, we represent the perturbed ray Hamiltonian as

$$H = H_0(I) + \varepsilon F(I, \vartheta, r), \tag{2.32}$$

where $\varepsilon \ll 1$ (perturbation is weak), the term $H_0(I)$ corresponds to the range-independent part of the Hamiltonian. The Hamiltonian ray equations in terms of the action-angle variables take the form

$$\frac{dI}{dr} = -\varepsilon\frac{\partial F}{\partial\vartheta}, \quad \frac{d\vartheta}{dr} = \omega(I) + \varepsilon\frac{\partial F}{\partial I}, \tag{2.33}$$

where $\omega = dH_0/dI$ is the spatial frequency of ray oscillations in the unperturbed waveguide. The function F can be represented by means of the Fourier series

$$F = \frac{1}{2}\sum_{l,m} F_{l,m}(I)e^{i(l\vartheta - mk_r r)} + \text{c.c.}, \qquad F^*_{l,m} = F_{-l,-m}, \tag{2.34}$$

where $k_r = 2\pi/\lambda_r$, and the symbol c.c. denotes complex conjugation. Then, the Hamiltonian equations (2.33) take the form

$$\frac{dI}{dr} = -\frac{i\varepsilon}{2}\sum_{l,m} lF_{l,m}(I)e^{i(l\vartheta - mk_r r)} + \text{c.c.} \tag{2.35}$$

and

$$\frac{d\vartheta}{dr} = \omega(I) + \frac{i\varepsilon}{2}\sum_{l,m}\frac{\partial F_{l,m}}{\partial\vartheta}e^{i(l\vartheta - mk_r r)} + \text{c.c.} \tag{2.36}$$

The ray-medium resonance condition implies that the following equation must be satisfied:

$$l\omega(I) - mk_r = 0, \tag{2.37}$$

where l and m are two integers. This means that we have to specify a pair of integers, (l_0, m_0), and the corresponding value, I_0, such that (2.37) converts into the following identity

$$l_0\omega(I_0) = m_0 k_r. \tag{2.38}$$

As a rule, the values $(l_0, m_0; I_0)$ can be found in abundance. This is due largely to the nonlinearity of ray oscillations, that is, to the dependence of ω on I. For brevity, we shall refer to the ray-medium resonance with some l_0 and m_0 as the $l_0 : m_0$ resonance.

The way to deal with Eqs. (2.35) and (2.36) is to analyze certain simplified situations. First, we examine an isolated resonance (2.38) and ignore all other possible resonances. This means that in the equations of motion

(2.35) and (2.36) only terms with $l = \pm l_0$, $m = \pm m_0$ should be retained which satisfy the resonance condition at $I = I_0$

$$\frac{dI}{dr} = \varepsilon l_0 F_0 \sin(l_0\vartheta - m_0 k_r r + \phi_0), \tag{2.39}$$

$$\frac{d\vartheta}{dr} = \omega(I) + \varepsilon \frac{\partial F_0}{\partial I} \cos(l_0\vartheta - m_0 k_r r + \phi_0), \tag{2.40}$$

where we set

$$F_{l,m} = |F_{l,m}| e^{i\phi_0} = F_0 e^{i\phi_0}. \tag{2.41}$$

It is also assumed that the deviation of the action from the resonance value

$$\Delta I = I - I_0 \tag{2.42}$$

is small, that is, Eq. (2.38) is examined in the vicinity of the resonance. In order to simplify Eqs. (2.39) and (2.40), the following approximations are used further:

(1) Set $F_0 = F_0(I_0)$ on the right-hand sides of Eqs. (2.39) and (2.40);
(2) Expand the spatial frequency $\omega(I)$ as follows:

$$\omega(I) = \omega_0 + \omega_I' \Delta I, \tag{2.43}$$

where $\omega_0 = \omega(I_0)$, $\omega_I' = d\omega(I_0)/dI$;
(3) Neglect the second term of the order of ε in Eq. (2.40).

Finally, we reduce Eqs. (2.39) and (2.40) to a simplified version

$$\begin{aligned} \frac{d}{dr}(\Delta I) &= -\varepsilon l_0 F_0 \sin\Psi, \\ \frac{d}{dr}\Psi &= l_0 \omega_I' \Delta I, \end{aligned} \tag{2.44}$$

where a new phase is introduced

$$\Psi = l\vartheta - mk_r r + \phi_0 - \pi. \tag{2.45}$$

Equations (2.44) can be presented in a Hamiltonian form:

$$\begin{aligned} \frac{d}{dr}(\Delta I) &= -\frac{\partial \tilde{H}}{\partial \Psi}, \\ \frac{d}{dr}\Psi &= \frac{\partial \tilde{H}}{\partial(\Delta I)}, \end{aligned} \tag{2.46}$$

where

$$\tilde{H} = \frac{1}{2} l_0 \omega_I' (\Delta I)^2 - \varepsilon l_0 F_0 \cos \Psi \tag{2.47}$$

is the so-called *universal Hamiltonian of nonlinear resonance.* Equations (2.46) can be rewritten as the pendulum-like equation

$$\frac{d^2\Psi}{dr^2} + \Omega_0^2 \sin \Psi = 0, \tag{2.48}$$

where the quantity Ω_0, being the analogue of the frequency of small-amplitude oscillations of a pendulum, is given by

$$\Omega_0 = l_0 \sqrt{\varepsilon \omega_I' F_0}. \tag{2.49}$$

The parameter Ω_0 is also referred to as a frequency of phase oscillations near resonance.

The three approximations mentioned above are valid if the following conditions are satisfied:

$$\varepsilon \ll \alpha \ll \varepsilon^{-1}, \tag{2.50}$$

where the dimensionless parameter α characterizes the degree of nonlinearity of oscillations

$$\alpha = \frac{I_0}{\omega_0} |\omega_I'|. \tag{2.51}$$

The first inequality in Eq. (2.50) means that the nonlinearity must be fairly strong, while the second one justifies neglecting of the non-resonant terms.

To determine the validity of the Hamiltonian (2.47), we need to estimate the amplitude of the phase oscillations

$$\frac{\max \Delta I}{I_0} \sim \left(\varepsilon \frac{F_0}{|\omega_I'|} \right)^{1/2} \frac{1}{I_0} \sim \left(\frac{\varepsilon}{\alpha} \right)^{1/2}, \tag{2.52}$$

where we put $F_0 \sim H_0 \sim \omega_0 I_0$. In the same way, we derive the frequency width of the resonance from the definition (2.49)

$$\frac{\max \Delta \omega}{\omega_0} = \frac{\Omega}{l_0 \omega_0} = (\varepsilon \alpha)^{1/2}. \tag{2.53}$$

Expression (2.52) provides the meaning for the first inequality in (2.50) and expression (2.53) — the meaning for the second, that is, the relative changes in action and spatial frequency of ray oscillations ω attributed to the ray-medium resonance must be small.

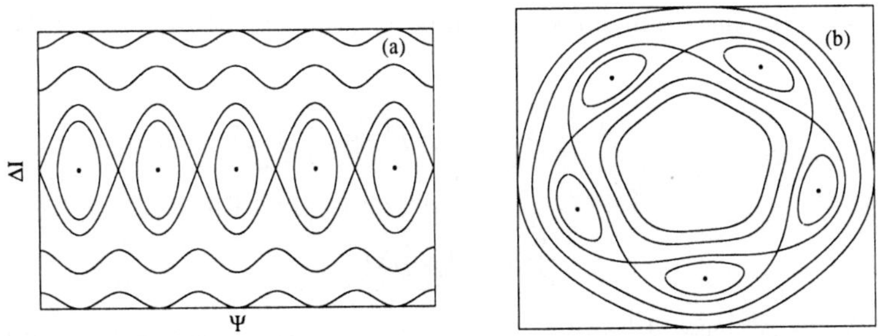

Fig. 2.4 (a) The phase portrait corresponding to Eq. (2.46) and describing ray oscillations near the ray-medium resonance with $l = 5$, $m = 1$. Bold points are the elliptic points of the resonance. (b) The same as in (a) but in the polar coordinates corresponding to the action and angle variables of the original ray equations (2.44). Here the action I is the radial distance from the center, and the angle variable ϑ determines the angular location of the islands.

The phase portrait of the system (2.46) for the 5 : 1 resonance is shown in Fig. 2.4(a). It coincides with the phase portrait of a nonlinear pendulum in Fig. 2.1 and consists of the areas of bounded and unbounded motion partitioned by the separatrix. The separatrix has two branches which intersect each other at fixed saddle points located at $\Delta I = 0$ and $\Psi = \pi \pm 2\pi n$, $n = 0, 1, 2, 3, \ldots$. Bounded motion is a rotation around the elliptic points that corresponds to oscillations near the exact resonance (2.38). Unbounded trajectories correspond to those rays which are not captured into the ray-medium resonance. In the phase space of the original ray equations (2.44), the same ray-medium resonance reveals itself as shown in Fig. 2.4(b). It forms a closed chain of five islands connected at the saddle points. Non-resonant rays now look as closed curves apart from the resonant chain.

2.4 Overlapping of Resonances

The approximations made above allows one to reduce the initial problem (2.33) with 3/2 degrees of freedom to the problem (2.44) with one degree of freedom. The cost of this reduction is omitting of the range-dependent terms in the Hamiltonian (2.47) which are mainly associated with non-resonant harmonics of the Fourier expansion (2.34). Are they really insignificant for ray motion?

To answer this question, we repeat the derivation of system (2.44) as in the preceding section retaining one of the non-resonant terms with $l = l'$

and $m = m'$ in the expansions (2.35) and (2.36). Then Eqs. (2.44) take the form

$$\frac{d}{dr}(\Delta I) = -\varepsilon l_0 F_0 \sin\Psi - \varepsilon l' F' \sin(\Psi + \omega' r + \Delta\phi), \tag{2.54}$$

$$\frac{d}{dr}\Psi = l_0 \omega_I' \Delta I, \tag{2.55}$$

where $\Delta\phi = \phi' - \phi_0$, F' and ϕ' are the modulo and phase of the complex Fourier amplitude $F_{l',m'}$, respectively, and

$$\omega' = (l' - l_0)\omega + (m' - m_0)k_r \tag{2.56}$$

is the relative frequency of the $l' : m'$ resonance with respect to the $l_0 : m_0$ resonance. The corresponding resonance Hamiltonian now reads as:

$$\tilde{H} = \frac{1}{2} l_0 \omega_I' (\Delta I)^2 - \varepsilon l_0 F_0 \cos\Psi - \varepsilon l' F' \cos(\Psi + \omega' r + \Delta\phi). \tag{2.57}$$

In contrast to (2.47), this Hamiltonian is range-dependent, therefore, equations (2.54) and (2.55) are, in general, non-integrable. Assume that the second and third terms in (2.57) are of the same order, i.e. the range dependence cannot be treated as a small perturbation. Then elimination of the range dependence is possible only if

$$\Omega_0 \ll |\omega'|. \tag{2.58}$$

Under this condition, one can average the right-hand side of Eq. (2.54) over the fast phase $\omega' r$. It is easy to see that this procedure reproduces Eqs. (2.44) and returns us to the Hamiltonian (2.47). However, the averaging technique fails in the vicinity of a separatrix, where even a weak perturbation may cause transition from the bounded regime to unbounded one, and vice versa.

Now it is worth to look at the motion near the separatrix. In the dynamical systems with one degree of freedom *a separatrix* is a special singular trajectory connecting hyperbolic *stationary points* (see the corresponding article in the Glossary) and separating topologically different regions of motion. Any range-dependent perturbation is sufficient to transform the unperturbed separatrix into a complicated pattern consisted of twisted and folded branches. The infinity of the period on the unperturbed separatrix induces an infinite number of oscillations as Ψ approaches a saddle point. This phenomenon is known as splitting of the separatrix (see the article "Melnikov method" in the Glossary). Different branches of the separatrix intersect each other creating new saddle points in the first approximation. For a neighboring pair of new saddles, the same consideration can be applied but at a higher approximation. A more complex picture will therefore emerge which points the way towards chaotic ray dynamics in a close vicinity of the unperturbed separatrix. Thus we arrive at the important point: even a well-isolated resonance has a narrow chaotic layer near the separatrix. In other words, a range-periodic sound-speed perturbation, just satisfying the condition (2.50), is sufficient for the onset of some amount of chaos in ray motion. As the relative frequency ω' approaches Ω_0, the width of this chaotic layer grows, until the domain inside the separatrix loop becomes completely chaotic [Chandre and Jauslin (2002)]. It reflects the increasing of influence of the $l' : m'$ resonance on the $l_0 : m_0$ resonance.

So, the following question arises: how to determine whether a resonance is well-isolated or not? In other words, we need a quantity which can be easily calculated and measures degree of isolation of resonance. Such a quantity was found by [Chirikov (1959, 1979)]. He considered locations of resonances in the phase space and introduced the following quantity:

$$K_{\mathrm{c}} = \frac{\Delta I}{\delta I}, \tag{2.59}$$

where ΔI is a maximal width of a chain of islands and δI is a distance between two neighboring chains.

Here the following point should be mentioned. Strictly speaking, resonances fill densely the whole phase space because the integers l and m in Eq. (2.37), can be chosen arbitrarily. However, the Fourier amplitudes $F_{l,m}$ in the expansion (2.34) decay rapidly with increasing $|l| + |m|$. In particular, if $F(\vartheta)$ is analytic, then

$$F_{l,m} \propto e^{-\varsigma(|l|+|m|)}, \tag{2.60}$$

where ς is a distance from the real axis to the nearest pole of the function $F(\vartheta)$. If the function $F(\vartheta)$ has poles on the real axis, the Fourier amplitudes decay in a power-like way. The function $F(\vartheta)$ is analytic for purely-water rays and has a singularity (or singularities) for rays reflecting off either the surface or the bottom. Both those cases imply decreasing of Fourier amplitudes and the respective resonance widths with increasing of the number. Thus, the resonances with high values of l_0 and m_0 don't affect rays significantly and can be disregarded.

If $K_c \ll 1$, then the resonances are far from each other in the phase space. The condition

$$K_c \gtrsim 1 \tag{2.61}$$

was proposed by Chirikov as a qualitative criterion in the occurrence of stronger chaos than the one which already exists. The criterion (2.61) infers the contact of separatrices of the neighboring resonances. If it appears, the chaotic layers situated near the separatrices merge and form the so-called chaotic sea.

We can reformulate the Chirikov criterion in terms of the spatial frequency of ray oscillations

$$K_c = \frac{\Delta\omega}{\delta\omega} \gtrsim 1, \tag{2.62}$$

where $\Delta\omega$ and $\delta\omega$ are the width of resonance and the distance between neighboring resonances, respectively. For small ΔI and δI, we can write down

$$\Delta\omega = |\omega_I'|\Delta I, \qquad \delta\omega = |\omega_I'|\delta I. \tag{2.63}$$

It is easy to estimate the value of K_c from the resonant condition (2.38) in the following two cases:

- Consider $k_r \ll \omega_0$ and, for the sake of simplicity, put $l_0 = 1$. The resonance besides (2.38) would then satisfy the condition

  $$\omega(I_0 + \delta I) = (m_0 \pm 1)k_r \tag{2.64}$$

  or

  $$\delta\omega = |\omega(I_0 + \delta I) - \omega(I_0)| = k_r. \tag{2.65}$$

  Thus, we derive

  $$K_c = \frac{\Delta\omega}{k_r} = (\varepsilon\alpha)^{1/2}\frac{\omega_0}{k_r} \gtrsim 1 \tag{2.66}$$

  from Eqs. (2.62), (2.65) and (2.53). Owing to condition (2.50), we have $\varepsilon\alpha \ll 1$ and $\omega_0/k_r \gg 1$, and expression (2.66) is the product of a small and large parameter, that is, condition (2.66) is non-trivial.
- Consider the reverse when $k_r \gg \omega_0$ and, for the sake of simplicity, put $m_0 = 1$. The resonance beside Eq. (2.38) would then satisfy the

condition

$$(l_0 \pm 1)\omega(I_0 + \delta I) = k_r \tag{2.67}$$

or

$$\delta\omega = |\omega(I_0 + \delta I) - \omega(I_0)| = \left|\frac{k_r}{l_0} - \frac{k_r}{l_0 \pm 1}\right| \sim \frac{\omega_0^2}{k_r} = \frac{\omega_0}{l_0}. \tag{2.68}$$

Once again, by applying Eqs. (2.68) and (2.53) in (2.59), we obtain

$$K_c = (\varepsilon\alpha)^{1/2}\frac{k_r}{\omega_0} \gtrsim 1. \tag{2.69}$$

This condition is similar to Eq. (2.66) since it is a product of both small ($\varepsilon\alpha \ll 1$) and large ($k_r/\omega_0 \gg 1$) parameters.

It is easy to see that both these cases correspond to the inequality

$$\Omega \geq |\omega'|. \tag{2.70}$$

To verify this, it is sufficient to substitute $\Omega = l_0\Delta\omega$ and rewrite ω' in terms of $\delta\omega$. Indeed, we have for $k_r \gg \omega_0$

$$l_0 = l' = 1, \quad |m' - m_0| = 1, \quad \omega' = k_r = \delta\omega, \tag{2.71}$$

and for $k_r \ll \omega_0$

$$m_0 = m' = 1, \quad |l' - l_0| = 1, \quad \omega' = \omega_0 = l\omega. \tag{2.72}$$

Under conditions (2.71) or (2.72), Eq. (2.70) transforms into (2.62).

According to Eqs. (2.66) and (2.69), the parameter K_c is proportional to the square root of the dimensionless parameter of nonlinearity α which depends only on the background sound-speed profile. Beron-Vera and Brown argued that this parameter can be regarded as the parameter of stability whose dependence on ray action controls ray instability [Beron-Vera and Brown (2003, 2004)]. In particular, it specifies the domains of the phase space where rays exhibit the strongest chaos.

Before we proceed with considering resonance overlapping, one point should be discussed. According to Eq. (2.50), the theory of the ray-medium resonance presented in Sec. 2.3 is valid when the parameter α is larger than the amplitude of a perturbation ε. It infers that the system of ray equations must be nondegenerate, i.e. dependence of the spatial frequency ω on action I must be monotonic. Nondegeneracy means that, given l_0 and m_0, the resonance condition (2.38) is satisfied for an unique value

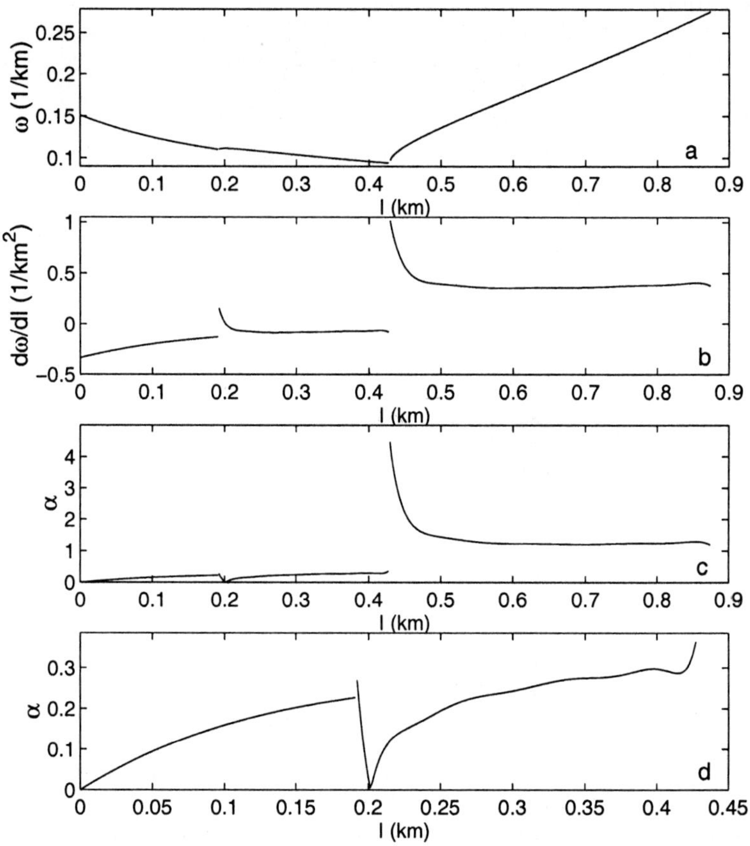

Fig. 2.5 Spatial frequency of ray oscillations (a), its derivative with respect to action I (b), and the nonlinearity parameter α (c) as functions of I for the canonical Munk waveguide. A closer view of the dependence $\alpha(I)$ for rays propagating without reflections off the bottom (d). The depth of the ocean bottom is taken of 5 km [Smirnov *et al.* (2001)].

of I_0. However, the monotonicity of the function $\omega(I)$ is violated due to reflections of rays off the surface and bottom. For example, let us consider the canonical Munk sound-speed profile (1.102). Assume that the bottom is located at 6 km depth. Figure 2.5 demonstrates the dependencies of ω, $d\omega/dI$ and α on action. The rays with actions in the interval from 0 to about I', where $I' \simeq 0.2$ km, don't reflect off neither the surface nor the bottom. The interval from I' to I'', where $I'' = 0.43$ km, corresponds to surface-bounce rays, but these rays don't reflect off the bottom. Larger

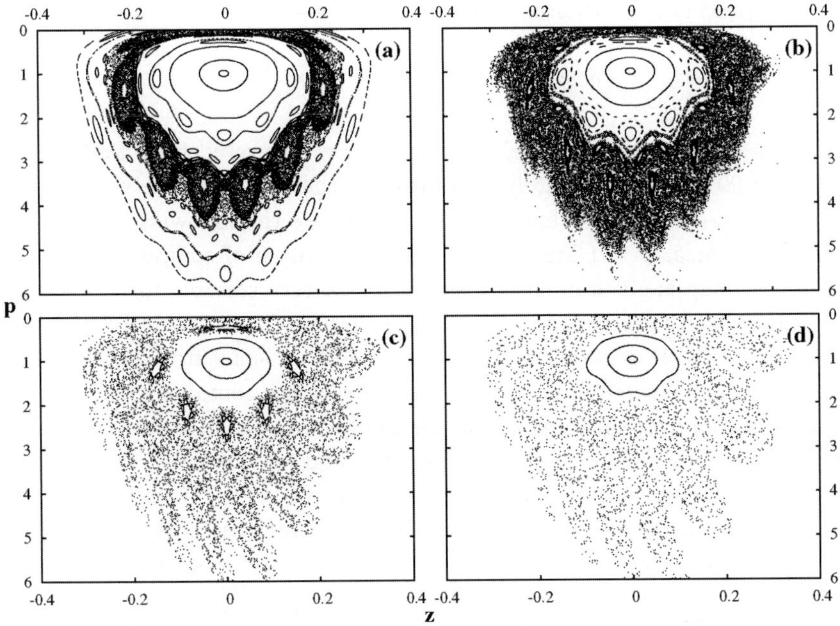

Fig. 2.6 Metamorphoses of the Poincaré map under increasing the amplitude of a sound-speed perturbation. (a) $\varepsilon = 0.005$, (b) $\varepsilon = 0.01$, (c) $\varepsilon = 0.02$, (d) $\varepsilon = 0.04$.

values of I correspond to rays reflecting off both the surface and bottom. The function $\omega(I)$ has two isolated extrema near $I = I'$ and one extremum near $I = I''$. At those extrema α vanishes and the condition (2.50) is not satisfied certainly. Degeneracies associated with isolated extrema of $\omega(I)$ are called *accidental* [Brown *et al.* (2003)]. They were studied earlier in the context of weakly-divergent beams [Petukhov (1994); Goncharov and Kurtepov (1994); Brekhovskikh *et al.* (1995)].

Ray motion near the accidental degeneracies is a very complicated issue. Studies of degenerate Hamiltonian systems show that dynamics near accidental degeneracies is very sensitive to values of perturbation's parameters. For instance, there may occur a barrier dividing a chaotic sea into two parts, and, thus, restricting chaotic transport. On the other hand, numerical simulations of ray dynamics in USCs expose remarkable enhancement of ray instability in the domains close to the degeneracies. An example of that is presented in Fig. 2.6(a), where we show the Poincaré map for the canonical Munk waveguide with the ocean bottom placed at $z = 6$ km. The sound-speed perturbation is taken in the form

$$\delta c(z, r) = 2\varepsilon c_0 \frac{z}{B} e^{-2z/B} \sin \frac{2\pi r}{\lambda_r}. \tag{2.73}$$

This strongly idealized model of perturbation, borrowed from Ref. [Smith *et al.* (1992)], approximately accounts for the first internal wave mode. The values of the perturbation's parameters are the following: $\varepsilon = 0.005$, $c_0 = 1500$ m/s, $B = 1$ km, $\lambda_r = 8$ km. Description of computation of the Poincaré map is given in the Glossary. It is a very suitable tool for studying periodically-perturbed dynamical systems and allows one to distinguish explicitly chaotic domains in the underlying phase space from regular ones. As it is demonstrated in the figure, the phase space contains several nested chains of islands of different width and number. All these chains correspond to ray-medium resonances with different l_0 and m_0. Notably, the width of a chain decreases with increasing the number of islands. This circumstance reflects decreasing of a resonance width with increasing order of resonance.

The chain corresponding to the $7:1$ resonance is submerged into a chaotic layer which looks in the figure as a self-closed band densely filled by points. All the other chains of resonances are placed within the regular domains where all the points form self-closed curves. Note that the chaotic layer divides the regular domain into two parts, the inner and outer ones. One can easily see that the chaotic layer contains the point $z = p = 0$, corresponding to the so-called surface-grazing ray, being the border between purely-water and surface-bounce rays. As we already saw from Fig. 2.5, the transition from purely-water to surface-bounce rays bears a pair of degeneracies near $I = I'$. The $7:1$ resonance overlaps with the degenerate domain that gives rise to the bounded chaotic layer.

The outer regular domain contains three chains, corresponding to the $22:3$, $15:2$ (little islands), and $8:1$ ray-medium resonances. Here we don't take into account the rays reaching the bottom because they are irrelevant for long-range sound propagation due to significant sound attenuation at the bottom. Since that the degeneracy about $I = I''$, associated with reflections off the bottom, doesn't affect ray motion.

Let's explore how the situation changes with increasing perturbation's amplitude ε. According to Eqs. (2.52) and (2.53), it should result in broadening of the ray-medium resonances. Figure 2.6(b) represents the Poincaré map computed with $\varepsilon = 0.01$. There one can see that the resonances in the outer regular domain appear to be completely overlapped. As a consequence of their overlapping, the outer regular domain is completely destroyed. The chaotic sea formed is not restricted by invariant curves from the outer side, therefore, there is no dynamical barriers preventing chaotic swinging of rays. As a result of that swinging, a ray may reach the bottom and lose the most of acoustic energy it carries. In other words, the chaotic layer is *open* and allows for rays to leave a waveguide. This phenomenon is referred to as ray escaping.

Increasing of the perturbation's amplitude also diminishes the area of the inner regular domain. Islands of the 7 : 1 resonance are still present in the map, but their area is significantly reduced. With further increasing of the amplitude to the level of $\varepsilon = 0.02$, the near-separatrix chaotic layer of the 6 : 1 resonance joins with the chaotic sea. Thus, the resonance chain transforms to the chain of stability islands in the chaotic sea. All other islands of stability vanish. Finally, with $\varepsilon = 0.04$ there is no stability islands in the chaotic sea at whole. Only the central domain, looking as an area with nested self-closed curves within, preserves stability.

To summarize the above analysis, the following points should be underlined. Chaos is always present near separatrices of a ray-medium resonance. However, until different ray-medium resonances are separated from each other, and from the regions where degeneracies occur, chaotic ray motion occurs only within narrow near-separatrix layers. The contact with a degeneracy leads to significant broadening of the near-separatrix chaotic layer. If the chaotic layers of neighboring resonances intersect each other, then they merge into a chaotic sea.

Now let's consider ray motion in a chaotic sea. Obviously, chaotic rays must diverge with each other in an exponential rate which tends to a global Lyapunov exponent in the long-range limit. This means that any compact ray bundle, forming a spot in the phase space, should spread rapidly until the chaotic sea will be covered completely. However, as it was shown in Sec. 2.2, the divergence of rays may be not monotonic due to intermittency. To examine peculiarities of ray diffusion in the chaotic sea, we consider once again the canonical Munk waveguide with the bottom at $z = 6$ km and the sound-speed perturbation (2.73). We just slightly modify the parameters of the perturbation, setting $\varepsilon = 0.01$ and $\lambda_r = 10$ km. In this case a wide bounded chaotic sea occurs. It is illustrated by means of the Poincaré map in Fig. 2.7. The sea is not uniform, and there are two chains of islands within, one chain relating to the 5 : 1 resonance and another one relating to the 6 : 1 resonance.

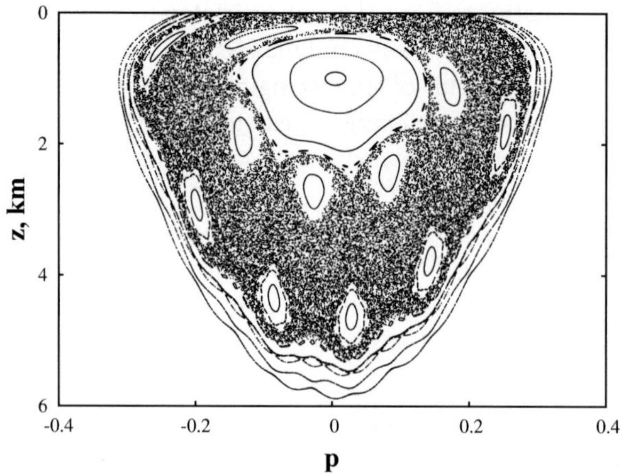

Fig. 2.7 Poincaré map calculated with $\varepsilon = 0.01$ and $\lambda_r = 10$ km.

Consider a bundle which forms a narrow strip in the phase space (see Fig. 2.8(a)). This strip can be used as a model of the so-called *Lagrangian manifold* — a manifold describing the surface of some propagating *wavefront* (for a brief review of manifolds see the corresponding article in the Glossary). Taking into account amplitude factors and phase shifts along the different rays of the manifold, one can compute the respective wavefield.

While moving in the chaotic sea, the strip undergoes a cascade of stretchings alternating with foldings, and the form of the strip becomes more and more complicated (see the article "Horseshoe mapping" in the Glossary). For example, at $r = 100$ km the strip transforms into a string with the z-like folding in the middle (see Fig. 2.8(b)). At $r = 500$ km the string appears significantly extended due to the numerous stretchings, and its middle part looks as a shrank accordion (see Fig. 2.8(c)), that is a result of numerous foldings. Thus, there are two concurrent tendencies: extending of the string and approach of different segments of the bundle. At $r = 1000$ km the string becomes long enough to surround the central domain of stability, as it is shown in Fig. 2.8(d). In this case the distance between rays starting from neighboring initial positions in the phase space becomes very large, so that we can see separate points instead of a continuous line corresponding to the Lagrangian manifold. It doesn't mean that the respective wavefront is broken and just reflects the deficiency of the density of initial conditions. This deficiency is an inevitable consequence

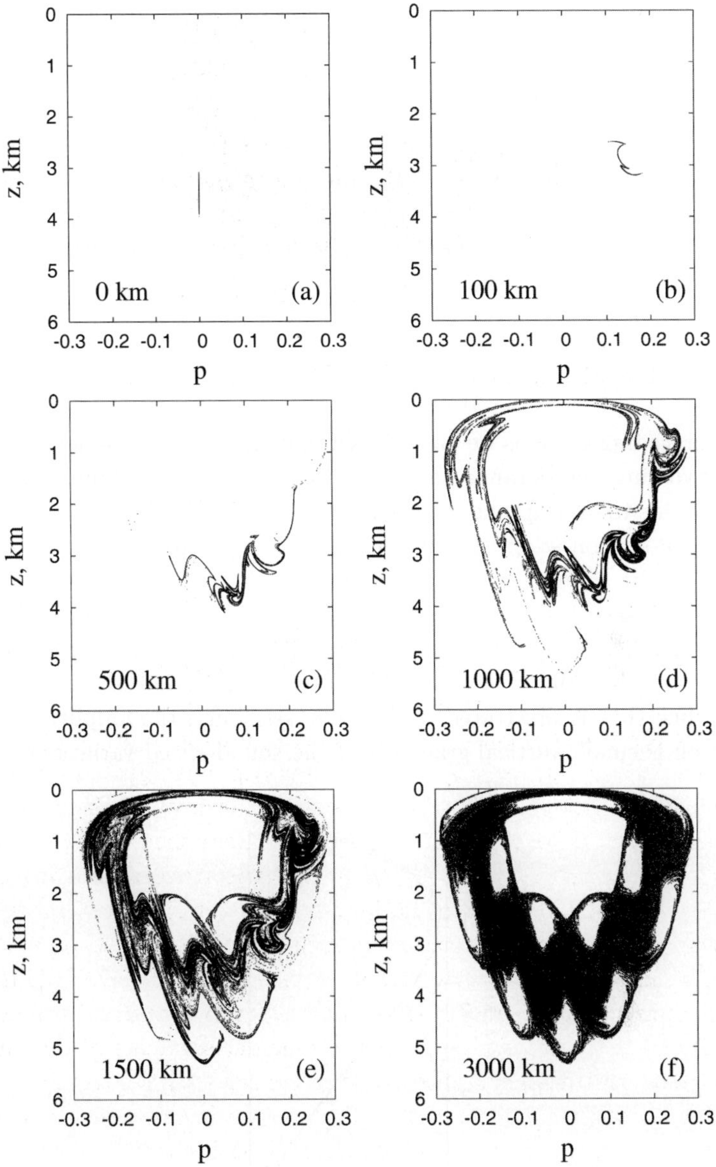

Fig. 2.8 Evolution of a Lagrangian manifold in chaotic sea. Each figure represents a snapshot of ray distribution in the phase space at the ranges of 0 (a), 100 (b), 500 (c), 1000 km (d), 1500 (e), and 3000 km (f).

of the Lyapunov instability, implying that the necessary number of rays for the accurate reproduction of a wavefront grows exponentially with range.

The further evolution of the ray ensemble looks rather as diffusion of separate points than some complicated transformation of the continuous line. A large number of rays reaches the area of the neighboring $6:1$ resonance (see Fig. 2.8(e)). At $r = 3000$ km, the chaotic layer becomes almost uniformly filled as it is demonstrated in Fig. 2.8(e). No continuous segments of the Lagrangian manifold can be resolved, and one can say about almost complete mixing of rays with close initial conditions.

2.5 Vertical Resonance

In the preceding sections we have shown that the simplest model of a waveguide involving a weak range-periodic inhomogeneity can exhibit chaotic ray behavior due to overlapping of resonances in the underlying phase space. In the present section we shall consider a more complicated case when a weak range-periodic sound-speed perturbation is a rapidly oscillating function of depth. Those fast depth oscillations mimic a fine structure produced by high-number modes of internal waves. The internal-wave fine structure results in an essentially non-smooth "chirped" distortion of a sound-speed profile, like that is illustrated in Fig. 2.9. Although the amplitude of the distortion is small, vertical gradients of the sound-speed variations caused

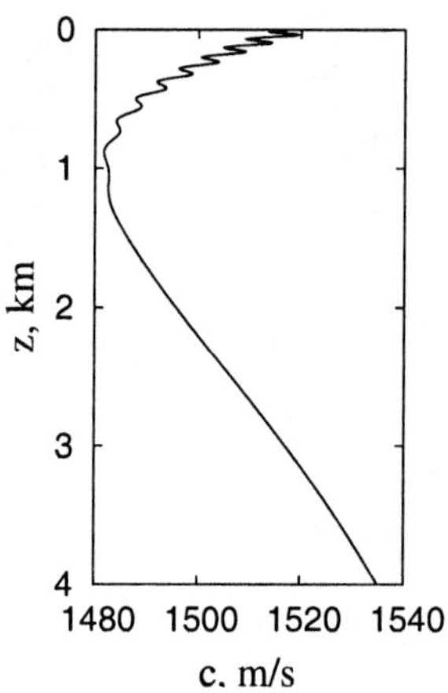

Fig. 2.9 "Chirped" distortion of a sound-speed profile due to an internal-wave fine structure.

by a fine structure are of the same order (or even larger) as the vertical gradient of the unperturbed sound-speed profile. It obliges one to take care when considering the effect of internal-wave fine structure on ray motion.

We begin with considering the following model of a sound-speed perturbation:

$$\delta c(z, r) = \varepsilon c_0 Y(z) \sin k_z z \sin k_r r, \tag{2.74}$$

where $\varepsilon \ll 1$ is a dimensionless constant, $Y(z)$ is a perturbation envelope being a smooth function of depth, k_z and k_r are vertical and horizontal wavenumbers of the perturbation. The vertical wavenumber is assumed to be large, so that

$$k_z \gg \left|\frac{dY}{dz}\right|. \tag{2.75}$$

The sound-speed perturbation (2.74) can be represented as a sum of upgoing and downgoing components

$$\delta c = \frac{\varepsilon}{2} c_0 Y \left[\cos(k_z z - k_r r) - \cos(k_z z + k_r r)\right]. \tag{2.76}$$

Then the ray equations can be cast in the following form:

$$\frac{dp}{dr} = \frac{\varepsilon}{2}\left[(k_z Y(\sin\Psi^- - \sin\Psi^+) - \frac{dY}{dz}(\cos\Psi^- - \cos\Psi^+)\right], \tag{2.77}$$

where $\Psi^{\pm} = k_z z \pm k_r r$. In accordance with Eq. (2.75), the range-dependent term oscillates rapidly along a ray path, except for the resonant regions, where one of the following conditions, either

$$\frac{d\Psi^+}{dr} = k_z p + k_r \simeq 0 \tag{2.78}$$

or

$$\frac{d\Psi^-}{dr} = k_z p - k_r \simeq 0, \tag{2.79}$$

is fulfilled. These resonances occur at the certain values of the ray momentum, when a ray is oriented parallel to either the upgoing or downgoing component of the sound-speed perturbation. Since these resonances are associated with ray motion with respect to depth oscillations of a sound-speed perturbation, we shall refer to the resonances of this sort as *vertical resonances*. In particular, we shall name resonance (2.78) as the upgoing vertical resonance and the resonance (2.79) as the downgoing vertical resonance.

This sort of resonances is known in the theory of classical Hamiltonian systems as resonances in slow-fast systems or resonances in systems with rapidly-varying phases. They have been firstly considered by Timofeev (1978), and then their theory has been developed by Itin *et al.* (2000); Vainchtein *et al.* (2006); Neishtadt and Vasiliev (2007). Let's discuss briefly their properties.

2.5.1 *Adiabatic approximation*

A.I. Neishtadt with his coworkers has studied resonances in slow-fast systems in the framework of the so-called adiabatic approximation [Neishtadt and Vasiliev (2005)]. It is based on a consideration of non-resonant and resonant motions individually. Far from the vertical resonances, ray dynamics can be reduced to an integrable one by averaging over the rapidly-varying phases Ψ^+ and Ψ^-. The averaging method yields an improved adiabatic invariant instead of the unperturbed action I. The reader who is interested in details in the averaging procedure is referred, for example, to the seminal monograph [Bogolyubov and Mitropolsky (1961)].

We intend here to concentrate attention at the motion near resonances. Let us consider one of the vertical resonances, for instance, the upgoing one. Firstly we simplify the equation (2.77). Since k_z is a large parameter, we can take into account only those terms in the right-hand side which are proportional to k_z. In addition, we can neglect the non-resonant term $\sim \sin\Psi^-$. After those simplifications, we obtain

$$\frac{dp}{dr} = -U_z - \frac{\varepsilon}{2}k_z Y(z)\sin\Psi^+, \tag{2.80}$$

where we used the shortnotation $U_z = dU/dz$. At the next step, we shall describe variations of the perturbation phase Ψ^+ along a ray path. For the brevity, we omit superscript "+". Twice differentiating the phase Ψ with respect to r and taking into account that

$$\frac{d}{dr} = \frac{\partial}{\partial r} + p\frac{\partial}{\partial z}, \tag{2.81}$$

we derive the pendulum-like equation

$$\frac{d^2\Psi}{dr^2} + \frac{\varepsilon}{2}k_z^2 Y(z)\sin\Psi + k_z U_z = 0. \tag{2.82}$$

This equation can be rewritten as a pair of the first-order equations in the

Hamiltonian form

$$\frac{d\Psi}{dr} = k_z y = \frac{\partial \tilde{H}(\Psi, y)}{\partial y},$$
$$\frac{dy}{dr} = -\frac{\varepsilon}{2} k_z Y(z) \sin\Psi - U_z = -\frac{\partial \tilde{H}(\Psi, y)}{\partial \Psi}. \tag{2.83}$$

Here y is the resonance detuning

$$y = p - p_{\text{res}} = p + \frac{k_r}{k_z}. \tag{2.84}$$

Equations (2.83) correspond to the resonant Hamiltonian $\tilde{H}$ of the following form:

$$\tilde{H}(\Psi, y) = \frac{k_z y^2}{2} - \frac{\varepsilon}{2} k_z Y \cos\Psi + U_z \Psi. \tag{2.85}$$

Expressions (2.82), (2.83), and (2.85) involve dependence on the ray depth z. In the adiabatic approach, one interprets z and p as slowly varying parameters specifying the structure of the phase space portrait in the (Ψ–y) plane.

Let us give an insight into the phase portraits in the (Ψ–y) space with fixed values of z. If

$$|U_z| < \frac{\varepsilon k_z Y}{2}, \tag{2.86}$$

then there arise pairs of fixed points on the phase portrait. Each pair consists of one saddle and one center. Saddle point lies on the separatrix which encloses a resonant domain with oscillating motion within (see Fig. 2.10).

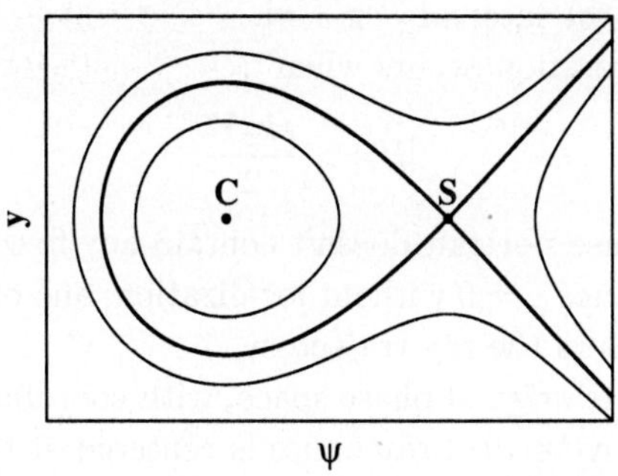

Fig. 2.10 The phase portrait with the resonant domain.

Coordinates of the fixed points can be found from the equation

$$U_z(z) + \frac{\varepsilon}{2} k_z Y(z) \sin \Psi = 0. \tag{2.87}$$

The center point is

$$y_{\mathrm{c}} = 0, \quad \Psi_{\mathrm{c}} = -\arcsin \frac{2U_z}{\varepsilon k_z Y} + 2\pi n, \tag{2.88}$$

and the saddle point is

$$y_{\mathrm{s}} = 0, \quad \Psi_{\mathrm{s}} = -\Psi_{\mathrm{c}} + \pi(2n+1), \tag{2.89}$$

where n is an integer.

Slow changes in z may move a ray, initially located outside the separatrix loop, into the resonant domain. When a ray enters the resonant domain, Ψ switches its behavior from rotational to an oscillating one, that is followed by localization of the ray momentum in a narrow interval near p_{res}. An exact criterion of falling into vertical resonance can be formulated as [Makarov and Uleysky (2007); Kon'kov *et al.* (2007); Makarov *et al.* (2008); Makarov and Uleysky (2008)]

$$\left.\tilde{H}\right|_{z(r)} \le \left.\tilde{H}_{\mathrm{s}}\right|_{z(r)}, \tag{2.90}$$

where H_{s} is a value of the Hamiltonian $\tilde{H}$ at the separatrix

$$\tilde{H}_{\mathrm{s}} = U_z \bar{\Psi}_{\mathrm{s}} - \frac{\varepsilon}{2} k_z Y \cos \bar{\Psi}_{\mathrm{s}}. \tag{2.91}$$

Subscript $z(r)$ in (2.90) indicates that $\tilde{H}$ and $\tilde{H}_{\mathrm{s}}$ are calculated along the ray trajectory, $\bar{\Psi}_{\mathrm{s}} = \Psi_{\mathrm{s}} + 2\pi n$, where n is chosen to satisfy the condition $-\pi \le \bar{\Psi}_{\mathrm{s}} \le \pi$, and Ψ is assumed to be a cyclic variable with a range of values confined by the interval $[-\pi : \pi]$.

The different situation occurs when

$$|U_z| > \frac{\varepsilon k_z Y}{2}. \tag{2.92}$$

In this case the phase portrait doesn't contain any fixed points, a ray goes by the exact resonance $y = 0$ without localization, and one shouldn't expect appreciable changes in the ray trajectory.

The region in the original phase space, with coordinates p and z, which is mainly affected by vertical resonance is centered at the point which can be found from the following arguments. The momentum coordinate of this point, p_{res}, corresponds to the exact resonance and is determined by

Eq. (2.78). The depth coordinate can be found from the condition of the largest area inside the separatrix loop. This condition corresponds to the maximal probability of entering the resonance domain. The respective area is determined by the integral

$$J = \frac{1}{2\pi} \oint y(\Psi, \tilde{H}_{\rm s})\, d\Psi, \tag{2.93}$$

where

$$y(\Psi, \tilde{H}_{\rm s}) = \sqrt{2\frac{\tilde{H}_{\rm s}}{k_z} - 2\frac{U_z}{k_z}\Psi + \varepsilon Y \cos\Psi}. \tag{2.94}$$

However, this integral is not suitable for an analytical treatment, therefore, it is reasonable to seek for a more convenient condition. A relevant candidate is the condition demanding the maximum of resonance width with respect to the resonance detuning y. According to Eq. (2.84), the maximal width with respect the the detuning is equal to the maximal width in the ray momentum space. The maximal detuning is expected when $\tilde{H} = \tilde{H}_{\rm s}$, i.e. the amplitude of phase oscillations inside the resonance domain is maximal, and the Ψ-dependent part of the Hamiltonian (2.85), acting as a "potential" for phase oscillations, takes on a minimal value. The latter condition yields $\Psi = \bar{\Psi}_{\rm c}$, where $\bar{\Psi}_{\rm c}$ is defined in the same way as $\bar{\Psi}_{\rm s}$, i.e., it is the value of $\Psi_{\rm c}$ in the frame $[-\pi : \pi]$. It follows that, given a depth z, the resonance width in the momentum space is

$$\Delta p(z) = 2y[(\Psi_{\rm c}(z), \tilde{H}_{\rm s}(z)] = 2\sqrt{\frac{2}{k_z}\left[\tilde{H}_{\rm s}(z) - \tilde{H}_{\rm min}(z)\right]}, \tag{2.95}$$

where

$$\tilde{H}_{\rm min} = U_z \bar{\Psi}_{\rm c} - \frac{\varepsilon}{2} k_z Y \cos\bar{\Psi}_{\rm c}. \tag{2.96}$$

So that, the region of the strongest influence of the upgoing vertical resonance is located around the point determined by the conditions

$$p = p_{\rm res} = -\frac{\lambda_z}{\lambda_r}, \quad \Delta p(z) = \max. \tag{2.97}$$

The respective formula for the downgoing vertical resonance differs only by the sign of the resonant momentum. It means that the upgoing and downgoing vertical resonances affect the same rays but at different phases of trajectories. The resonance (2.78) arises only on upgoing part of a ray

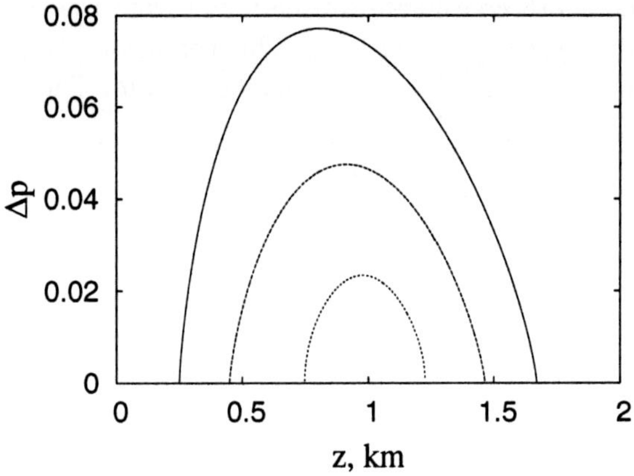

Fig. 2.11 Width of vertical resonance Δp vs depth for the waveguide with the sound-speed profile given by Eq. (2.27). Value of k_z are: $2\pi/200$ m^{-1} (solid), $2\pi/500$ m^{-1} (dashed), and $2\pi/2000$ m^{-1} (dotted). The depth of the waveguide axis is $z_{\rm a} = 1$ km. The value of the perturbation's amplitude is ε is taken to be 0.005.

trajectory ($p < 0$), whereas the resonance (2.79) arises on the downgoing part ($p > 0$). Therefore, one can formulate a common condition for the upgoing and downgoing resonances as

$$|p| = |p_{\rm res}| = \frac{\lambda_z}{\lambda_r}, \quad \Delta p(z = z_{\rm res}) = \max. \tag{2.98}$$

If variations of the function $Y(z)$ are weak, then Δp achieves its maximum close to the waveguide axis $z_{\rm a}$, where ray momentum takes on the maximal value. It is demonstrated in Fig. 2.11 which presents the dependencies $\Delta p(z)$ for the waveguide described by Eq. (2.27) with different values of k_z. Thus, one can approximately set $z_{\rm res} = z_{\rm a}$, and the common condition (2.98) now reads

$$p_{\max}(H) = \sqrt{2(H - U(z_{\rm a}))} \simeq \frac{\lambda_z}{\lambda_r}. \tag{2.99}$$

Inverting this formula, we obtain the value of H corresponding to the region affected by vertical resonance

$$H_{\rm res} = \frac{k_r^2}{2k_z^2} + U(z_{\rm a}). \tag{2.100}$$

2.5.2 *Passage through a resonance*

The inequality (2.92) holds only for a finite piece of a ray trajectory. Thus, each ray, after being fall into a resonance, for certain will exit the resonant domain at some range. Variations of a ray trajectory after a passage through a resonance depends qualitatively on how long this ray has been delayed in the resonant domain. Two regimes of the passage should be distinguished: *scattering on resonance* and *capturing into resonance*. Let's consider each of them in detail.

2.5.2.1 *Scattering on a resonance*

In this case a ray is not delayed for long ranges in the resonant area and leaves it already during the first cycle around the center point (2.88). Nevertheless, crossing the resonant area is followed by a jump-like variation of the action variable δI. An approximate formula for the action jump can be derived using the procedure described by [Vainchtein *et al.* (2006)]. As a starting point, we have

$$\delta I = \int_{r_1}^{r_2} \dot{I}\, dr, \tag{2.101}$$

where r_1 and r_2 are the values of range at which a ray enters and exits the resonant domain, respectively, and the dot denotes differentiation with respect to r. Far from resonances, the action I is near a constant, therefore, we can replace the limits $r_{1,2}$ by $\mp\infty$. Then we transform the integration over r to the integration over Ψ

$$\delta I = \frac{1}{k_z} \int_{-\infty}^{\infty} \frac{\dot{I}}{y(\Psi)}\, d\Psi, \tag{2.102}$$

where $y(\Psi)$ is given by Eq. (2.94). Range evolution of the action is governed by the equation

$$\dot{I} = -\frac{\partial H}{\partial \vartheta} = -\frac{\partial V}{\partial \vartheta}. \tag{2.103}$$

Substituting Eq. (2.76) into Eq. (2.103) and taking into account only the leading resonant term, we obtain

$$\dot{I} \simeq -\frac{\varepsilon k_z p Y \sin\Psi}{2\omega}. \tag{2.104}$$

When a ray moves inside the resonant domain, variations of the ray depth z and momentum p are slow compared with variations of Ψ. Since that we can fix values of z and p at the values at separatrix crossing. It yields

$$\delta I = \frac{\sqrt{k_z}\varepsilon Y^* p^*}{2\omega^*} \int\limits_{\Psi^*}^{\infty} \frac{\sin\Psi\, d\Psi}{\sqrt{2\tilde{H}_{\mathrm{s}}^* - 2U_z'^*\Psi + \varepsilon k_z Y^* \cos\Psi}}, \qquad (2.105)$$

where the superscript $*$ means that the respective values are taken at the moment of separatrix crossing. The respective formula for the downgoing vertical resonance (2.79) differ only by the sign of δI. The emphatic point is that the action jump depends extremely on initial conditions. Since the action jumps in different passages through a resonant area are not correlated, rays exhibit chaotic diffusion. Neishtadt and Vasiliev (2007) have rigorously established that there are no stable periodic trajectories in the domain of separatrix crossings, except, maybe, for periodic trajectories passing anomalously close to the saddle point. So, the vertical resonance leads to onset of a chaotic layer in the phase space. This issue will be addressed in detail in subsection 2.5.4.

2.5.2.2 *Capturing into a resonance*

In the case of large k_z or ε, there occurs a possibility of capturing into a resonance, when a ray undergoes a cascade of fast oscillations inside the resonant domain, whereas the resonant domain drifts slowly with depth. For example, consider the resonance (2.78). By definition, we have

$$z(r) = \frac{1}{k_z}\left(\Psi(r) - k_r r\right) = \frac{\Psi(r)}{k_z} + p_{\mathrm{res}} r. \qquad (2.106)$$

Here the last term corresponds to depth drift of the resonant domain. Denote as L the depth interval at which the resonant domain exists. L is given by the formula

$$L = |z_2 - z_1|, \qquad (2.107)$$

where z_1 and z_2 are roots of the equation

$$|U_z| = \varepsilon k_z Y. \qquad (2.108)$$

Period of ray oscillations inside the resonant domain is given by the formula

$$L_\Psi = \sqrt{\frac{2}{k_z}} \int\limits_{\Psi_1}^{\Psi_2} \left[\tilde{H} - U_z\Psi + (\varepsilon k_z Y/2)\cos\Psi\right]^{-1/2} d\Psi. \qquad (2.109)$$

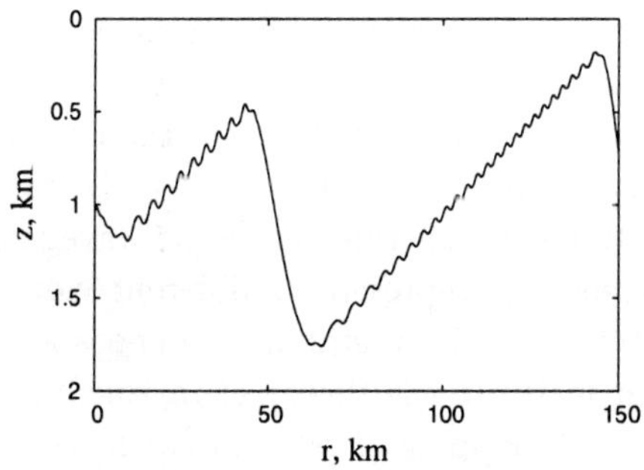

Fig. 2.12 Ray trajectory under capturing into a vertical resonance.

Then the criterion of capturing into resonance can be expressed as

$$\frac{L}{|p_{\rm res}|} \gg L_\Psi. \tag{2.110}$$

If this criterion is satisfied, we can average (2.106) over fast oscillations of Ψ, that leads to the formula

$$\bar{z}(r) = p_{\rm res} r + z^*, \tag{2.111}$$

where z^* is a depth at the moment of entering the resonant domain.

Capturing into a resonance can be interpreted in a simple way. A sound-speed perturbation poses a wave-like distortion onto the background sound-speed profile, so that there occur multiple local minima associated with resonant domains. Some rays appear to be trapped near those minima. The minima move upwards or downwards (or in the both directions) with range carrying the trapped rays. The trajectory of a trapped ray acquires the specific "elevator"-like forms shown in Fig. 2.12.

A ray captured into a resonance deviates strongly from the ray with the same initial conditions in the unperturbed waveguide. Each capturing, as well as each scattering, is followed by a jump of the ray action. Magnitude of the jump depends on how long the ray is delayed in the resonance domain, and can be of a large value. Thus, capturing into a vertical resonance can produce fast chaotic diffusion occupying a large domain in the phase space.

2.5.3 *Vertical resonance versus ray-medium resonance*

Before we proceed with illustrations of ray instability caused by vertical resonance, one important issue should be discussed. Theory of vertical resonance treats ray dynamics in the different manner than the theory of ray-medium resonance described in the preceding Section. It looks somewhat surprising, because the theory of ray-medium resonance, being based on the KAM theory, seems to provide a very general tool for studying ray motion in weakly-perturbed range-periodic waveguides. However, the theory of vertical resonance represents a different view of dynamics. The mechanism of the transition from stability to chaos also looks differently, at the first glance at least. In the theory of ray-medium resonance, global chaos emerges under overlapping of resonances. In vertical resonance, the route to appearance of wide chaotic layers is crossing of the "breathing" separatrix of a resonant domain. Actually, one can object that the origin of instability in the theory of ray-medium resonance is almost the same, i.e. transport through the separatrix due to its destruction. But in the case of vertical resonance, the transition to global chaos arises without influence of any neighboring resonances, and the reason, which makes the crossing possible, is the evolution of "slow" variables due to ray oscillations in a waveguide. So, it is worthwhile to understand how these two approaches interrelate, and where the boundaries beyond which one of them becomes inapplicable while another one remains valid.

Formally, the only distinction between the two approaches is the presence of depth-dependent term $k_z z$ in the expression for the resonance phase under vertical resonance. In the theory of ray-medium resonance, this term is "hidden" in Fourier-amplitudes of the perturbation. In order to examine how it reveals itself in Fourier spectra, we consider the ray Hamiltonian (2.32) and add the factor $\sin k_z z$ to the perturbation term. Now Eq. (2.32) reads

$$H = H_0(I) + \varepsilon F(I,\, \vartheta,\, r) \sin k_z z(I, \vartheta), \tag{2.112}$$

where $F(r) = F(r + \lambda_r)$. If the function F has more than one range harmonic, then condition (2.99) looks as

$$p_{\max}^{(m)} = \frac{m k_r}{k_z}, \tag{2.113}$$

where m is a number of a harmonic. Consider a single ray-medium resonance

$$l\omega = m k_r = \frac{2\pi m}{\lambda_r}. \tag{2.114}$$

The Fourier-amplitudes of the perturbation $F \sin k_z z$ are given by the formula

$$H_{l,m} = \frac{1}{8\pi i \lambda_r} \int\limits_{-\lambda_r/2}^{\lambda_r/2} dr \exp\left(i\frac{2\pi m r}{\lambda_r}\right) W_l(I,r) + \text{c. c.}, \tag{2.115}$$

$$W_l(I,r) = \int\limits_{-\pi}^{\pi} F(I,\vartheta,r)\, e^{il\vartheta + ik_z z}\, d\vartheta - \int\limits_{-\pi}^{\pi} F(I,\vartheta,r)\, e^{il\vartheta - ik_z z}\, d\vartheta. \tag{2.116}$$

With large k_z, the integrals in Eq. (2.116) can be evaluated using the stationary phase method (see Appendix D). Denoting the respective phases as

$$\Psi = l\vartheta \pm k_z z(\vartheta), \tag{2.117}$$

we find the stationary phase conditions

$$\frac{d\Psi}{d\vartheta} = l \pm k_z \frac{dz}{d\vartheta} = l \pm \frac{k_z p}{\omega} = 0. \tag{2.118}$$

According to Eq. (2.114), ω can be replaced by mk_r/l. Then the stationary phase conditions become

$$mk_r \pm k_z p(I,\vartheta) = 0. \tag{2.119}$$

In order to estimate the strength of resonance, we need the second derivative

$$\frac{d^2\Psi}{d\vartheta^2} = \pm \frac{k_z}{\omega^2} \frac{dp}{dr}. \tag{2.120}$$

Carrying out the integration and assuming

$$\frac{dp}{dr} \approx -U_z \tag{2.121}$$

we find the solution as a sum of contributions given by the stationary points

$$W(I,r) \approx \frac{1}{2i} \sum_j \left(D\sqrt{k_z\,|U_z(\vartheta_j)|} \right)^{-1} F(\vartheta_j, r) e^{i\Psi_j + i(\pi/4)\operatorname{sgn}(d\Psi_j^2/d\vartheta^2)} + \text{c. c.}, \tag{2.122}$$

where we dropped the indice l, ϑ_j is the j-th solution of Eq. (2.119), and Ψ_j is the corresponding value of the phase. It follows from Eq. (2.122) that the function $H_{l,m}(I)$ has a singularity at the waveguide axis, where $U_z = 0$ and $|p| = p_{\max}$. Being combined with Eq. (2.119), it yields condition of

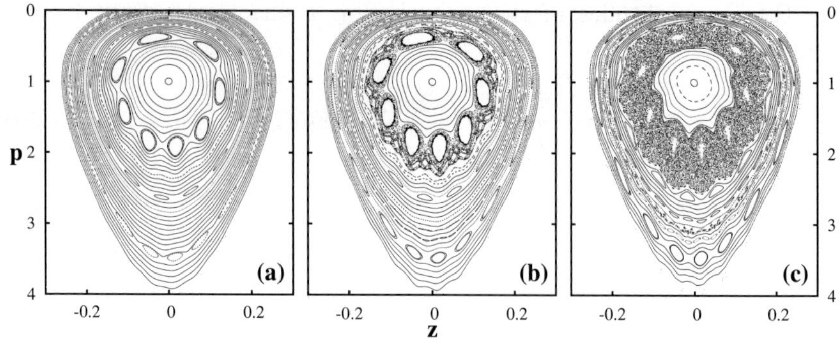

Fig. 2.13 Broadening of the resonance 8 : 1 and the birth of a chaotic layer under vertical resonance with $\lambda_z = 0.6$ km and $\lambda_r = 5$ km. (a) $\varepsilon = 0.0005$, (b) $\varepsilon = 0.001$, (c) $\varepsilon = 0.002$.

the vertical resonance (2.113). This infers strong broadening of the respective ray-medium resonance when it is placed close to vertical resonance in the phase space. Apparently, this broadening may result in overlapping of ray-medium resonances and global chaos. On the contrary, ray-medium resonances are expected to be very weak if they are located far from vertical resonance in the phase space, and Eq. (2.119) doesn't have any roots. This is because the respective Fourier-amplitudes appear to be strongly suppressed by fast oscillations of the integrands in Eq. (2.116).

To illustrate that, we return to the model of a waveguide described by Eq. (2.27) and set $\lambda_z = 600$ m. The horizontal period of the perturbation is taken to be $\lambda_z = 5$ km. Under this choice, vertical resonance with $p_{\max} = \lambda_z/\lambda_r = 0.12$ influences mainly on the ray-medium resonance $8:1$. Figure 2.13 represents the Poincaré maps constructed with various values of the perturbation's amplitude ε. At $\varepsilon = 0.0005$, the resonance $8:1$ is remarkably enhanced as compared with the other resonances, but the respective Poincaré map shown in Fig. 2.13(a) reveals predominantly regular pattern. With increasing ε, the resonance $8:1$ overlaps with neighboring resonances having fractional order $l : m$. Subsequently, a chaotic layer appears in the vicinity of the separatrix of the resonance $8:1$ (see Fig. 2.13(b)). With further growth of the perturbation's strength the chaotic layer broadens and transforms into a wide chaotic sea occupying large domain in the phase space, as it is shown in Fig. 2.13(c).

However, there can be much more intricate scenario of global chaos. Such a selective enhancement of resonance Fourier-amplitudes in the vicinity of vertical resonance leads to growth of the derivatives $dH_{l,m}(I)/dI$

and $d^2H_{l,m}(I)/dI^2$. Consequences of that growth can be illustrated in the following way. Let's rewrite Eq. (2.40) as

$$\frac{d\vartheta}{dr} = \frac{\partial H_0(I)}{\partial I} - \varepsilon\frac{\partial H_{l,m}(I)}{\partial I}\cos\Psi_{l,m}, \tag{2.123}$$

where $\Psi_{l,m} = l\vartheta - mk_r r + \phi_0 - \pi$, and consider its behavior near elliptic (or center) fixed points of resonance. They are determined by conditions $\Psi_{l,m} = 0$ (this follows from Eq. (2.39)) and $ld\vartheta/dr = mk_r$. Taking that into account, Eq. (2.123) transforms to

$$\frac{m}{l}k_r = \frac{\partial H_0(I)}{\partial I} - \varepsilon\frac{\partial H_{l,m}(I)}{\partial I}. \tag{2.124}$$

Unless

$$\left|\frac{\partial^2 H_0}{\partial I^2}\right| < \varepsilon\left|\frac{\partial^2 H_{l,m}}{\partial I^2}\right|, \tag{2.125}$$

Eq. (2.124) has an unique root, determining the location of the elliptic point in the action space. However, if inequality (2.125) holds, then a pair of additional roots of Eq. (2.124) may occur. Each novel root corresponds to an elliptic or a hyperbolic fixed point. This is either a saddle-center or a pitchfork bifurcation. The same situation happens with hyperbolic fixed points located at $\Psi_{l,m} = \pi$. The bifurcations can appear pairwise: if the saddle-center bifurcation emerges near elliptic fixed points, then the pitchfork bifurcation almost simultaneously arises with hyperbolic ones, and vice versa. The elliptic points formed correspond to distinct chains with l islands, therefore, the resonance $l:m$ acquires two satellites of the same order.

Number of roots grows with increasing of ε, therefore, more and more periodic ray orbits appear. This process we shall refer to as the *proliferation of periodic ray orbits*. Each hyperbolic orbit belongs to a chaotic layer, and the proliferation results in merging of these layers into a chaotic sea. So that, global ray chaos arises.

To illustrate this situation, we invoke once again the model (2.27) and put $\lambda_z = 0.2$ m. Vertical resonance with $p_{\max} = 0.04$ affects the area of the ray-medium resonance 15 : 2. With $\varepsilon = 0.0001$, there is an unique chain of 15 islands (see Fig. 2.14(a)). Twice increasing of ε doesn't change neither the location of the chain nor the number of islands, but the islands become stretched in the direction of increasing action. It is demonstrated in Fig. 2.14(b). This is just an initial stage before the bifurcation, after which the phase space patterns looks as in Fig. 2.14(c), where the case of

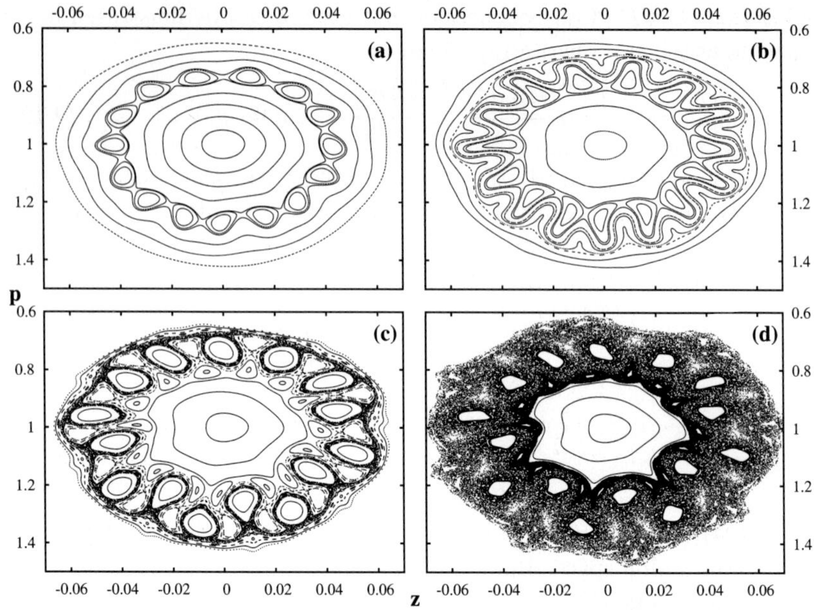

Fig. 2.14 The birth of a chaotic layer in the vicinity of vertical resonance with $\lambda_z = 0.2$ km and $\lambda_r = 5$ km. (a) $\varepsilon = 0.0001$, (b) $\varepsilon = 0.0002$, (c) $\varepsilon = 0.00035$, (d) $\varepsilon = 0.0005$.

$\varepsilon = 0.00035$ is presented. One chain transforms into three, each has the same number of islands as the initial one. Such bifurcations are peculiar for degenerate dynamical systems or non-twist maps [Van der Weele and Valkering (1990); Budyansky *et al.* (2009)]. With further increasing ε, there emerges a chaotic layer which devours one of the chains (see Fig. 2.14(d)). The other two survive as islands of stability in the chaotic environment.

Thus, we considered two possible scenarios of the transition to global chaos under vertical resonances. The first one is overlapping of resonances due to selective growth of the respective Fourier-amplitudes. The second one is associated with the proliferation of periodic ray orbits due to bifurcations.

2.5.4 *Resonance-induced chaotic layer*

As it follows from preceding subsections vertical resonance induces a chaotic layer in the phase-space region specified by the formula (2.99). Here we proceed with studying the phase space under vertical resonance and give insight onto metamorphoses of the phase-space pattern which happen as λ_z varies and λ_r is a constant.

As it is realized from the above analysis, a variation of the ratio λ_z/λ_r changes the location of a vertical resonance in the phase space, but doesn't change the size of the separatrix loop. However, it should be emphasized that the vertical resonance actually affects much larger region than one enclosed by the separatrix. To demonstrate it, we rewrite the Hamiltonian (2.85) as

$$\tilde{H} = \frac{k_z y^2}{2} - \frac{\varepsilon}{2} k_z Y\big(z(r)\big)\cos\Psi + \Psi U_z'\big(z(r)\big). \tag{2.126}$$

The only meaningful difference of this formula as compared with Eq. (2.85) is that we underlined explicitly the range dependence of the depth variable z. If we assume that $z(r)$ is some known periodic function of range, the Hamiltonian (2.126) becomes a Hamiltonian of a simple nonlinear pendulum with the frequency given by

$$\Omega = k_z\sqrt{\varepsilon Y(r)}. \tag{2.127}$$

Generally speaking, this assumption is not valid because we cannot decouple a ray trajectory $z(r)$ and dynamics near resonance, but it provides a way to understanding some important properties of vertical resonance.

For simplicity, we set $Y = 1$. Then the third term in the Hamiltonian (2.126) serves as an external perturbation, which leads to destruction of the separatrix of the pendulum and chaos, as we convinced ourselves in Sec. 2.3. The emphatic point is that the width of this chaotic layer depends on the ratio of frequencies of the perturbation and the unperturbed pendulum. In our case the frequency of the perturbation is ω, the spatial frequency of ray oscillations in a waveguide. So, the most wide chaotic layer is expected in the "adiabatic" regime [Soskin *et al.* (2005)]

$$\frac{\omega}{\Omega} = \frac{\omega}{k_z\sqrt{\varepsilon}} \ll 1, \tag{2.128}$$

i.e. in the limit of large values of k_z. Thus, the larger k_z, the larger the chaotic layer induced by vertical resonance.

We present some numerical illustrations, using the model of a waveguide described by Eq. (2.27). Now we fix the amplitude of the sound-speed perturbation at $\varepsilon = 0.005$ and consider four values of λ_z: 2000 m, 1000 m, 500 m and 200m. In each case we construct a Poincaré map. The horizontal wavelength of the perturbation λ_r is taken to be 5000 m for all the cases.

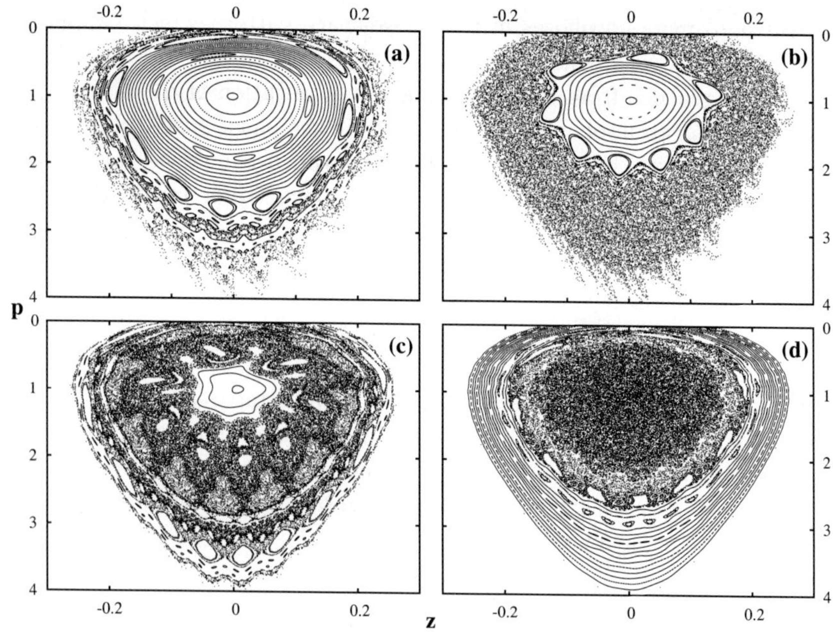

Fig. 2.15 Poincaré maps computed with $\lambda_z = 2000$ m (a), $\lambda_z = 1000$ m (b), $\lambda_z = 500$ m (c), and $\lambda_z = 200$ m (d).

The results of our calculations is presented in Fig. 2.15. Let us discuss them in the order of decreasing λ_z. In the case of $\lambda_z = 2000$ m, the zone of vertical resonance, determined by the condition $p_{\max}(E) = |p(H, z = z_a)| = 0.4$, lies outside the interval corresponding to rays propagating without reflections off the lossy bottom. Since that ray motion is predominantly stable, as it is shown in the Poincaré map depicted in Fig. 2.15(a). The phase space of the ray equations contains a large central island of stability surrounded by the chaotic sea. There are several chains of ray-medium resonances inside the central island. The most prominent one corresponds to the ray-medium resonance $9:1$ ($m = 9$ and $n = 1$). Other chains have remarkably smaller widths.

With decreasing λ_z twice, the condition (2.99) is fulfilled for steep rays with $p_{\max} \simeq 0.2$. This reduces considerably the area of the central island and destroys the ray-medium resonance $9:1$ (see Fig. 2.15(b)). The ray-medium resonance $8:1$, on the contrary, is enhanced. The respective chaotic layer is open, i.e. there are no barriers preventing rays from chaotic swinging with an eventual reaching the attenuating bottom. This is an effect of ray escaping causing additional transmission losses in a long-range sound-propagation. As it was shown by Makarov and Uleysky (2007), vertical resonance affecting steep rays causes especially strong losses due to ray escaping.

With $\lambda_z = 500$ m, the phase space pattern becomes especially non-trivial, as one can see in Fig. 2.15(c). There are two nested chaotic layers separated from each other by a narrow strip of stability. The inner layer, occurred due to the vertical resonance with $p_{\max} \simeq 0.1$, contains numerous islands of stability within. An attentive view reveals that these islands belong to three chains, each of them consists of 8 islands. Such chains arise after the bifurcation discussed in subsection 2.5.3. The outer layer is comparatively narrow and mainly caused by the contacts of ray-medium resonances with the accidental degeneracy associated with reflections off the ocean surface.

The case of $\lambda_z = 200$ m is depicted in Fig. 2.15(d). The main peculiarity of this Poincaré map is the presence of a wide and almost uniform chaotic layer occupying the central domain of the phase space and being surrounded by invariant curves. Another detail to be noted is that there is no remarkable islands inside the layer, except for the chain near the outer boundary. It implies almost ergodic ray diffusion inside the chaotic layer. To verify that, let us consider a narrow-angle ray bundle initially belonged to the chaotic layer in the phase space. Let the bundle be emitted from the waveguide axis. The phase-space pattern corresponding to the initial bundle is shown in Fig. 2.16(a). In numerical simulation, the bundle was modeled as an ensemble of 200000 rays with random initial values of depth and momentum, concentrated about $z = z_{\mathrm{a}} = 1$ km and $p = 0$ with very small variances. High density of rays allows one to interpret the bundle as an example of a Lagrangian manifold.

Evolution of the bundle in the phase space is presented in Fig. 2.16. At the first stage, the initial spot is stretching and transforms into a weakly-folded string. It should be noted that the vertical period of the string coincides with λ_z. Then, after a sequence of stretchings and foldings, different segments of the Lagrangian manifold begin to intersect each other. It leads to a transformation of the initial spot into a fuzzy cloud.

Mechanism of the transformation becomes clear if we give insight to single ray trajectories involved into the bundle. Figure 2.17 illustrates three rays with initial momenta of -0.0001, 0, and 0.0001. Since the rays are very close, they simultaneously fall into the vertical resonance at the first

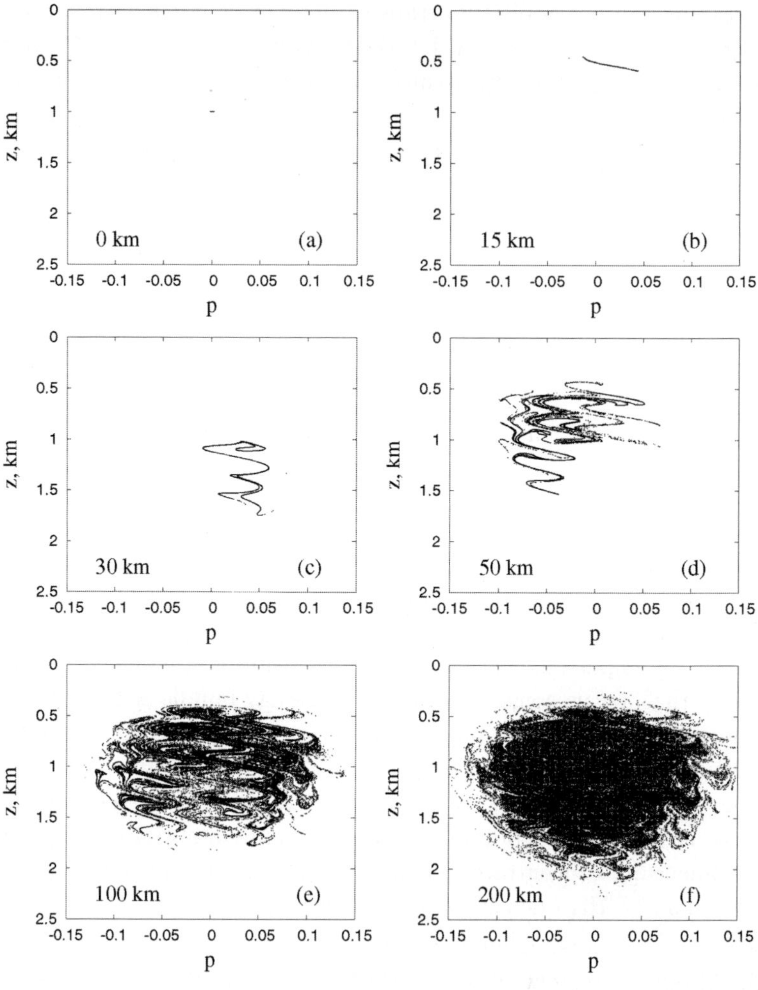

Fig. 2.16 Evolution of a Lagrangian manifold corresponding to an initially narrow bundle of chaotic rays. Each figure represents a snapshot of ray distribution in the phase space at the ranges of 0 (a), 15 km (b), 30 km (c), 50 km (d), 100 km (e), and 200 km (f).

half-cycle. Then two of them, after being scattered on the resonance, propagate like rays in an unperturbed waveguide, weakly diverging. The third ray falls into the resonance again and appears to be delayed in the upper ocean layer. Consider a Lagrangian manifold embedded into the original one and containing those three rays. Assume that the first and third rays

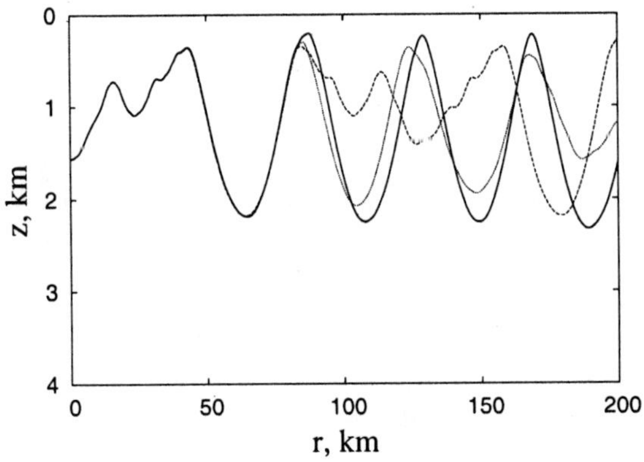

Fig. 2.17 Three close rays from the ray bundle shown in Fig. 2.16 [Makarov and Kon'kov (2007)].

are the ends of the embedded manifold, and the second ray lies amid them. Then the observed delay of the third ray at small depths means fast increasing of the distance between the ends of the manifold, and the embedded manifold begins to stretch fast. When the third ray leaves the resonance and intersects the other two rays in the opposite phase of the trajectory, the embedded manifold undergoes folding and self-crossing.

Now let's give insight on the phase space region corresponding to steep rays. There is no any signatures of chaos, despite of the presence of local degeneracy associated with reflections off the ocean surface. It is not surprising, because the inhomogeneity acts on steep rays in a non-resonant way. As it was pointed out in the beginning of this Section, far from vertical resonance, range-dependent terms in the ray Hamiltonian can be removed using the averaging method, therefore, ray equations are reduced to the integrable ones with solutions stable by Lyapunov. Moreover, as it was shown subsection (2.5.3), far from vertical resonance ray-medium resonances appear suppressed, and there is no route to chaos by means of their overlapping.

2.5.5 *Influence of vertical resonance on a timefront of a received pulse*

In a range-dependent waveguide, rays with different values of action arrive to a receiver at different travel times. It was discussed in detail in Chapter 1. The steeper ray, the smaller the travel time. In experiments, on long-range

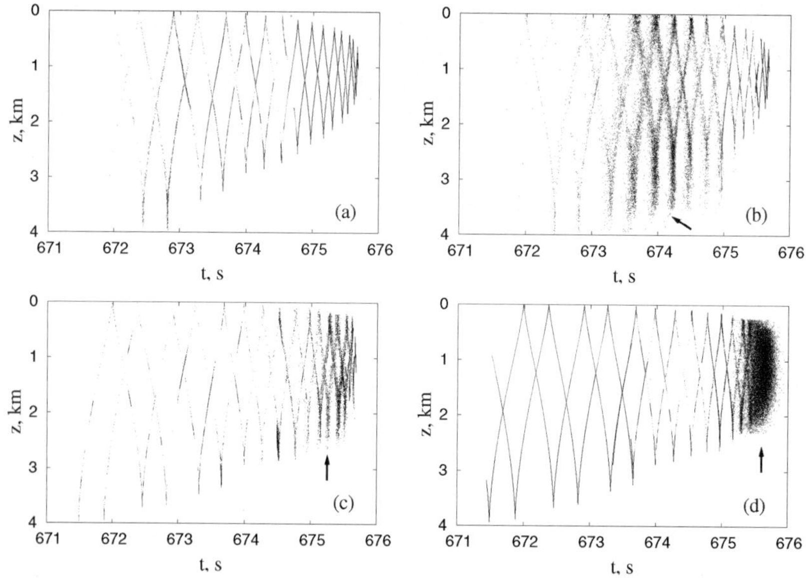

Fig. 2.18 Timefronts of a received pulse in an underwater sound channel described by Eq. (2.27). The distance between the source and receiver is 1000 km. Amplitude of the sound-speed perturbation ε is taken to be 0.005. The values of λ_z are: (a) 2000 m, (b) 1000 m, (c) 500 m, and (d) 200 m. Arrows indicate on chaotic segments of the timefronts.

sound propagation, the main measurable characteristics is a timefront, i.e. a wavefield represented in the plane arrival time–arrival depth (see, for instance, Fig. 1.18). Vertical resonance causes chaos of rays nearby a certain value of action and, therefore, affects a segment of timefront, corresponding to chaotic rays. Figure 2.18 represents timefronts ensuing from the ray calculations with the same values of λ_z as in Fig. 2.15. The source is placed at the channel axis $z_{\mathrm{a}} = 1$ km. In the case of $\lambda_z = 2000$ m, there is no remarkable signatures of irregularity in the timefront because ray motion is stable. With $\lambda_z = 1000$ m, vertical resonance causes strong escaping and, thus, filtering of early arrivals. As a consequence, the total duration of the pulse decreases. When λ_z decreases to 500 m, the most steepest rays become stable, and escaping ceases. Therefore, the total duration of the pulse is restored. On the other hand, chaos penetrates into the middle part of the timefront, causing fuzzy irregular pattern. Finally, in the case of $\lambda_z = 200$ m, vertical resonance affects near-axial rays and gives rise to the unresolved irregular pattern in the arrival finale. On the contrary, steep rays, being insensitive to vertical resonance, exhibit remarkable stability. A bare comparison of Fig. 2.18(d) with Fig. 1.18 suggests that vertical oscillations of an internal-wave field should play an essential role in stability of early arrivals and chaoticity of late ones. More detailed discussion of travel time distributions under chaotic conditions will be presented in the next Section.

2.6 Manifestation of Regular and Chaotic Ray Motion in Distributions of Ray Travel Times

The properties of the ray travel times in a range-independent waveguide were examined in Sec. 1.4. In the present section, we will see how they change in the presence of a perturbation with periodic range dependence. Our attention will be focused on the following two issues. The first of them is a remarkable relationship between the ray travel time and the identifier (topology) of the ray trajectory. It turns out that the rays (regular or chaotic) with equal identifiers have close travel times. This is a general property which remains valid, as we will see in Chapter 4, in the presence of a random perturbation. The second issue considered here is specific for a range-periodic waveguide. A typical feature of this environmental model is the coexistence of regular and chaotic rays. In the present section, we will see how this fact manifests itself in the distribution of ray travel times and in the distribution of sound energy in the time-depth plane at finite frequencies [Smirnov *et al.* (2001, 2002, 2004, 2005b)].

2.6.1 *Perturbed waveguide*

The presence of a perturbation δc, giving rise to ray chaos, makes the distribution of ray travel times much more complicated compared with what we have seen in a range-independent waveguide (Sec. 1.4). The general properties of the travel times under conditions of ray chaos are illustrated below with an example of a perturbed waveguide with the sound speed field $c(r,z) = \bar{c}(z) + \delta c(r,z)$, where $\bar{c}(z)$ is the Munk profile (1.102), and

$$\delta c(r,z) = -2\varepsilon c_0 \frac{z}{B} e^{-2z/B} \sin(2\pi r/\lambda_r), \tag{2.129}$$

with $\varepsilon = 0.01$ and $\lambda_r = 10$ km, is a periodic perturbation that generates the ray chaos. It is assumed that a perfectly reflecting bottom is set at depth $z = 5$ km. The phase portrait of the ray system in this waveguide is

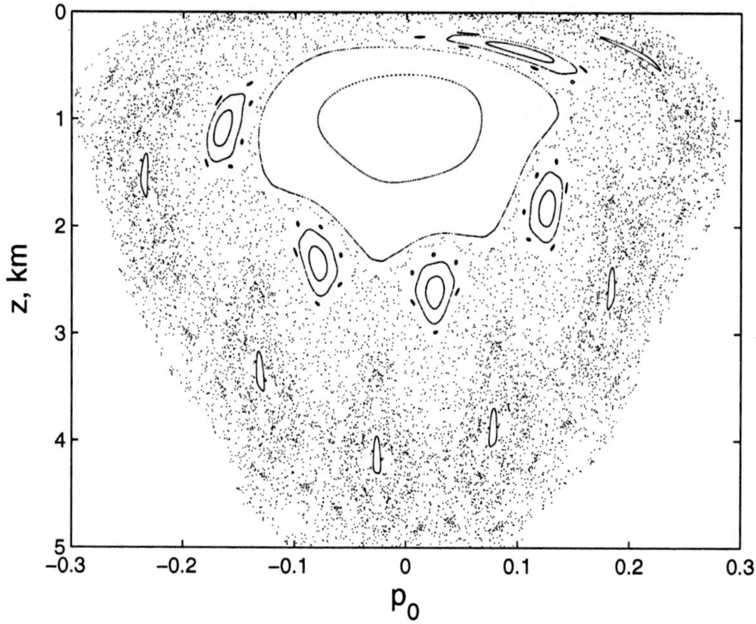

Fig. 2.19 Phase portrait of the ray system constructed using the Poincaré map.

shown in Fig. 2.19. It was plotted using the Poincaré map described in the Glossary. This is almost the same map as depicted in Fig. 2.7, except for the difference in the depth of the ocean bottom. In Sec. 2.3 we have emphasized the role of ray-medium resonances in the emergence of ray chaos. In the environmental model under consideration, the dependence of the cycle length on the magnitude of the ray momentum p_{a} at the sound channel axis $z = z_{\mathrm{a}}$ in the absence of perturbation ($\delta c = 0$) is shown in Fig. 2.20. The rays with $p_{\mathrm{a}} < 0.2193$ propagate without reflection from the surface and bottom. The ray with p_{a} from an interval $0.2193 < p_a < 0.274$ are surface reflected, and the rays with $p_a > 0.274$ reflect off both the surface and the bottom. The period of the perturbation in our environmental model is $\lambda_r = 10$ km. For the rays with not large values of p_a shown in Fig. 2.20, there exist only two primary resonances with $D = 5\lambda_r$ (for $p_{\mathrm{a}} = 0.1558$) and $D = 6\lambda_r$ (for $p_{\mathrm{a}} = 0.2465$). A five-fold island chain surrounding a central big island in the phase portrait in Fig. 2.19 is associated with the first of these resonances. Each of the five islands is surrounded by six small satellite islands associated with one of the secondary resonances. The second

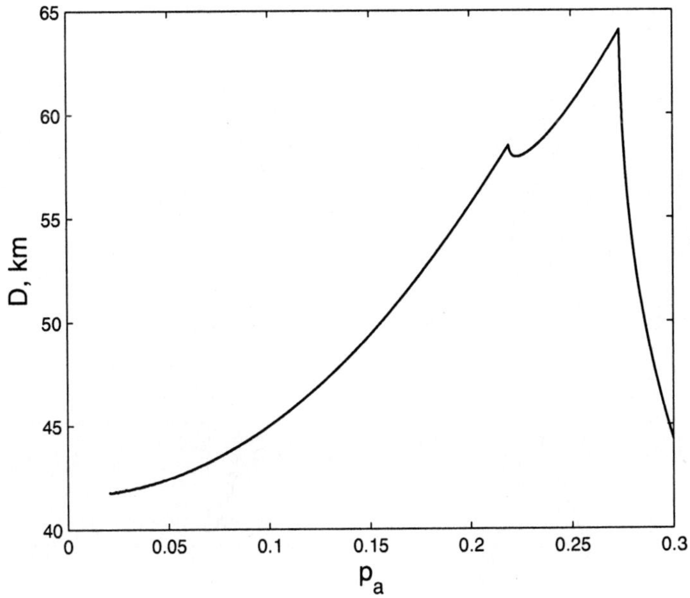

Fig. 2.20 Ray cycle length D as a function of ray momentum p at the sound channel axis $z_{\mathrm{a}} = 1$ km in the unperturbed (range-independent) waveguide.

primary resonance is associated with a similar chain of six islands located closer to the outer border of the chaotic sea. All other islands (the total number of these islands is infinite) are indiscernible in the phase portrait. In this section we consider the rays escaping a point source set at the sound channel axis $z_{\mathrm{a}} = 1$ km. Therefore, starting points (in the phase plane) of the rays leaving the source lie on the straight line $z = z_{\mathrm{a}}$. According to the phase portrait shown in Fig. 2.19, the central part of this line corresponding to an interval of starting momenta

$$-0.131 < p_0 < 0.133 \tag{2.130}$$

belongs to the central stable island. This means that flat rays are predominantly regular, while steep rays with p_0 outside the above interval are predominantly chaotic. In Chapter 4 we will see that in a more realistic environmental model with a perturbation δc induced by random internal waves, on the contrary, the steep rays are much more regular than the flat ones [Simmen *et al.* (1997); Brown and Viechnicki (1998); Beron-Vera *et al.* (2003)].

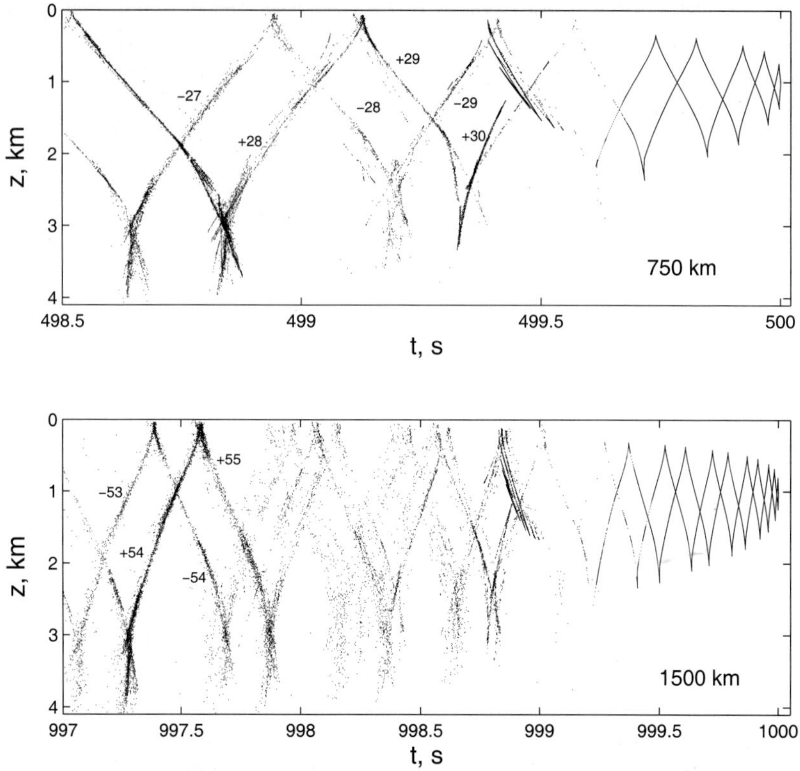

Fig. 2.21 Timefronts at 750 km (upper panel) and 1500 km (lower panel) in the presence of a perturbation giving rise to ray chaos. The identifiers of rays forming some resolved fuzzy segments are shown next to these segments.

2.6.2 *Timefront*

Figure 2.21 presents timefronts at ranges 750 km (upper panel) and 1500 km (lower panel) formed by rays with the same starting parameters as in Fig. 1.4. Note that the dependence of the ray depth at the range of observation on the ray travel time is always expressed by a continuous function. However, under conditions of ray chaos this function is so complicated that our fan of rays used in the evaluation of the timefront is not enough dense for visualization of this continuity. The presence of randomly scattered points in the $t - z$ plane is an indicator of chaos.

Comparison of Figs. 1.4 and 2.21 shows that the segments corresponding to arrival times $t > 499.7$ s at 750 km and $t > 999.4$ s at 1500 km in the presence of a perturbation retain their shapes and almost do not change. The point is that in the case of a periodic perturbation some rays remain regular. Effects caused by the coexistence of chaotic and regular rays will be considered below. In the environmental model determined by Eqs. (1.102) and (2.129) considered in this section, regular rays propagate at small grazing angles and form stable segments in the late portion of the timefront. The rays with steeper grazing angles are chaotic.

Arrival of chaotic rays with a given identifier form in the $t - z$ plane what we call the fuzzy segment. Some fuzzy segments in Fig. 2.21 are resolved in the sense that they do not overlap with their neighbors. It turns out that each fuzzy segment (no matter resolved or not) is localized in the vicinity of a segment with the same identifier in the unperturbed waveguide. This fact expresses a remarkable effect that travel times of the rays with equal identifiers are close not only in the unperturbed waveguide, but in the perturbed waveguide as well [Palmer *et al.* (1991); Tappert and Tang (1996); Virovlyansky (2001); Beron-Vera *et al.* (2003)]. Figure 2.22 shows fuzzy segments associated with the identifiers +58 (points) and +59 (circles) at 1500 km that are not resolved in Fig. 2.21. We see that these segments

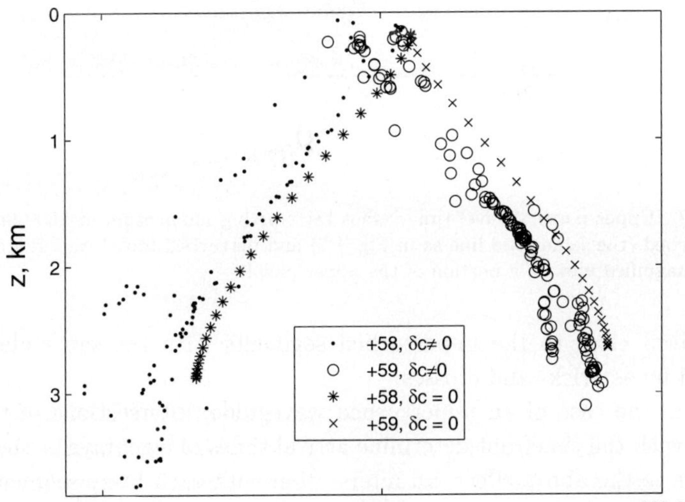

Fig. 2.22 Fragments of the timefronts at 1500 km range formed by rays with the identifiers equal to +58 and +59 in the perturbed (points and circles) and unperturbed (asterisks and crosses) waveguide.

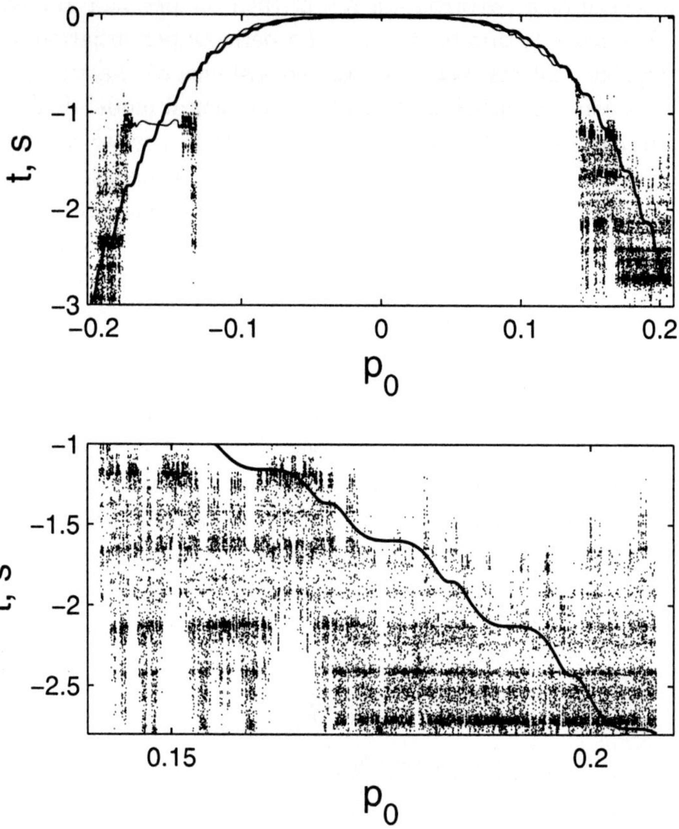

Fig. 2.23 Upper panel: Travel time versus ray starting momentum at 1500 km in the unperturbed (the same solid line as in Fig. 1.7) and perturbed (dots) waveguide. Lower panel: magnified view of a portion of the upper panel.

are indeed close to the unperturbed segments with the same identifiers marked by asterisks and crosses.

As in the case of an unperturbed waveguide, intersections of the line $z = z_r$ with the timefront determine arrival times of eigenrays at the depth z_r. Due to the above effect, an intersection with each fuzzy segment gives a group of eigenrays with comparatively close travel times. It means that that the ray travel times come in compact clusters [Palmer *et al.* (1991); Tappert and Tang (1996); Virovlyansky (2001); Beron-Vera *et al.* (2003)]. Each cluster is formed by eigenrays with equal identifiers. The center of a cluster is close to the arrival time of an unperturbed ray with the same identifier. In Chapter 4 (Sec. 4.2) we will see that the effect of clustering of ray travel times is observed in an environmental model with a random perturbation δc, as well. Its quantitative explanation will be given for a case when the perturbation is induced by random internal waves.

Figure 2.23 shows the ray arrivals in the $t - p_0$ plane at 1500 km in the presence of a perturbation (points). Solid line (the same as in Fig. 1.7) graphs the $t(p_0)$ curve in the unperturbed waveguide. It is clearly seen that the perturbation only weakly affects the distribution of arrivals of regular rays with $|p_0| < 0.12$. In contrast, the arrivals of chaotic rays with starting momenta $p_0 > 0.12$ are depicted by randomly scattered points. However, the density of these points has maxima at regular dark horizontal stripes. It should be emphasized that the stripes are located at travel times close to that of "shelves" on the unperturbed $t(p_0)$ curve [Smirnov *et al.* (2002)]. Since starting momenta p_0 of rays whose arrivals are presented in Figs. 1.7 and 2.23 are uniformly distributed within a fixed interval, the unperturbed "shelves", where the derivative $|dt/dp_0|$ is small, may be interpreted as parts of the $t(p_0)$ curve with the highest density of points. So, we can state that under conditions of ray chaos positions of maxima of the density of ray travel times remain relatively stable. In Chapter 4 we will see that a similar effect is observed in the presence of a random perturbation, as well. Its qualitative explanation will be done in Sec. 4.2.

2.6.3 *Amplitude of a pulse signal in plane "travel time–depth"*

The upper panel in Fig. 2.24 shows the amplitude of a short pulse as a function of time and depth in the unperturbed waveguide with $\delta c = 0$. This function has been computed at a range of 3000 km for a case when an initial pulse emitted by the point source has a shape

$$s(t) = \exp\left(-\frac{t^2}{2\tau^2}\right)\cos(\omega_0 t) \qquad (2.131)$$

determined by two parameters: a central frequency $\Omega/2\pi = 75$ Hz and an effective pulse duration $\tau = 0.017$ s. The lower panel presents a timefront computed by numerical ray tracing in the same (unperturbed) waveguide. It is seen that the timefront properly predicts distinctive features in the distribution of acoustic energy in the time-depth plane. Figure 2.25 presents results of similar computations performed in the perturbed waveguide. Consistent with our intuition, travel times of regular rays form

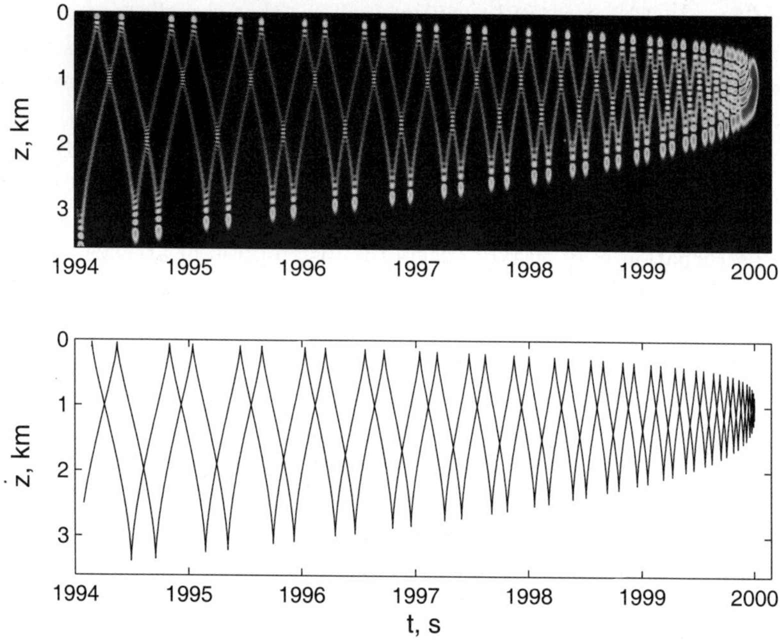

Fig. 2.24 Upper panel: Pulse signal in the unperturbed waveguide at the range 3000 km. Amplitude of the acoustic wave field is shown as a function of travel time and depth. Lower panel: Timefront in the unperturbed waveguide at the range 3000 km. The source is set at the sound channel axis $z_a = 1$ km.

regular curves on the plot depicting the timefront. In contrast, arrivals of chaotic rays at the same plot are presented as randomly scattered points (see comment to Fig. 2.21). A regular part of the perturbed timefront (a part of the lower panel corresponding to times $t > 1998.5$ s) is formed by rays propagating at small grazing angles and belonging to the big central island in the phase portrait. Fuzzy segments formed by rays with travel times of about 1995 are resolved comparatively well. The upper panel in Fig. 2.25 demonstrates that there are fuzzy segments formed by both regular and chaotic rays which clearly manifest themselves in the distribution of the sound field amplitude in the $t - z$ plane.

2.6.4 *Gap between travel times of chaotic and regular rays*

In Refs. [Smirnov *et al.* (2002, 2004, 2005b)] it has been shown that in our environmental model the coexistence of chaotic and regular rays causes an

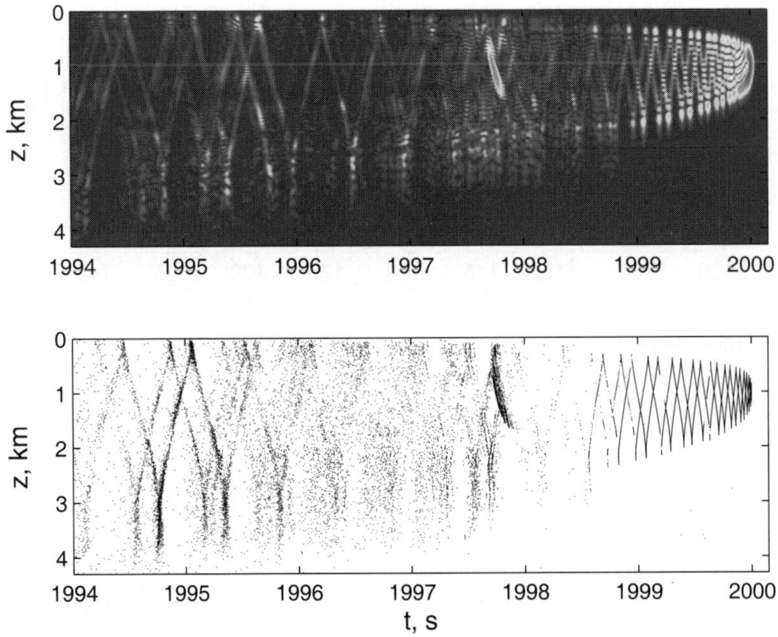

Fig. 2.25 The same as in Fig. 2.24 but for the perturbed waveguide.

appearance of a gap in distribution of ray travel times. In the lower panel of Fig. 2.25 this gap is observed in the interval of times between 1998 s and 1998.5 s. It separates arrivals of regular (belonging to the central island) and chaotic rays. In Fig. 2.21 similar gaps are seen in the timefronts at ranges 750 and 1500 km. In order to explain this phenomenon turn our attention to Fig. 2.26 where a distribution of numbers of turning points M at range 3000 km is shown for rays with negative (upper panel) and positive (lower panel) starting momenta. Both distributions have gaps at $M = 121$, 122, and 123. So the percentage of rays with the identifiers ± 121, ± 122, and ± 123 is comparatively small. Earlier it has been noted that the ray travel time to a significant extent is determined by the identifier (that is, by the topology) of a corresponding ray path. Therefore it is not surprising that a gap in the distribution of ray turning points corresponds to a gap in the distribution of travel times. Equation (1.115) allows one to establish a quantitative connection between these two gaps. This formula may be considered as an estimate of the travel time spread for rays with the same identifier in the unperturbed waveguide. In Sec. 4.2 we will see that in the

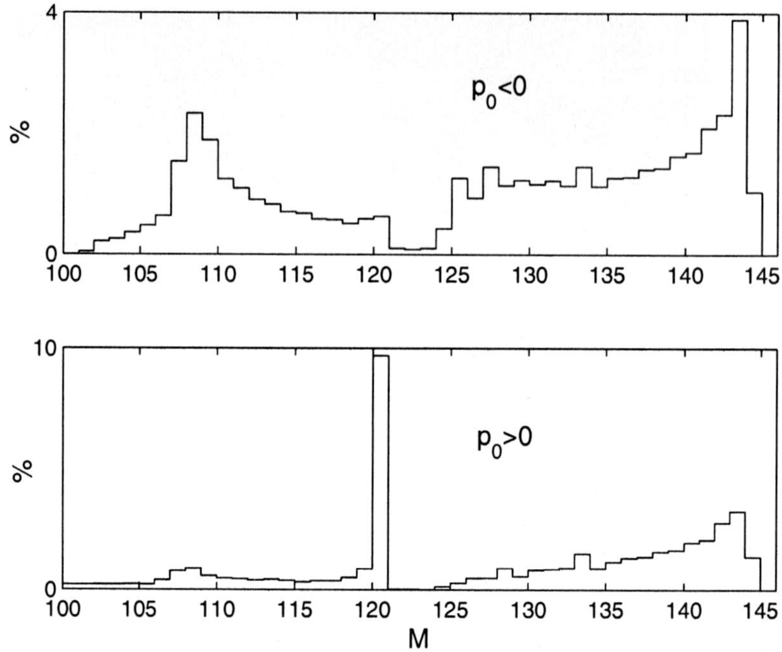

Fig. 2.26 Percentage of rays as a function of M, the number of turning points, at the range of 3000 km. Upper and lower panels present results for rays with negative and positive starting momenta, respectively.

presence of a weak perturbation this formula may also be applied as a rough estimate. Then the quantity $\bar{I}$ should be considered as the mean value of starting actions of rays which have the given identifier at the observation range. Thus, the width of the gap in the travel time distribution can be roughly estimated as

$$\Delta T_{\text{gap}} = \pi \bar{I} \Delta M / c_0, \tag{2.132}$$

where ΔM is the gap in the distribution of ray turning points, and $\bar{I}$ is the mean starting action of rays whose identifiers at the observation range in the unperturbed waveguide belong to our gap. In our case $\Delta M = 3$, $\bar{I} = 0.08$ km and according to Eq. (2.132) we find $\Delta T_{\text{gap}} = 0.5$ s. This is consistent with the results of numerical simulation presented in Fig. 2.25. Similarly, Eq. (2.132) gives correct order-of-magnitude estimates relating the gaps in travel times observed in Fig. 2.21 to the corresponding gaps in distributions of ray turning points at 750 and 1500 km (not shown).

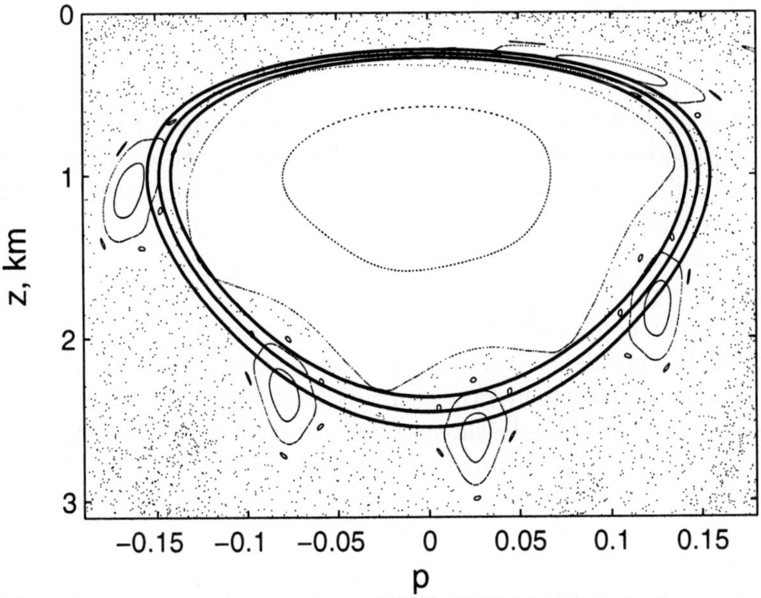

Fig. 2.27 Trajectories in phase space of unperturbed rays with the identifiers +121, +122, and +123 (thick solid curves) superimposed on the phase portrait.

The presence of the gap in the distributions of ray turning points may be qualitatively explained by the presence of the five stable islands surrounding the central island in the phase portrait (Fig. 2.19) [Smirnov *et al.* (2002)]. The point is that chaotic and regular (unperturbed) rays with equal identifiers have equal mean cycle lengths and, as a consequence, close values of mean actions. Therefore trajectories of these rays in the phase space should be more or less close. Thick solid curves in Fig. 2.27 depict trajectories of unperturbed rays with the identifiers +121, +122, and +123 in the unperturbed waveguide. It is seen that they intersect areas occupied by the five stable islands present in the phase portrait. The same is true of trajectories with the identifiers −121, −122, and −123 (not shown). Since chaotic rays cannot "penetrate" into the islands this fact qualitatively explains the lack of rays with the aforementioned identifiers in the presence of perturbation. Comparing the upper and lower panels in Fig. 2.25 we see that due to effects of diffraction the gap, so clearly seen in the timefront, to a considerable extent is "filled" (in both space and time) in the arrival pattern computed at a finite wavelength.

2.6.5 *"Focusing" of ray travel times*

Numerical simulation demonstrates another interesting feature in the distribution of sound energy in the time-depth plane [Smirnov *et al.* (2002, 2005b)]. We mean a bright spot clearly seen in the upper panel of Fig. 2.25 at times between 1997.7 s and 1997.8 s. Its appearance is another manifestation of the coexistence of chaotic and regular rays. In order to examine this phenomenon, first note that the density of dots in the timefront is approximately proportional to the acoustic field intensity. In the lower panel of Fig. 2.25 presenting the timefront at the range 3000 km we see an area with abnormally high density of points. It is located at the same part of the time-depth plane as the bright spot in the upper panel and is formed predominantly by rays with the identifier −120. Figure 2.28 shows a magnified view of the bright spot (left panel) and depicts arrivals of rays with the identifier −120 in the time-depth plane (right panel). Recall that according to the distribution presented in Fig. 2.26 the percentage of rays with identifier −120 is much greater than that for any other identifier.

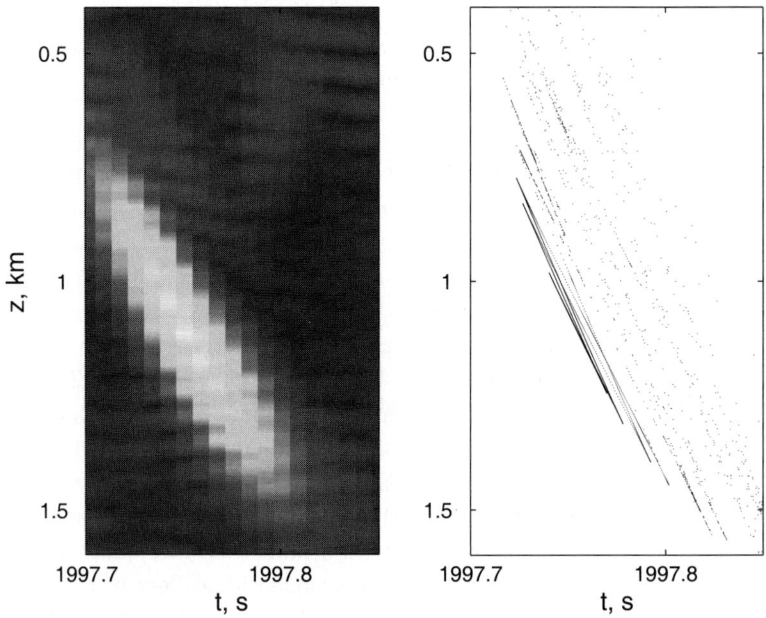

Fig. 2.28 Left panel: A magnified view of the portion of the arrival pattern containing the bright spot. Right panel: A part of the time front in the perturbed waveguide formed by rays with the identifier −120.

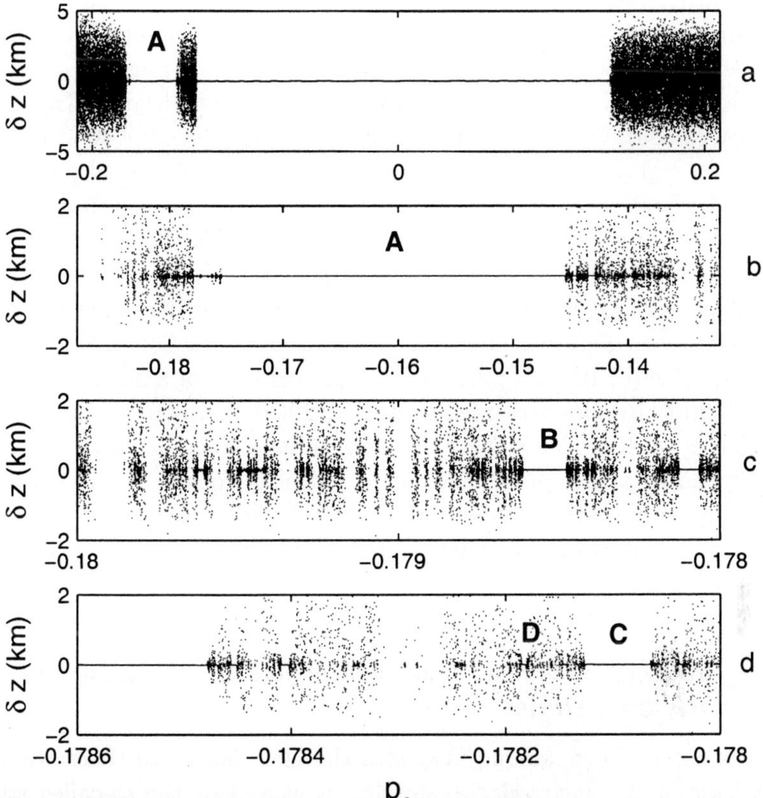

Fig. 2.29 Depth differences at 3000 km range for pairs of rays with infinitesimally close starting momenta. Each point represents the absolute difference in arrival depth for two rays: a ray with original momentum and one with a slightly perturbed momentum (which is the original plus 10^{-7}). The differences are shown for all fan rays whose starting momenta belong to the interval $-0.21 < p_0 < 0.21$ (a) and for rays with the identifier -120 and starting momenta from the intervals $-0.19 < p_0 < -0.13$ (b), $-0.18 < p_0 < -0.178$ (c), and $-0.1786 < p_0 < -0.178$ (d).

Another important thing about the rays with the identifier -120 is that they are relatively stable. In order to characterize quantitatively the sensitivity of ray paths to small variations in initial conditions we have followed Ref. [Simmen *et al.* (1997)] and computed the absolute difference, $\delta z = |z_1 - z_2|$, in arrival depths, z_1 and z_2, of two rays with starting momenta that differ by $\delta p_0 = 10^{-7}$. In Fig. 2.29(a) this difference is shown as a function of starting momentum p_0 at the range 3000 km. The plot has been constructed by tracing numerically 100 000 pairs of fan rays escaping the point source set at depth z_a. In this plot we see two intervals of starting momenta corresponding to very small δz. One of them presents a segment of straight line $z = z_a$ that lies within the big central island. This is the interval (2.130). Another interval — it is marked by symbol **A** — includes p_0 from a segment of line $z = z_a$ within a smaller stable island (one of the five-fold island chain) to the left of the central one. A detailed analysis shows that all rays belonging to the interval **A** have the identifier -120. In Fig. 2.29(b) we present δz for all pairs corresponding to rays with the identifier -120. It is seen that a lot of rays with this identifier belong to the chaotic sea. Figure 2.29(c) is a zoom of a small fraction of Fig. 2.29(b), and Fig. 2.29(d) in turn provides a magnified view of a small fraction of Fig. 2.29(c). Results presented in Figs. 2.29(b), (c), and (d) look similar suggesting that the dependence of δz on p_0 has a fractal structure (see the corresponding article in the Glossary). All plots consist of intervals filled with randomly scattered points interspersed with intervals of very small δz. At such intervals, the dependencies of δz on p_0 look like segments of straight lines.

2.6.6 *Role of stickiness and chaotic jets in "focusing" of ray travel times*

As was noted above, among rays with the identifier -120 there are a lot of chaotic ones. Their relative stability is caused by the so-called stickiness, i.e., the presence of such parts in chaotic trajectories where the latter exhibit an almost regular behavior (see the article "Dynamical traps" in the Glossary). This occurs when after wandering in the phase space the trajectory approaches a stable island and "sticks" to its border for some time that may be fairly long [Zaslavsky *et al.* (1997)]. The point is that regular rays belonging to any of the islands forming the five-fold island chain at the range 3000 km have the identifier -120. A significant part of rays starting from points in the phase plane close to these islands stick to their borders and keep near the borders at a long enough range interval. At this interval the sticky rays exhibit almost regular behavior mimicking that of regular rays forming the islands. Therefore, the sticky rays have the same identifiers as their regular counterparts from the islands.

This statement is illustrated in Fig. 2.30. In the left column we see two Poincaré maps for a typical ray with starting momentum from the interval **B** in Fig. 2.29(c). The trajectory has been computed up to a

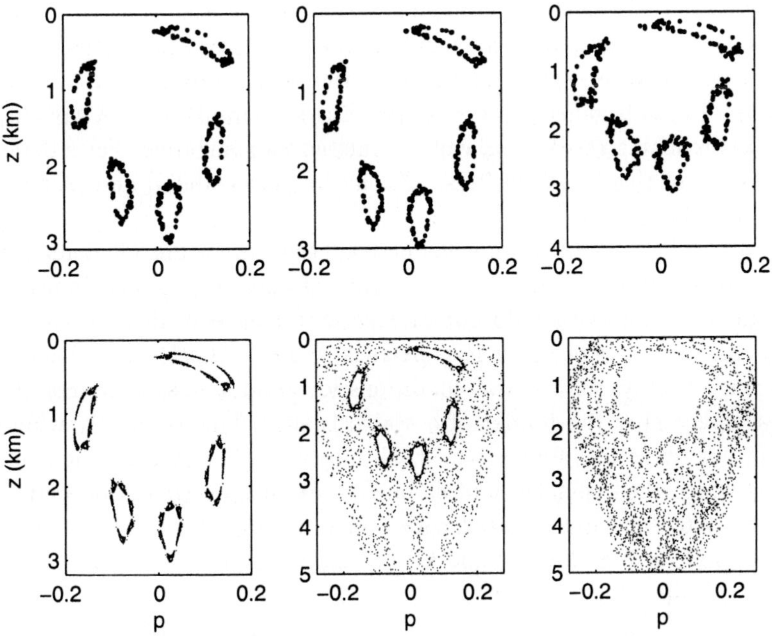

Fig. 2.30 Poincaré maps constructed using ray trajectories with 3000 km lengths (upper row of plots) and 30 000 km lengths (lower row of plots).

range of 30 000 km. In the upper plot we use only the first 3000 km of this trajectory. The lower plot has been constructed using the whole ray path. By comparing the upper and lower panels we conclude that while up to 3000 km the trajectory sticks to the island boundaries at longer ranges it begins to break away. In the central column of Fig. 2.30, we see two analogous Poincaré maps computed for a ray with the starting momentum from the interval **C** in Fig. 2.29(d). Although up to 3000 km the ray adheres to the islands, the remaining part of its 30 000 km trajectory passes through different portions of the chaotic sea. In the right column we have presented Poincaré maps constructed for a ray from an interval **D** (see Fig. 2.29(d)) filled with randomly scattered points. Although at long ranges it covers almost all the chaotic sea, at first 3000 km it does not deviate significantly from the stable islands.

The phenomenon of stickiness is a consequence of an extremely complicated structure of the phase space in the neighborhood of the island (see the article "Cantorus" in the Glossary). Besides comparatively big islands seen in Fig. 2.19 there exist an infinite number of smaller islands indiscernible in our plots. Imbedding of such islands in the stochastic sea creates a fractal pattern of the phase space (its manifestation we see in Fig. 2.29). A fractal system of islands generates a corresponding system of traps. A ray trajectory approaching this system will be trapped for some time. For a detailed discussion of this issue see Refs. [Zaslavsky *et al.* (1997)] and [Afanasiev *et al.* (1991)].

Beside the stickiness, we encounter another important phenomenon of the chaotic motion. This is an existence of chaotic jets, i.e., bunches of chaotic trajectories that do not diverge at a long enough range interval. As it has been predicted in Ref. [Afanasiev *et al.* (1991)], the chaotic jet is formed by rays with initial conditions localized at a special area of the phase space near the border of a stable island. This is exactly what we see in our numerical simulation. Each interval of the starting momentum p_0, at which the dependence of δz on p_0 looks like a segment of a straight line $\delta z = 0$, can be associated with a chaotic jets. Typical examples are intervals **B** (in Fig. 2.29(c)) and **C** (in Fig. 2.29(d)).

2.6.7 *Smoothed intensity of pulse signal*

It is clear that the extreme sensitivity of chaotic trajectories to initial conditions and variations of environmental parameters should impose severe limitations on the predictability of underwater sound fields in the scope of the geometrical optics approximation (see Eq. (2.10)). On the other hand, it is known that even under condition of ray chaos the geometrical optics can properly predict intensities of some large-scale features of the wave field [Virovlyansky and Zaslavsky (2000)]. We have seen that the large bright spot in the arrival pattern is formed by rays with relatively stable trajectories. This suggests that the ray theory may predict a coarse-grained intensity in a corresponding part of the received signal.

In order to confirm this conjecture we consider a smoothed intensity defined as

$$J(r,t) = \frac{1}{\sqrt{2\pi}\Delta_t} \int dz'dt' \left|u(r,z',t')\right|^2 \exp\left(-\frac{(t-t')^2}{2\Delta_t^2}\right), \tag{2.133}$$

where Δ_t is a smoothing scale and the integration over the vertical coordinate z' goes along the whole cross-section of the waveguide. It follows from Eqs. (1.40) and (1.42) that in the geometrical optics approximation the wave field at a given observation point (r, z) is presented in the form

$$u(r,z,t) = \sum_j \sqrt{\frac{k_0}{2\pi i|\partial z/\partial p_0|}}\, s\left(t - t(r, p_0, z_s)\right)\Bigg|_{p_0 = p_{0,j}}, \tag{2.134}$$

where the symbol $t(r, p_0, z_0)$ denotes the travel time of a ray with starting momentum and coordinate equal to p_0 and z_0, respectively, at range r, and index j numbers eigenrays arriving at (r, z). Using incoherent summation of rays, we present the field intensity as

$$\begin{aligned} |u(r,z,t)|^2 &= \sum_j \frac{k_0}{2\pi|\partial z/\partial p_0|}\left|s\left(t - t(r,p_0,z_s)\right)\right|^2_{p_0=p_{0,j}} \\ &= \frac{k_0}{2\pi}\int dp_0\, \delta\left(z - z(r,p_0,z_s)\right)\left|s\left(t - t(r,p_0,z_s)\right)\right|^2. \end{aligned} \tag{2.135}$$

Assuming that the smoothing scale is greater than the duration of an initial pulse τ, i.e.

$$\Delta_t > \tau, \tag{2.136}$$

substituting Eq. (2.135) into Eq. (2.133), and neglecting small contributions from rapidly oscillating terms representing products of pulses coming to the observation point through different eigenrays, we arrive at

$$J(r,z,t) = \frac{k_0\tau}{2^{3/2}\pi\Delta_t}\int dp_0 \exp\left(-\frac{\left[t - t(r,p_0,z_0)\right]^2}{2\Delta_t^2}\right). \tag{2.137}$$

The last integral can be evaluated as a sum over a dense enough fan of rays escaping a point source. In our calculation we have considered a fan of 100 000 rays with initial grazing angles within the interval $\pm 12.5°$. In Fig. 2.31 we compare a ray-based prediction of the smoothed intensity predicted by Eq. (2.137) with the smoothing scale $\Delta_t = 0.04$ s to a result obtained by direct solving the parabolic equation (the same data has been used when constructing plots presented in Fig. 2.25). The comparison demonstrates that the ray-based approach provides a good prediction for both maxima of the smoothed intensity formed by contributions from predominantly stable rays. On the other hand, there is a significant discrepancy between the ray-based and PE-based results in the temporal interval from 1998 s to 1998.5 s. As it has been already noticed at the end of Sec. 2.6.4 the ray-based approach is unable to describe "filling" the gap in the timefront due to effects of diffraction.

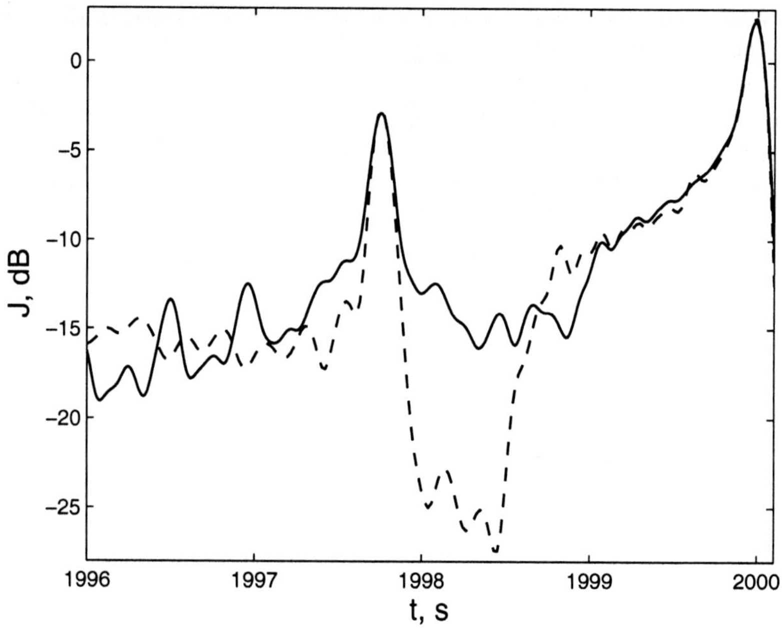

Fig. 2.31 Smoothed pulse intensity $J(r,t)$ defined by Eq. (2.133). Solid line: result obtained by direct solution of the parabolic equation. Dashed line: ray-based prediction provided by Eq. (2.137).

2.7 Summary

We described briefly in Sec. 2.1 the basic concepts and approaches in the theory of dynamical chaos in Hamiltonian systems. A notion of Lyapunov instability was described in more details in Sec. 2.2.

A simple model of the USC with a periodic range dependence, producing a ray-medium resonance, was considered in Sec. 2.3. As a result of the resonance, there appear chains of islands in the phase space. In Sec. 2.4 we show that overlapping of the neighboring chains produces stochastic layers within which ray motion is chaotic. Rays in any compact bundle rapidly diverge in that layer due to the Lyapunov instability.

In Sec. 2.5 we considered a more complicated USC model with a periodic range dependence taking into consideration oscillations with depth. It was shown that the vertical oscillations change qualitatively the dynamics of rays under conditions of a vertical resonance producing chaos in a region in the phase space, determined by the ratio of vertical and horizontal length scales of a sound-speed perturbation. It is shown that vertical resonance is accompanied by bifurcations of periodic ray orbits and leads to smearing of a specific segment of the timefront representing ray arrivals in the time–depth plane. In the case of small-scale vertical oscillations of a perturbation vertical resonance mainly affects near-axial rays forming the late portion of a received pulse. It is worth mentioning that ray motion far from vertical resonance is regular. This circumstance can explain the surprising stability of early arrivals observed in experiments.

In Sec. 2.6 we examined the distribution of ray travel times in a waveguide with a weak range-periodic perturbation of the sound speed field. The most important property exhibited by both chaotic and regular rays is a close connection between the travel time and the identifier of the ray path. Rays with equal identifiers have close travel times. Ray tracing demonstrated that the coexistence of chaotic and regular rays causes an appearance of a gap in the timefront. Besides the gap, a "focusing" of ray travel times was found within a comparatively small temporal interval preceding the gap. In numerical simulation of a sound pulse propagation, this phenomenon reveals itself in the appearance of a bright spot in the distribution of acoustic energy in the time–depth plane. It has been shown that the effect of focusing is a manifestation of the so-called stickiness, i.e., the presence of such parts of the chaotic trajectory where the latter exhibits an almost regular behavior.

be applied with a caution in the geophysical context because the theory is asymptotic in time but all underwater sound characteristics are finite-time ones.

3.1.1 *Ehrenfest time*

There are three levels of description of dynamics: classical (ray), semiclassical, and quantum (wave). In the semiclassical approach, we try to obtain approximate solutions to a wave equation in terms of wave amplitudes and phases using only classical notion of rays (trajectories). Classical and semiclassical approximate solutions follow to an "exact" wave solution for a time scale, known as a break time (which may strongly differ for classical-wave and semiclassical-wave correspondence), then begin to deviate from that. It is important to distinguish between the correspondence at classical and quantum levels and validity of the semiclassical approximation. *The Ehrenfest break-time* T_E is defined as the time scale for which propagation of the centroid of a wave packet can be approximately described by solving the classical Hamiltonian equations of motion without using the semiclassical approximation for the wave packet. If the classical motion is chaotic this time scale can be estimated as follows. The width of the one-dimensional wave packet Δx typically grows in the unstable direction as $\Delta x_0 \exp(\nu t)$, where ν is a Lyapunov exponent of the classical dynamics. The Ehrenfest break-time is defined as the time for Δx to reach the characteristic scale of the nonlinear confining potential l

$$T_E = \frac{1}{\nu} \ln \frac{l}{\Delta x_0}. \tag{3.1}$$

This is a time of correspondence between the wave and ray descriptions for waves of any nature. In the classically chaotic regime, T_E tends to infinity in the limit $\Delta x_0 \to 0$ only logarithmically. The other name of the time scale (3.1) is "a log time".

In quantum mechanics the volume in the phase space covered by a wave packet with the minimal uncertainty is $\Delta x_0 \Delta p_0 \simeq \hbar$, and the quantum-classical correspondence in the classically chaotic regime breaks down at the time [Berman and Zaslavsky (1978); Zaslavsky (1981)]

$$T_E^{\hbar} = \frac{1}{\nu} \ln \frac{l \Delta p_0}{\hbar} \simeq \frac{1}{\nu} \ln \frac{I}{\hbar}, \tag{3.2}$$

where $\hbar$ is the Plank constant, I is a characteristic classical action, and $I/\hbar$ is a quasiclassical parameter. In the regular regime $T_E^{\hbar}$ scales as a small power of $\hbar$.

Ray chaos in long-range sound propagation in the ocean is quantified by the exponential (in average) divergence of rays with initially close values of the depth z and momentum p

$$\Delta z = \Delta z_0 e^{\nu r}, \quad \Delta p = \Delta p_0 e^{\nu r}, \tag{3.3}$$

where the Lyapunov exponent ν is a measure of divergence of rays over a distance, not time. The finite-size maximal Lyapunov exponent was estimated in realistic models of underwater sound channels to be of the order of $(1/300 \div 1/100)\text{km}^{-1}$ [Simmen *et al.* (1997); Beron-Vera *et al.* (2003)]. The volume in the phase space covered by a sound wave packet (pulse signal) is $\Delta z \Delta p \simeq k_0^{-1}$, where the wave number k_0 plays the role of $\hbar^{-1}$. In analogy with (3.2) we obtain *the Ehrenfest break-distance* under ray chaos

$$R_E = \frac{1}{\nu} \ln(l \Delta p k_0) = \frac{1}{\nu} \ln(I k_0), \tag{3.4}$$

beyond which the wave-ray correspondence breaks down. This spatial scale should not be confused with the so-called predictability horizon scale (2.10) which limits the predictability of ray trajectories in the chaotic regime due to inevitable uncertainties in initial ray variables.

At frequencies of the order of 100 Hz, the Ehrenfest break-distance R_E for long-range sound propagation does not exceed a few hundreds of kilometers. It does not mean that at greater distances the ray approximation fully fails to describe adequately the wave field. Numerous comparisons of the results, based on computing ray and wave equations, and those with experimental data [Worcester *et al.* (1999); Colosi *et al.* (1999)] have shown that ray theory is not able to predict fine details of the wave pattern but adequately describes the large-scale characteristics of the wave field at distances up to $3000 \div 5000$ km.

3.1.2 *Semiclassical propagator*

In this subsection we represent a somewhat different ray-based approach to description of wave motion. That approach is borrowed from quantum mechanics and seems to be the most suitable way to compare results of ray and wave modeling. Assume that the wavefield evolution is fairly described by the standard parabolic equation and rewrite it in the operator form

$$i k_0 \frac{\partial u}{\partial r} = \hat{H}(z, r) u, \tag{3.5}$$

where

$$\hat{H}(z,r) = -\frac{1}{2k_0^2}\frac{\partial^2}{\partial z^2} + \frac{\Delta c(z) + \delta c(z,r)}{c_0}. \tag{3.6}$$

The solution of (3.5) can be formally written as

$$u(z,\, r + \Delta r) = \hat{K}_{\mathrm{D}}(r,\, r + \Delta r)u(z,\, r), \tag{3.7}$$

where $\hat{K}_{\mathrm{D}}$ is the Dyson evolution operator

$$\hat{K}_{\mathrm{D}}(r,\, \Delta r) = \hat{\exp}\left(-ik_0 \int\limits_{r}^{r+\Delta r} \hat{H}(r')\, dr'\right). \tag{3.8}$$

This exponential operator can be expanded into the Dyson series

$$\hat{\exp}\left(-ik_0 \int\limits_{r}^{r+\Delta r} dr'\, \hat{H}(r')\right)$$

$$= 1 - ik_0 \int\limits_{r}^{r+\Delta r} dr'\, \hat{H}(r') - k_0^2 \int\limits_{r}^{r+\Delta r} dr' \int\limits_{r}^{r'} dr''\, \hat{H}(r'')\hat{H}(r') + \dots \tag{3.9}$$

In order to compare ray- and wave-based descriptions, we need in some representation of wave evolution in terms of deterministic trajectories. It can be obtained in the following way. Assume that a wavefield is concentrated in $z = z'$ at range r. Then its value in $z = z''$ at range $r + \Delta r$ is given by

$$K(z'',\, z',\, r + \Delta r,\, r) = \int \delta(z'' - z)\hat{K}_{\mathrm{D}}\delta(z' - z)\, dz. \tag{3.10}$$

Here $\delta(z)$ is the Dirac delta function. The quantity K is termed as the *propagator* and acts as a Green function, i.e.,

$$u(z'', r + \Delta r) = \int K(z'',\, z',\, r + \Delta r,\, r)u(z', r)dz'. \tag{3.11}$$

The propagator obeys the following conditions [Gutzwiller (1990)]:

$$\lim_{\Delta r \to 0} K(z'',\, z',\, r + \Delta r,\, r) = \delta(z'' - z'), \tag{3.12}$$

$$\frac{i}{k_0}\frac{\partial K}{\partial \Delta r} = -\frac{1}{2k_0^2}\frac{\partial^2 K}{\partial z''^2} + \frac{\Delta c(z) + \delta c(z'', r + \Delta r)}{c_0}K. \tag{3.13}$$

The well-known Huygens principle, stating that each point of an incoming wavefront is in fact a source of an outgoing wavefront, can be written in terms of the propagator as

$$K(z'',\, z',\, r+\Delta r,\, r) = \int K(z'',\, \tilde{z},\, r+\Delta r,\, \tilde{r}) K(\tilde{z},\, z',\, \tilde{r},\, r)\, d\tilde{z}. \tag{3.14}$$

Feynman represented the propagator as a *path integral* [Feynman (1948)]

$$K(z'',\, z',\, r+\Delta r,\, r) = \int \mathcal{D}(z) \exp\left(ik_0 \int\limits_{r}^{r+\Delta r} L\, dr \right), \tag{3.15}$$

where the differential $\mathcal{D}(z)$ means that integration is carried over *all paths* connecting points (z', r') and (z'', r''), and L is the Lagrangian

$$L = 1 + \frac{p^2}{2} - \frac{\Delta c(z) + \delta c(z, r)}{c_0}. \tag{3.16}$$

Evaluation of the Feynman path integral in the limit $k_0 \to \infty$, using the stationary phase method, yields the so-called *semiclassical propagator* (also frequently referred to as the Van Vleck or Van Vleck-Gutzwiller propagator), which is a sum of contributions from different eigenrays [Gutzwiller (1967)]:

$$K_{sc} = \left(\frac{k_0}{2\pi i} \right)^{1/2} \sum_j \left| \frac{\partial^2 S}{\partial z' \partial z''} \right|^{1/2} e^{iS(z'', z', r+\Delta r, r) - i\mu_j \pi/2}, \tag{3.17}$$

where j labels ray trajectories connecting points (z', r) and $(z'', r+\Delta r)$, μ_j is the Maslov index, and S is the Hamilton's principal function or ray eikonal given by

$$S(z'', z',\, r+\Delta r, r) = \int\limits_{r}^{r+\Delta r} L\, dr. \tag{3.18}$$

Here the integration is carried out along a ray path. Details of the derivation of the semiclassical propagator from the Feynman path integral can be found, for instance, in [Stöckmann (1999)].

Since the eikonal serves as a generating function for ray variables (see Chap. 1), we have

$$\frac{\partial S}{\partial z'} = -p', \quad \frac{\partial S}{\partial z''} = p'', \tag{3.19}$$

where p' and p'' are initial and final ray momentum, respectively. Taking Eq. (3.19) into account, it is easy to see that K_{sc} coincides with the ray-based wavefield given by Eq. (1.41).

Replacing K by K_{sc} in (3.11), we arrive at the expression for a wavefield by means of the semiclassical propagator

$$u(z'', r + \Delta r) = \int K_{sc}(z'', z', r + \Delta r, r) u(z', r) dz'. \tag{3.20}$$

Despite of severe limitations of the ray approximation, especially under chaotic conditions, it was recognized that the formula (3.20) appears to be remarkably accurate far beyond the Ehrenfest break-distance [Tomsovic and Heller (1991); Sepulveda *et al.* (1992); Tomsovic and Heller (1993)]. The cost one pays for this accuracy is the chaos-induced exponential growth of a number of eigenrays with range. Indeed, in the ray limit, any wavepacket is described by a Lagrangian manifold. As it was mentioned above, under chaotic conditions, a Lagrangian manifold undergoes a cascade of stretchings and foldings which result in its transformation into a complicated pattern. At relatively long ranges, the number of rays needed for reproducing the manifold appears to be excessively large. Thus, under chaotic conditions the ray method can preserve enough accuracy at long ranges but requires not less computational efforts than for modeling a wavefield by solving the parabolic equation.

3.1.3 *Fidelity or overlap of wave fields*

One of the contradictions of the wave-ray correspondence is the following. A small change in the initial conditions of the ray equations may cause an exponential divergence of chaotic rays with the rate given by the maximal Lyapunov exponent. A small change in the control parameters of the Hamiltonian with fixed initial conditions also may cause an exponential divergence of chaotic rays with the same Lyapunov exponent. In the same system, overlap of two wave fields with different initial conditions is constant in time, whereas this overlap decays in time under a small change in the system's parameters.

It is easily demonstrated in quantum mechanics that because of the unitarity of quantum evolution changes in the initial conditions do not grow in time. Let $|\Psi(0)\rangle$ and $|\Psi_\delta(0)\rangle$ be two very close quantum states (in the Dirac notations) with the initial distance between them measuring by the overlap $\delta(0) = |\langle\Psi_\delta(0)|\Psi(0)\rangle|^2$. Any pure quantum state evolves in time as $|\Psi(t)\rangle = \hat{U}(t)|\Psi(0)\rangle$ with $\hat{U}(t)$ being a unitary evolution operator, i.e., $\hat{U}(t)\hat{U}^+(t) = I$, where I is an identity operator. The distance

$$\delta(t) = |\langle\Psi_\delta(t)|\Psi(t)\rangle|^2 = \left|\langle\Psi_\delta(0)|\hat{U}^+(t)\hat{U}(t)|\Psi(0)\rangle\right|^2 = \delta(0) \tag{3.21}$$

does not change in time. A useful measure to quantify the stability of quantum systems, called "fidelity", has been introduced by Asher Peres [Peres (1984)]. Let the same initial states evolves under the action of two slightly different Hamiltonians H and H'. The fidelity is a squared overlap of the state $|\Psi(t)\rangle$ obtained by the unperturbed evolution $\hat{U}(t)$ and the state $|\Psi'(t)\rangle$ obtained by the perturbed evolution $\hat{U}'(t)$, i.e.

$$F(t) = |\langle\Psi(t)|\Psi'(t)\rangle|^2 = \left|\langle\Psi(0)|[\hat{U}'(t)]^{+}\hat{U}(t)|\Psi(0)\rangle\right|^2. \tag{3.22}$$

It can be also interpreted as an overlap of the initial state $|\Psi(0)\rangle$ and this state evolved forward in time with one Hamiltonian and then backward in time with a slightly different one. That is why the fidelity is called sometimes "Loschmidt echo".

It follows from Eq. (3.22) that $F = 1$ only if quantum states $|\Psi(t)\rangle$ and $|\Psi'(t)\rangle$ are identical. The greater the deviation from unity, the greater the difference between the two states and the lower the fidelity of the quantum state under consideration. The rate of the fidelity decay gives an estimate for how much the wave field can change propagating in a slightly different environment. This rate depends on initial conditions, the type of perturbation, and the type of underlying classical dynamics. Peres [Peres (1984)] has shown numerically that the long-time behavior of the fidelity for classically chaotic and regular systems was clearly distinct. In a chaotic regime $F(t)$ decays rapidly to a small constant, whereas in a regular regime it demonstrates strong oscillations and recurrences. In Ref. [Jalabert and Pastawski (2001)] it has been shown that for a single quantum particle with a classically chaotic Hamiltonian $F(t)$ decays exponentially with a rate given by the maximal classical Lyapunov exponent. Then it has been found that for sufficiently small perturbations one gets a Gaussian decay in strongly chaotic regime, whereas for stronger perturbations the decay is exponential [Cerruti and Tomsovic (2002)]. In the first experiments with nuclear spins weakly coupled to the environment [Levstein *et al.* (1998)], the decay of the Loschmidt echo was found to be Gaussian.

An analogue of fidelity in quantifying the stability of any wave field under perturbations is the modulus square of the overlap of two wave fields. In underwater acoustics it is the quantity

$$C(r) = \left|\int dz\Psi^*(z,r)\Psi'(z,r)\right|^2, \tag{3.23}$$

where $\Psi(z,r)$ is a wave field propagated to range r in a given sound speed profile and $\Psi'(z,r)$ the wave field at r propagated from the same initial state in a slightly different environment.

Being motivated by the problem of the ray-wave correspondence, the authors [Hegewisch *et al.* (2005)] have studied the convergence of the wave field propagating in an underwater sound channel with the Garrett-Munk spectrum of internal waves. To answer the question: "how adding more vertical modes in the internal wave expansion alters the propagation beyond some maximum mode number", they solved the parabolic equation with the initial wave field in the form of a normalized Gaussian wave packet and computed $C(r)$ for various values of mode number cutoff J. The overlap $C(r)$ was computed as a function of J for the source frequencies 25, 75, 150, and 250 Hz at a fixed range and for the ranges 1000, 2000, 3000, and 4000 km at a fixed frequency. It was found that, for example, for 75 Hz and $r = 1000$ km the overlap $C(r)$ is greater than 0.99 above $J = 50$. So $J = 50$ is the minimum number such that by including higher vertical modes there is a little change in the wave propagation. Increasing the range from 1000 km to 4000 km decreases C very little, from 0.99 to 0.96. So, $J = 50$ is a good choice for 75 Hz even for $r = 4000$ km. So, at low frequencies, the influence of weak small-scale inhomogeneities can be considered to be negligibly small in the wave propagation. On the other hand, such inhomogeneities may increase ray chaos in the ray approach. In Sec.3.4 we discuss this contradiction in a more detail.

3.1.4 *Dynamical localization*

Diffusive-like motion in the phase space is a hallmark of chaotic deterministic systems (see Chapter 2 and the Glossary). One of the simplest toy models with chaotic behavior is a kicked rotor, i.e., a free pendulum with the angular momentum L subjected to time periodic kicks. Its equations of motion in the limit of delta-like kicks can be reduced to the well-known standard map with a single control parameter K, the normalized kicking strength. If K is sufficiently large, the motion of the kicked rotor is fully chaotic [Chirikov (1979)]. The sequence of discrete values of L is a random walking which satisfies to the normal-diffusion law $\langle(\Delta L)^2\rangle = Bt$, where $B \equiv K^2/2$ is the diffusion coefficient (see the dashed line in Fig. 3.1).

The behavior of the corresponding quantum kicked rotor is different and is schematically shown in Fig. 3.1 by the solid line [Casati *et al.* (1979)]. Classical diffusion grows linearly in time whereas quantum diffusion grows

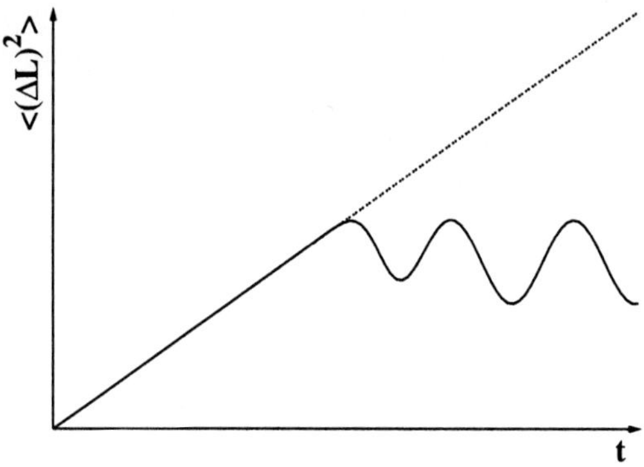

Fig. 3.1 Suppression of diffusion in a quantum kicked rotor (solid line) as compared with the classical one (dashed line).

only up to a break time. Why does quantum diffusion stops after that time? The answer to this question has been done in Refs. [Fishman *et al.* (1982); Grempel *et al.* (1984)]. The spectrum of the quantum kicked rotor is discrete and the solution of the corresponding Scrödinger equation in the momentum representation may be written as a discrete sum

$$\Psi(L) = \sum_j a_j u_j(L) \exp(i\omega_j t), \tag{3.24}$$

where the functions $u_j(L)$ form a complete orthonormal basis in L. The authors [Fishman *et al.* (1982)] proved that the eigenfunctions are localized in the momentum space, i.e., for a given j the function $u_j(L)$ decays on average with L roughly exponentially away from a center value $\bar{L}_j$

$$u_j(L) \sim \exp\left(-\frac{L - \bar{L}_j}{L_s}\right), \tag{3.25}$$

where L_s is the localization length of the eigenfunction. The effect is called "dynamical quantum localization". The similar effect, known as Anderson localization [Anderson (1958, 1978)], exists for electrons in a random potential, for example, in disordered solids, and implies an exponential localization of the electron wave function not in time but in configuration space. Manifestations of quantum localization, suppression of diffusion (Fig. 3.1), and exponential decay (3.25) have been observed with microwave-driven Rydberg atoms [Bayfield (1974)] and with cold atoms in a modulated standing light wave [Moore *et al.* (1994)].

An example of wave localization in a range-dependent waveguide was given by Iomin and Bliokh (2003). In that work a two-dimensional waveguide bounded by the ocean surface $z = 0$ and a corrugated bottom at $z = h_0 + \varepsilon\tilde{h}(r)$ was considered. The sound speed is uniform within the waveguide. A corrugation function $\tilde{h}(r)$ is chosen as

$$\tilde{h}(r) = \cos(k_r r). \tag{3.26}$$

Ray equation can be reduced to the following area-preserving map

$$\begin{aligned} \chi_{n+1} &= \chi_n + 2\varepsilon k_r \sin k_r r_n^*, \\ r_{n+1}^* &= r_n^* + \frac{2h_0}{\chi_{n+1}}, \end{aligned} \tag{3.27}$$

where χ_n is the ray grazing angle, and r_n^* is the range of n-th ray reflection off the bottom. This is the so-called Ulam map [Lichtenberg and Liberman (1983)]. It exhibits chaotic behavior in the range of low values of χ. The condition of the validity of the map (3.27) is

$$\frac{\varepsilon}{\lambda} < \frac{\pi}{2k_r h_0}, \tag{3.28}$$

where λ is an acoustic wavelength. This condition implies weakness of the corrugation of the waveguide. The maximal size of the chaotic region in the angle space is estimated as

$$\chi_{\max} \approx 2k_r\sqrt{\varepsilon h_0}. \tag{3.29}$$

The full-wave analysis yields that wave dynamics follows its ray counterpart only within finite ranges. Beyond that ranges, wave localization occurs. For instance, the localization length in terms of the ray grazing angle is asymptotically estimated as

$$\Delta\chi \approx \begin{cases} k_r\varepsilon\chi_0, & \lambda \ll \varepsilon, \\ \frac{1}{2}k_r\chi_0\lambda\left[\ln\frac{\lambda}{2\varepsilon}\right]^{-1}, & \lambda \gg \varepsilon, \end{cases} \tag{3.30}$$

where χ_0 can be thought of as an initial inclination of a wavepacket.

3.2 Normal Modes

This section presents main formulas expressing the modal representation of the wave field in the WKB approximation. It is explained how the modal representation, initially introduced in a range-independent waveguide, can be applied in the presence of a range-dependent perturbation giving rise to mode coupling.

Significant attention is paid to a discussion of the ray-mode relations. A ray-based method that allows one to express the mode amplitude in a range-dependent waveguide through solutions of the ray equations is derived.

It is shown that the description of the wavefield in a waveguide with periodic range-dependence can be simplified using the Floquet theory. In this case, the solution to the parabolic equation (1.17) can be presented as a sum of terms, analogous to the Floquet states in quantum mechanics.

3.2.1 *Range-independent waveguide*

3.2.1.1 *Eigenfunctions and eigenvalues in the WKB approximation*

In the range-independent waveguide ($V = 0$), the parabolic equation (1.17) is readily solved by the method of separation of variables. Then the wave field is expressed in the form of a decomposition

$$u(r,z) = \sum_{m=1}^{\infty} a_m(r)\varphi_m(z), \tag{3.31}$$

where $\varphi_m(z)$ are solutions to the Sturm-Liouville eigenvalue problem

$$\frac{1}{2}\frac{d^2\varphi_m}{dz^2} + k_0^2(E_m - U)\varphi_m = 0 \tag{3.32}$$

with appropriate boundary conditions. The eigenfuntions $\varphi_m(z)$ and eigenvalues E_m (hereafter numbered from $m = 1$) are complete analogues of the wave functions and energy levels, respectively, of a mechanical particle oscillating in a potential well $U(z)$ [Landau and Lifshitz (1977)]. Pushing further this analogy, the free water surface at $z = 0$ can be interpreted as an infinitely high potential barrier. For simplicity, we will assume that $U(z)$ tends to infinity as $z \to \infty$. Then the wave field is formed by the modes of a discrete spectrum. The eigenfunctions $\varphi_m(z)$ are orthogonal and normalized in such a way that

$$\int dz\ \varphi_m(z)\varphi_{m_1}(z) = \delta_{mm_1}, \tag{3.33}$$

where the integration goes over the whole waveguide vertical section. In a waveguide model that accounts for the sound penetration into the bottom, the modal representation becomes much more complicated and involves the modes of a continuous spectrum [Brekhovskikh and Godin (1999)].

In the limit of small wavelength, the eigenvalues E_m and eigenfunctions $\varphi_m(z)$ can be evaluated using the WKB approximation [Landau and Lifshitz (1977)] Below we will consider only modes whose turning points (or turning depths) are located within the water bulk. In the WKB approximation, the m-th mode is associated with the ray path whose action variable I_m satisfies the quantization rule

$$k_0 I_m = m - 1/2. \tag{3.34}$$

Then the eigenvalue of the m-th mode is

$$E_m = H_0(I_m). \tag{3.35}$$

A "differential" form of the quantization rule can be obtained by combining Eqs. (3.34), (3.35), and (1.53). This yields

$$\frac{dE_m}{dm} \equiv E_{m+1} - E_m = \frac{2\pi}{k_0 D(I_m)}. \tag{3.36}$$

For short, the quantity $D(I_m)$, or the cycle length of a ray with action I_m, sometimes will be denoted by D_m.

The WKB approximation of the m-th eigenfunction $\varphi_m(z)$ between its turning points is represented in the form [Landau and Lifshitz (1977); Brekhovskikh and Lysanov (1991)]

$$\varphi_m(z) = \varphi_m^+(z) + \varphi_m^-(z), \tag{3.37}$$

with

$$\varphi_m^\pm(z) = Q_m e^{\pm i[k_0 g(I_m,z)+\phi_b/2]}, \tag{3.38}$$

where

$$g(I,z) = \int\limits_{z_{\min}}^{z} dz \sqrt{2\left[H_0(I) - U(z)\right]} \tag{3.39}$$

$$Q_m = \frac{1}{\sqrt{D(I_m)p_m(z)}}, \tag{3.40}$$

$$p_m(z) = \sqrt{2\left[E_m - U(z)\right]}. \tag{3.41}$$

Note that $g(I,z)$ coincides with the generating function $G(I,z)$ determined by Eq. (1.46) for positive p.

3.2.1.2 *Mode amplitudes*

Substituting Eq. (3.31) into Eq. (1.17) yields

$$a_m(r) = b_m e^{-ik_0 E_m r}, \tag{3.42}$$

where the constant b_m, which is the mode amplitude at $r = 0$, is determined by the starting wave field as

$$b_m = \int dz\ u(0,z)\varphi_m(z). \tag{3.43}$$

In the case of a point source set at depth z_s (see Eq. (1.20)), we have

$$a_m(r) = \varphi_m(z_\mathrm{s})e^{-ik_0 E_m r}. \tag{3.44}$$

3.2.1.3 *Brillouin waves*

Each solution to the Sturm-Liuoville eigenvalue problem defines a solution to the parabolic equation

$$u(r,z) = \varphi_m(z)e^{-ik_0 E_m r}, \tag{3.45}$$

which is called the *normal mode* or simply *mode.* Using Eq. (3.37) allows one to represent the mode as a superposition of two terms

$$\varphi_m^{\pm}(z)e^{-ik_0 E_m r} = Q_m e^{\pm i[k_0 g(I_m,z)+\phi_b/2]-ik_0 E_m r} \tag{3.46}$$

describing the so-called Brillouin waves. Each of these waves (approximately) satisfies the parabolic equation, but not the boundary conditions. Therefore, the Brillouin waves can propagate only in pairs. In the limit of high frequency, each of them has a slowly varying amplitude and a rapidly varying phase.. Therefore, the Brillouin wave can be associated with a congruence of rays. Considering an argument of the exponential on the right-hand side of Eq. (3.46) as an action integral (multiplied by k_0) and using Eq. (1.38) connecting the action integral to the ray momentum, we can easily describe these congruences. It turns out that both of them include rays whose action variables are equal to I_m. Turning depths of these rays coincide with the turning depths of the m-th mode. At depth z, each ray corresponding to the Brillouin wave $\varphi_m^{\pm}(z)e^{-ik_0 E_m r}$ has a momentum $p = \pm p_m(z)$. Each congruence is formed by shifting the same ray path along the r axis.

3.2.1.4 *Matrix elements*

When dealing with normal modes, it is often necessary to evaluate the so-called matrix elements defined as

$$F_{mm'} = \int dz\ \varphi_m(z) F(z) \varphi_{m'}(z), \tag{3.47}$$

where $F(z)$ is a given function. If this function is smooth enough, then in the WKB approximation, the matrix elements with $m, m' \gg 1$ can be expressed through coefficients of the expansion (1.67). Indeed, take $F(z)$ whose characteristic scale is much greater than the wavelength $2\pi/k_0$. Then the matrix elements are very small unless $|m - m'| \ll m$. Making use of Eqs. (3.37) and (3.38), we rewrite Eq. (3.47) as

$$F_{m,m+\Delta m} = 2 \int_{z_{\min}}^{z_{\max}} dz\ F(z) Q_m^2(z) \cos\left[k_0 \left(g(I_{m+\Delta m}, z) - g(I_m, z)\right)\right]. \tag{3.48}$$

From Eq. (3.34) we find

$$k\left[g(I_{m+\Delta m}, z) - g(I_m, z)\right] = \left.\frac{\partial g(I, z)}{\partial I}\right|_{I=I_m} \Delta m. \tag{3.49}$$

According to Eqs. (1.52), (1.35) and (3.40), integration over z in Eq. (3.48) can be replaced with integration over the angle variable θ along the unperturbed ray with the action $I = I_m$. Then

$$dz\ Q_m^2 = \frac{d\theta}{2\pi}. \tag{3.50}$$

Replacing $\partial g/\partial I$ in Eq. (3.49) by θ (see Eqs. (1.47)), we obtain the desired connection between the matrix elements and the coefficients of Eq. (1.67)

$$F_{m,m+\Delta m} = \frac{1}{\pi} \int_0^{\pi} d\theta\ F(z(I_m, \theta)) \cos(\Delta m\ \theta) = \frac{\epsilon_{\Delta m}}{2} F_{|\Delta m|}(I_m), \tag{3.51}$$

where $\epsilon_{\Delta m} = 1$ if $\Delta m = 0$ and $\epsilon_{\Delta m} = 2$ otherwise.

3.2.1.5 *Ray-mode relations*

The complex amplitudes of the wave field can be represented in two alternative forms expressed by Eqs. (1.41) (ray representation) and (3.31) (modal representation). Then a natural question arises: "How these sums can be transformed into each other?" This issue has been addressed in many publications (see, e.g., [Batorsky and Felsen (1971); Tindle and Guthrie (1974); Ahluwalia and Keller (1977); Brekhovskikh and Lysanov (1991)] and references therein). In application to a range-independent waveguide, a comprehensive answer to this question was known already in the early 1980s. Main results obtained in this area were summarized in two excellent papers [Felsen (1981)] and [Gao and Shang (1982)].

In most works, the connection between Eqs. (1.41) and (3.31) is established using the Poisson sum formula. We will illustrate the application of this formula with an example of a point source. According to Eqs. (3.31) and (3.44), in this case the modal representation of the wave field is

$$u(r,z) = \sum_m \varphi_m(z_s)\varphi_m(z)e^{-ik_0 E_m r}. \tag{3.52}$$

The ray representation is given by Eqs. (1.40) and (1.41) as

$$u(r,z) = \sqrt{\frac{k_0}{2\pi i}} \sum_j \left| \frac{\partial z}{\partial p_0} \right|_{p_0=p_{0,j}}^{-1/2} e^{ik_0 S_j - i\mu_j \pi/2}, \tag{3.53}$$

where the subscript j numbers eigenrays arriving at the observation point (r,z). In Refs. [Felsen (1981); Kamel and Felsen (1982)] it is shown that a deep insight into the ray-mode relations can be obtained using the Poisson sum formula taken in the form

$$\sum_{m=-M_1}^{M_2} f(m) = \frac{1}{2}\left[f(M_1) + f(M_2)\right] + \sum_{N=-\infty}^{\infty} \int_{M_1}^{M_2} dm\, f(m) e^{2\pi i N m}. \tag{3.54}$$

This relation follows from a well-known result of Fourier series theory [Titchmarsh (1948)]. For a function $F(x)$, continuous and of bounded variation on the interval $0 < x < 1$, we have

$$\frac{1}{2}\left[F(0) + F(1)\right] = \sum_{N=-\infty}^{\infty} \int_0^1 dx\, F(x) e^{2\pi i N x}. \tag{3.55}$$

Applying Eq. (3.55) to the function $F(x) = f(m+x)$ yields Eq. (3.54) [Arnold and Felsen (1983)].

Making use of Eqs. (3.37)–(3.41), we rewrite Eq. (3.52) in the form

$$u(r,z) = \sum_{\sigma_1,\sigma_2} u_{\sigma_1,\sigma_2}(r,z), \tag{3.56}$$

where both σ_1 and σ_2 take the values -1 and 1 (so that the above sum consists of four terms),

$$u_{\sigma_1,\sigma_2}(r,z) = \sum_m A(I_m) e^{ik_0\Phi(I_m,\sigma_1,\sigma_2)}, \tag{3.57}$$

$$A = D^{-1}(I_m)\left[2\left(H_0(I) - U(z_s)\right)\right]^{-1/4}\left[2\left(H_0(I) - U(z)\right)\right]^{-1/4}, \tag{3.58}$$

$$\Phi(I,\sigma_1,\sigma_2) = \sigma_1 g(I_m, z_s) + \sigma_2 g(I_m, z) - H_0(I)r - (\sigma_1+\sigma_2)\pi/4. \tag{3.59}$$

Now transform the right-hand side of Eq. (3.57) using the Poisson sum formula (3.54) with $M_1 = 1$ and $M_2 = \infty$ and neglecting the term $[f(0) + f(\infty)]/2$. This yields

$$u_{\sigma_1,\sigma_2}(r,z) = \sum_{N=-\infty}^{\infty} u^{(N)}_{\sigma_1,\sigma_2}(r,z), \tag{3.60}$$

where

$$u^{(N)}_{\sigma_1,\sigma_2}(r,z) = \int_1^{\infty} dm\ A(I_m) e^{ik_0\Phi(I_m,\sigma_1,\sigma_2)+2\pi i N m}. \tag{3.61}$$

We will evaluate this integral using the method of stationary phase and neglecting a contribution from the endpoint $m = 1$. A more rigorous and detailed analysis is given in Refs. [Felsen (1981)] and [Kamel and Felsen (1982)].

The quantities m and I_m are related by the quantization rule (3.34). Making use of this relation, change the integration variable in Eq. (3.61) replacing m by $kI - 1/2$. This yields

$$u^{(N)}_{\sigma_1,\sigma_2}(r,z) = k_0 e^{-i\pi N - i(\sigma_1+\sigma_2)\pi/4} \int dI\ A(I) e^{ik_0\Psi(I)}, \tag{3.62}$$

where

$$\Psi(I) = \sigma_1 g(I, z_s) + 2\pi N I + \sigma_2 g(I,z) - H_0(I) r. \tag{3.63}$$

In order to evaluate the integral on the right-hand side of Eq. (3.62), we will exploit the stationary phase technique (Appendix D). Using Eqs. (1.47),

(1.53), (1.54), and (1.57) it is not difficult to show that

$$\frac{\partial \Psi}{\partial I} = \frac{2\pi}{D(I)} \left(R(I,z) - r \right), \tag{3.64}$$

where

$$R(I,z) = \sigma_1 L(I,z_0) + N D(I) + \sigma_2 L(I,z). \tag{3.65}$$

Equation (3.65) coincides with Eq. (1.61), that is, the function $R(I,z)$ describes a ray path. The stationary value of action I satisfying the condition $\partial\Psi/\partial I = 0$ corresponds to an eigenray arriving at the observation point (r,z). To evaluate the contribution from this eigenray to the total wave field, we should find the second derivative

$$\frac{\partial^2 \Psi}{\partial I^2} = \frac{2\pi}{D(I)} \frac{\partial R}{\partial I}. \tag{3.66}$$

Since we consider a point source, all the paths escape the same depth with different starting momenta p_0. The action variable of the ray path can be considered as a function of p_0 determined via

$$H_0(I) = \frac{p_0^2}{2} + U(z_s). \tag{3.67}$$

Making use of Eq. (1.53), we obtain $\partial I/\partial p_0 = p_0/\omega(I)$. This relation allows one to replace differentiation with respect to I by differentiation with respect to p_0:

$$\frac{\partial R}{\partial I} = \frac{\partial R/\partial p_0}{\partial I/\partial p_0} = \frac{2\pi}{D(I)} \frac{\partial R}{\partial p_0}. \tag{3.68}$$

It is easy to see that

$$\frac{\partial z}{\partial p_0} = \tan\chi \frac{\partial R}{\partial p_0}, \tag{3.69}$$

where χ is the ray grazing angle at the observation point and $\partial z/\partial p_0$ is a derivative of the ray depth at a fixed range r with respect to the starting momentum. The same derivative is present in Eq. (3.53). Since $\tan\chi$ is equal to the momentum p at the observation point (Eq. (1.29)), we have

$$\frac{\partial^2 \Psi}{\partial I^2} = -\frac{(2\pi)^2}{D^2(I) p p_0} \frac{\partial z}{\partial p_0}. \tag{3.70}$$

Note that

$$p_0 = \left[2\left(H_0(I) - U(z_s)\right)\right]^{-1/4}, \quad p = \left[2\left(H_0(I) - U(z)\right)\right]^{-1/4}. \qquad (3.71)$$

Comparison of Eqs. (3.63) and (1.63) shows that Ψ gives the eikonal of the eigenray. Applying Eq. (D.4), we obtain

$$u^{(N)}_{\sigma_1,\sigma_2}(r,z) = \sqrt{\frac{k_0}{2\pi}} \left|\frac{\partial z}{\partial p_0}\right|^{-1/2} e^{ik_0 S - i\pi N + i(\gamma - \sigma_1 - \sigma_2)\pi/4}, \qquad (3.72)$$

where $\gamma = -\,\mathrm{sgn}(pp_0 \partial z/\partial p_0)$. The function $\mathrm{sgn}(x)$ gives the sign of its argument. A set of three numbers (N, σ_1, σ_2) determines the topology of a ray path (number of cycles and signs of grazing angle at end points) whose contribution is given by Eq. (3.72). Note that a stationary phase point, typically, exists for not all the integrals (3.62). Lack of the stationary phase point means lack of an eigenray with the corresponding topology.

From the comparison of Eqs. (3.72) and (3.53) we find an expression for the Maslov index $\mu = 2N + 1 - \gamma + \sigma_1 + \sigma_2$. Since N indicates the number of cycles of the ray path, this expression is consistent with the known fact that at each cycle a ray path in a range-independent waveguide twice passes through caustics [Watson *et al.* (1981)].

An interesting and comprehensive discussion of different aspects of the ray-mode relations in a range-independent waveguide is given in Refs. [Felsen (1981)] and [Gao and Shang (1982)]. It turns out that at short ranges normal modes with close numbers constructively interfere along the trajectory of a ray path. Moreover, a contribution from a single ray, sometimes, is formed by a group of constructively interfering modes, that is, by a part of the sum (3.52). And vice versa, the sum of constructively interfering rays (a part of the sum (3.53)) may form a contribution to the wave field (at the observation point) from a single mode. Usually, this occurs at long ranges.

Although most results on studying parallels between rays and modes have been obtained for range-independent waveguides or waveguides slowly varying with range, the ray-mode analogy goes beyond these simple models. In Sec. 3.2.4, we present a method that allows one to express the mode amplitude through parameters of the ray paths in a range-dependent environment. In Sec. 4.3, this method is generalized to the case of a transient wave field. In Sec. 4.4, the ray-mode relations are considered in the case where the wave field is excited by a distributed source.

Concluding this subsection, note that the ray-mode interrelation can be used even for analysis of diffraction effects in waveguide media. In

Ref. [Virovlyansky *et al.* (1997)] it is shown that the Fresnel zones for modes can be introduced, which are analogues of the usual Fresnel zones for rays [Born and Wolf (1968); Kravtsov and Orlov (1990); Rypina and Brown (2007)]. However, this issue will not be broached here.

3.2.2 *Range-dependent waveguide*

In order to introduce normal modes in a range-dependent waveguide, one should associate each range point with a reference range-independent waveguide. Then, the wave field at any particular cross section can be decomposed into eigenfunctions of the corresponding reference waveguide. A similar technique was discussed in Sec. 1.3.4 when introducing the action-angle variables. In the present subsection we consider the same two types of reference waveguide as in Sec. 1.3.4.

In the range-dependent environment, the wave equation no longer yields to a separation of variables and normal modes, are coupled by the perturbation δc. The mode amplitudes become range-dependent and they are governed by the mode coupling equations presented below.

3.2.2.1 *Coupled mode equations: slow and strong range dependence*

In this case, it is convenient to use the same choice of a reference waveguide as in Sec. 1.3.4.1: a waveguide associated with a range point r has the cross-section coinciding with that of the actual waveguide at this range. Normal modes defined in this way are called the *local* modes. In the scope of this approach, called "the reference waveguide method" [Brekhovskikh and Godin (1999)], the wave field at range r is decomposed into a sum over eigenfunctions of the corresponding reference waveguide. Modal representation (3.31) translates to

$$u(r,z) = \sum_m a_m(r)\varphi_m(r,z), \tag{3.73}$$

where we have emphasized the range dependence of the eigenfunctions. The eigenvalues E_m also become functions of r. By substituting the above expression into Eq. (1.17) and using the orthonormality condition (3.33), we find the mode coupling equations

$$\frac{da_m}{dr} + ik_0 E_m a_m = -\sum_{m_1} a_{m_1} \int dz\, \frac{\partial \varphi_{m_1}}{\partial r} \varphi_m. \tag{3.74}$$

Thus, the solution of the parabolic equation (1.17) reduces to the solution of a system of ordinary differential equations (3.74). For a point source set at a depth z_s, the system should be solved with the initial conditions

$$a_m(0) = \varphi_m(0, z_s). \tag{3.75}$$

The eigenfunctions $\varphi_m(r, z)$ form a complete function system, and the derivative $\partial\varphi_m/\partial r$ can be presented as an expansion

$$\frac{\partial \varphi_m}{\partial r} = \sum_{j\neq 0} K_{mj}\varphi_{m+j}. \tag{3.76}$$

Substituting this in Eq. (3.74) yields

$$\frac{da_m}{dr} + ik_0 E_m a_m = \sum_{j\neq 0} K_{mj} a_{m+j}. \tag{3.77}$$

It can be shown that

$$K_{mj} = \frac{1}{E_m - E_{m+j}} \int dz\ \varphi_m \frac{\partial V}{\partial r} \varphi_{m+j}. \tag{3.78}$$

This formula is derived by differentiating Eq. (3.32) in the reference waveguide (where φ_m, E_m, and U acquire an additional argument r) with respect to r and exploiting the normalization condition (3.33).

If characteristic scales of a medium range variation are large, compared with the cycle length, then the adiabatic approximation, which consists in neglecting the right-hand side of the mode coupling equation (3.77), can be employed [Brekhovskikh and Lysanov (1991); Brekhovskikh and Godin (1999)]. Then Eqs. (3.74) and (3.77) are readily solved to give $a_m(r) = a_m(0)\exp(-ik_0E_mr)$.

3.2.2.2 *Coupled mode equations: weak range dependence*

In the case of weak perturbation, it is convenient to use the decomposition of the wave field into eigenfunctions $\varphi_m(z)$ of an unperturbed waveguide. Then, as in Sec. 1.3.4.2, the unperturbed waveguide is used as a reference one at any range point. Let us seek the solution to the parabolic equation (1.17) with $\delta c \neq 0$ in the form

$$u(r, z) = \sum_m a_m(r) e^{-ik_0 E_m r} \varphi_m(z), \tag{3.79}$$

where, in contrast to Eq. (3.73), $\varphi_m(z)$ are the eigenfunctions of the unperturbed Sturm-Liouville eigenvalue problem (3.32). Substituting this into Eq. (1.17) yields

$$\sum_m \left[\frac{da_m}{dr} + ik_0 E_m a_m + ik_0 V(r,z) \right] \varphi_m(z) = 0. \tag{3.80}$$

Presenting a_m as

$$a_m = b_m(r) e^{-ik_0 E_m r} \tag{3.81}$$

and using the orthonormality condition (3.33), we arrive at the mode coupling equation

$$\frac{db_m}{dr} = -ik_0 \sum_n V_{mn} b_n e^{ik_0 (E_m - E_n) r}, \tag{3.82}$$

where

$$V_{mn}(r) = \int dz\ \varphi_m(z) V(r,z) \varphi_n(z). \tag{3.83}$$

3.2.3 *Normal modes corresponding to the Helmholtz equation*

Here we briefly outline how the above formulas, expressing the modal representation of the wave field, change when the analysis is based on the Helmholtz equation.

3.2.3.1 *Range-independent waveguide*

A solution to the Helmholtz equation (1.10) in the unperturbed (range-independent) waveguide, obtained using separation of variables, has the form (cf. Eq. (3.31))

$$v(r,z) = \frac{1}{\sqrt{r}} \sum_m a_m(r)\, \varphi_m(z), \tag{3.84}$$

where $\varphi_m(z)$ are eigenfunctions of the Sturm-Liouville problem

$$\frac{d^2\varphi_m}{dz^2} + \left(k_0^2 n^2(z) - k_m^2\right) \varphi_m = 0, \tag{3.85}$$

where k_0 is a wavenumber in the reference medium, with appropriate boundary conditions. As earlier, we will restrict our attention to normal modes with turning points within the water bulk. The eigenfunctions are orthogonal and normalized according to Eq. (3.33). Eigenvalues of the Sturm-Liouville problem k_m are called the horizontal wavenumbers of normal modes.

In the WKB approximation, the m-th mode is once again associated with the ray whose action variable satisfies the quantization rule (3.34). Then, the corresponding horizontal wavenumber is

$$k_m = -k_0 H_0(I_m), \tag{3.86}$$

where $H(I)$ is the Hamilton function determined by Eq. (1.99). If $1/k$ is much less than the characteristic scales of $H(I)$, then according to Eqs. (3.34), (3.86), and (1.53), we find

$$\frac{dk_m}{dm} \equiv k_{m+1} - k_m = -\omega(I_m) = -\frac{2\pi}{D(I_m)}. \tag{3.87}$$

This is a "differential" form of the quantization rule corresponding to the Helmholtz equation (cf. Eq. (3.36)).

In the WKB approximation, the eigenfunction of the m-th mode between its turning depths can be presented in the same form as in the parabolic approximation (cf. Eqs. (3.37) and (3.38)) [Ahluwalia and Keller (1977); Brekhovskikh and Lysanov (1991)]

$$\varphi_m(z) = \varphi_m^+(z) + \varphi_m^-(z), \tag{3.88}$$

with

$$\varphi_m^{\pm}(z) = Q_m e^{\pm i[k_0 g(I_m, z) + \phi_b/2]}, \tag{3.89}$$

where

$$g(I, z) = \int_{z_{\min}}^{z_{\max}} dz\, \sqrt{\bar{n}^2(z) - H_0^2(I)}, \tag{3.90}$$

$$Q_m = \sqrt{\frac{|H_0(I_m)|}{p_m(z)\, D(I_m)}}, \tag{3.91}$$

$$p_m(z) = \sqrt{n^2(z) - H_0^2(I_m)}. \tag{3.92}$$

3.2.3.2 *Mode amplitudes*

Substituting Eq. (3.84) in Helmholtz equation (1.10) with $n(r,z) = \bar{n}(z)$ and exploiting Eq. (3.33), it is easy to show that

$$a_m(r) = \text{const}\sqrt{r}H_0^{(1)}(k_m r), \tag{3.93}$$

where $H_0^{(1)}$ is the Hankel function, and factor const is determined by the source exciting the wave field. Already at ranges exceeding a few wavelengths, the Hankel function can be approximated by its large-argument asymptotic representation [Abramovitz and Stegun (1964)]

$$H_0^{(1)}(x) \simeq \sqrt{\frac{2}{\pi x}} e^{ix - i\pi/4}. \tag{3.94}$$

Then we get

$$a_m(r) = b_m e^{ik_m r}. \tag{3.95}$$

To find an explicit expression for b_m, assume that the wave field is governed by a forced Helmholtz equation

$$\frac{1}{r}\frac{\partial}{\partial r} r \frac{\partial v}{\partial r} + \frac{\partial^2 v}{\partial z^2} + k_0^2 n^2(r,z) v = -2\frac{\delta(r)}{r} q(z), \tag{3.96}$$

that generalizes Eq. (1.12). The function $q(z)$ determines the source distributed along the line $r = 0$. If $n(r,z) = \bar{n}(z)$, it is easy to show that

$$b_m = \sqrt{\frac{2\pi}{k_m}} e^{i\pi/4} \int dz\, q(z) \varphi_m(z). \tag{3.97}$$

In the important case of a point source

$$q(z) = \delta(z - z_s) \tag{3.98}$$

(cf. Eq. (1.12)) we get

$$a_m(r) = \pi i \varphi_m(z_0) \sqrt{r} H_0^{(1)}(k_m r) \simeq \sqrt{\frac{2\pi}{k_m}}\, \varphi_m(z_0)\, e^{ik_m r + i\pi/4}. \tag{3.99}$$

3.2.3.3 *Brillouin waves*

The wave field of a single mode is

$$v(r,z) = \varphi_m(z) H_0^{(1)}(k_m r) \simeq \sqrt{\frac{2}{\pi k_m r}} \varphi_m(z) e^{ik_m r + i\pi/4} \tag{3.100}$$

As in the case of a parabolic equation, it can be (approximately) presented as a superposition of two quasi-plane Brillouin waves

$$v(r,z) = \sqrt{\frac{2}{\pi k_m r}} Q_m \exp\left[\pm i k_0 g(I_m, z) + i k_m r \pm i\phi_b + i\pi/4\right]. \tag{3.101}$$

3.2.3.4 *Range-dependent waveguide*

Derivations and discussions of the mode coupling equations following from the Helmholtz equations have been presented in many publications (see, for example, [Dozier and Tappert (1978); Brekhovskikh and Lysanov (1991); Jensen *et al.* (1994)]). Without going into detail, we only note that in the case of a weak perturbation it is convenient to present the mode amplitude in the same form as in Eq. (3.81)

$$a_m(r) = b_m(r) e^{ik_m r}, \tag{3.102}$$

where $b_m(r)$ is an envelope function. It is assumed that $b_m(r)$ is a slow function of range, such that

$$\left| \frac{d^2 b_m}{dr^2} \right| \leq k_0 \left| \frac{db_m}{dr} \right|. \tag{3.103}$$

Substituting Eq. (3.84) into Eq. (1.17), neglecting the second derivative $d^2 b_m / dr^2$, and exploiting the orthonormality of eigenfunctions φ_m, we find the mode coupling equation in the form

$$\frac{db_m}{dr} = \frac{ik_0^2}{k_m} \sum_{m'} (\delta n)_{mm'} \, e^{i(k_{m'} - k_m) r}. \tag{3.104}$$

A comprehensive analysis of the reference waveguide method, in which the wave field is decomposed into a sum of local modes, is given in Ref. [Brekhovskikh and Godin (1999)].

3.2.4 *Ray-based description of mode amplitudes*

In Sec. 3.2.1.5, it was indicated that the ray and modal representations of the wave field in a range-independent waveguide can be transformed into each other using the Poisson sum formula. In the present subsection, we will consider an alternative approach for establishing a link between rays and modes that remains valid in a range-dependent environment [Berman and Zaslavsky (1978, 1979); Virovlyansky and Zaslavsky (2000); Virovlyansky (2000); Virovlyansky *et al.* (2005)]. It consists in projecting the ray representation of the wave field onto normal modes and allows one to express (approximately) the mode amplitude through parameters of the ray paths, that is, through solutions of the Hamilton (ray) equations.

3.2.4.1 *Parabolic equation*

Take a solution of parabolic equation (1.17) in a range-dependent waveguide in the form of a decomposition into a sum of local modes (3.73). Our starting point is the equation

$$a_m(r) = \int dz\; u(r,z)\varphi_m(r,z) \tag{3.105}$$

that follows from the orthonormality condition (3.33).

Substituting $u(r,z)$ from Eq. (1.23) and using the WKB approximation for $\varphi_m(r,z)$, we obtain $a_m = a_m^+ + a_m^-$ with

$$a_m^\sigma = \int dz\; \tilde{A}(r,z,\sigma)e^{ik_0\Phi(r,z,\sigma)}, \tag{3.106}$$

where σ denotes plus or minus,

$$\tilde{A} = e^{-i\mu\pi/2 + i\sigma\phi_b/2}\, AQ_m, \tag{3.107}$$

and

$$\Phi = S + \sigma g(I_m, z, r). \tag{3.108}$$

In the high frequency approximation (large k_0), integral (3.106) can be evaluated using a stationary phase technique. Making use of Eq. (D.4) yields

$$a_m^\sigma = \sqrt{\frac{2\pi}{k_0\,|\partial^2\Phi/\partial z^2|}}\,\tilde{A}\exp\left[ik_0\Phi + i\frac{\pi}{4}\,\mathrm{sgn}\left(\partial^2\Phi/\partial z^2\right)\right]\Bigg|_{z=Z_m}, \tag{3.109}$$

where Z_m is a stationary phase point for which the condition

$$\frac{\partial\Phi}{\partial z} = 0 \tag{3.110}$$

is met. Consider the eikonal S as a function of range r, starting coordinate z_0, and final coordinate z. According to the second of Eqs. (1.38), we have

$$\frac{\partial S}{\partial z} = p, \tag{3.111}$$

where p is the momentum at range r. Recall that that the function $g(I_m, z, r)$ in Eq. (3.108) is a generating function of the canonical transformation (for positive p) to action-angle variables in a reference waveguide associated with the range r. From the first of Eqs. (1.47), we find

$$\frac{\partial g(I_m, z, r)}{\partial z} = p_m, \tag{3.112}$$

where p_m is determined by Eq. (3.41) in the reference waveguide. From Eqs. (3.111) and (3.112) it follows

$$\frac{\partial \Phi}{\partial z} = p + \sigma p_m. \tag{3.113}$$

This derivative vanishes at point z, such that the ray arriving at this point has an action

$$I = I_m \tag{3.114}$$

and

$$\sigma = -\,\mathrm{sgn}(p). \tag{3.115}$$

Condition (3.114) singles out the rays whose actions at a given range r are equal to the action I_m associated with the m-th mode. In what follows, these rays will be called the *mode rays*. The quantity Z_m in Eq. (3.109) is the depth of a mode ray at the observation range.

For the second derivative of Φ we have an expression

$$\frac{\partial^2 \Phi}{\partial z^2} = \frac{\partial p}{\partial z} - \sigma \frac{\partial U/\partial z}{p_m} = \frac{1}{p}\frac{\partial H}{\partial z} = \frac{\omega}{p}\frac{\partial I}{\partial z}, \tag{3.116}$$

where H and I are considered functions of the ray arrival depth at the given range r. From Eq. (3.109) it follows that

$$a_m^\sigma = A\left|k_0 \partial I/\partial z\right|^{-1/2} \exp\left(ik_0\Phi - i\pi/4\sigma\,\mathrm{sgn}(\partial I/\partial z)\right)\Big|_{z=Z_m}. \tag{3.117}$$

Now we should express the derivative $\partial I/\partial z$ through parameters of the mode rays. This problem is easily solved for sources of two types considered below.

Point source. The starting wave field is given by Eq. (1.20). In this case, we deal with the congruence of rays leaving the point $(0, z_s)$. Labeling each ray by its starting momentum p_0, we represent the derivative $\partial I/\partial z$ as $(\partial I/\partial p_0)/(\partial z/\partial p_0)$. Then, insertion of Eq. (1.40) into Eq. (3.117) yields

$$a_m^\sigma = \frac{1}{\sqrt{2\pi\left|\partial I/\partial p_0\right|}} \times \exp\left\{k_0\left[S + \sigma g(I_m, Z_m, r)\right] - i\left[\frac{\mu\pi}{2} + \frac{\sigma\phi_b}{2} + \frac{(\beta-1)\pi}{4}\right]\right\}, \tag{3.118}$$

with

$$\beta \equiv -\sigma \,\mathrm{sgn}(\partial I/\partial p_0)\,\mathrm{sgn}(\partial z/\partial p_0). \tag{3.119}$$

Equation (3.118) accounts for the contribution from a ray satisfying condition (3.114), that is, from a mode ray. The mode amplitude is evaluated by summing up contributions from all the mode rays.

In a range-independent waveguide, for any mode there are exactly two rays satisfying Eq. (3.114) at any range point. They escape the source at grazing angles (equal in absolute value and opposite in sign) equal to grazing angles of the Brillouin waves $\varphi_m^+ e^{-ik_0 E_m r}$ and $\varphi_m^- e^{-ik_0 E_m r}$ at the source depth z_s. This statement remains valid in an adiabatic waveguide, as well.

Single mode source. A similar result can be derived with a source exciting a single mode with $m = m_0$. In this case, we have two congruences of rays associated with the Brillouin waves. Initial actions of all rays are equal to I_{m_0}, and their starting depths are located between the mode turning points. Now the rays are labeled by their starting depths z_0. There are two rays escaping each point z_0 with the starting momenta

$$p_0 = \pm\sqrt{2(E_{m_0}(0) - U(0, z_0))} \tag{3.120}$$

and complex amplitudes (due to Eq. (1.39))

$$A = \frac{\mathrm{const}}{\sqrt{|\partial z/\partial z_0|}}. \tag{3.121}$$

At $r = 0$, we have $\partial z/\partial z_0 = 1$ and A should be equal to the amplitude of the Brillouin wave at $z = z_0$. Taking this into account, we find

$$A = \frac{Q_{m_0}(z_0)}{\sqrt{|\partial z/\partial z_0|}} e^{\pm i[k_0 g(I_{m_0}, z_0, 0) + \phi_b/2]}. \tag{3.122}$$

Substituting Eq. (3.122) into Eq. (3.117) and representing $\partial I/\partial z$ as $(\partial I/\partial z_0)/(\partial z/\partial z_0)$ yields

$$a_m^\sigma = \frac{\varphi_m^{\sigma_0}(r, z_0)}{\sqrt{k\,|\partial I/\partial z_0|}} \exp\left[k_0\left(S + \sigma g(I_m, Z_m, r)\right) - i\mu\pi/2 + i\sigma\phi_b/2 + i\beta\pi/4\right], \tag{3.123}$$

with

$$\beta \equiv -\sigma \,\mathrm{sgn}(\partial I/\partial z_0)\,\mathrm{sgn}(\partial z/\partial z_0), \quad \sigma_0 = \mathrm{sgn}(p_0). \tag{3.124}$$

An important cautionary remark should be made. When using the stationary phase technique, we assume that the integration in Eq. (3.106) goes over an interval exceeding

$$\delta z = \sqrt{\frac{2\pi}{k_0 \left| \frac{\partial^2 \Phi}{\partial z^2} \right|}}. \tag{3.125}$$

Therefore, our approach requires that

$$\delta z \leq \Delta z, \tag{3.126}$$

where Δz is the difference between the mode turning depths. Denote by ΔI a spread of the ray action at the range of observation. The corresponding spread of the "energy" H in accord with Eqs. (1.53) and (1.54) is $\Delta H = 2\pi \Delta I / D$. Assuming that the spread of ray depths at the range of observation has the same order of magnitude as Δz and using Eq. (3.116), we obtain an order-of-magnitude estimate

$$\frac{\partial^2 \Phi}{\partial z^2} \approx \frac{1}{p} \frac{\Delta H}{\Delta z} = \frac{1}{p} \frac{2\pi}{D} \frac{\Delta I}{\Delta z}, \tag{3.127}$$

where p is the rms value of momentum at range r. As follows from Eq. (3.114), an initially excited mode breaks up into a group of approximately

$$\Delta m = k_0 \Delta I \tag{3.128}$$

modes. Using a rough estimate $pD/2 \approx \Delta z$ and combining Eqs. (3.125), (3.127), and (3.128), we obtain

$$\delta z \approx \frac{\Delta z}{\sqrt{\Delta m / 2}}. \tag{3.129}$$

Condition (3.126) is met only if $\Delta m \geq 1$, that is, if many modes are excited. Thus, Eq. (3.123) is valid only in a strongly nonadiabatic environment and at sufficiently long ranges where an initially excited mode breaks up into many modes. In contrast, a point source usually excites a large number of modes and, therefore, condition (3.126) practically does not restrict the applicability of Eq. (3.118).

The main results of our approach are given by Eqs. (3.118) and (3.123). They express the amplitudes of the local modes through solutions of the Hamilton (ray) equations. If the modal representation of the wave field is taken in the form of Eq. (3.79), that is, if $u(r, z)$ is decomposed into eigenfunctions of an unperturbed waveguide, then the ray-based description of the mode amplitudes a_m can be derived along exactly the same lines. The mode amplitudes once again are expressed by Eqs. (3.118) and (3.123), but the functions g and $\varphi_m^{\sigma_0}$ on the right-hand sides of these equations should be evaluated in an unperturbed waveguide. Correspondingly, all parameters of the ray paths should be evaluated using the formulas from Sec. 1.3.4.2.

3.2.4.2 *Helmholtz equation*

In the case of the Helmholtz equation, the above results require only minor modifications. Omitting the derivation, we present the final expressions for the mode amplitudes.

In the case of a point source determined by Eq. (1.12), the expression for the mode amplitude (analogue of Eq. (3.118)) is

$$a_m^\sigma = \frac{1}{\sqrt{k_0 \left| H \partial I / \partial p_0 \right|}} \exp\left(k_0 \left(S + \sigma S_m\right) - i\mu\pi/2 + i\sigma\phi_b/2 + i\beta\pi/4\right), \tag{3.130}$$

with

$$\beta \equiv -\sigma \operatorname{sgn}(\partial I/\partial p_0) \operatorname{sgn}(\partial z/\partial p_0). \tag{3.131}$$

In the case of a single mode source, the wave field in the unperturbed waveguide has the form

$$v(r,z) = \frac{1}{\sqrt{r}} \varphi_m(z) e^{ik_m r + i\pi/4}. \tag{3.132}$$

Then we have

$$a_m^\sigma = \sqrt{\frac{k_m}{\left| H \partial I / \partial z_0 \right|} \frac{\varphi_m^{\sigma_0}(z_0)}{k}} \times \exp\left\{k_0 \left[S + \sigma g(I_m, z)\right] - i \left[\frac{\mu\pi}{2} + \frac{\sigma\phi_b}{2} + \frac{\beta\pi}{4}\right]\right\}, \tag{3.133}$$

with

$$\beta \equiv -\sigma \operatorname{sgn}(\partial I/\partial z_0) \operatorname{sgn}(\partial z/\partial z_0), \quad \sigma_0 = \operatorname{sgn}(p_0). \tag{3.134}$$

3.2.5 *Floquet modes of a periodic waveguide*

The modal representation of the wave field, remarkably simple and physically clear in a range-independent environment, becomes much more complicated in the presence of a range-dependent perturbation. But if the range-dependence is periodic, the description of the wave field may be greatly simplified by applying the Floquet theory [Reichl (1992); Stöckmann (1999)]. We will consider this issue in the scope of the parabolic approximation.

Take a waveguide with the sound speed field (1.1) where $\delta c(r, z)$ is a periodic function of range r with period λ_r. Then the perturbation function $V(r, z)$ defined by Eq. (1.19) is also periodic so that

$$V(r + \lambda_r, z) = V(r, z). \tag{3.135}$$

Then the solution of Eq. (1.17) can be presented as a sum of Floquet modes

$$u_m(r, z) = e^{i\alpha_m r/\lambda_r} \Psi_m(r, z), \tag{3.136}$$

where $\Psi_m(r+\lambda_r, z) = \Psi_m(r, z)$, α_m are real constants, and $m = 1, 2, \ldots$. In quantum mechanics analogous terms describe the so-called Floquet states, and constants α_m are proportional to their quasi-energies [Grempel *et al.* (1984); Reichl (1992); Stöckmann (1999)]. The functions $u_m(r, z)$ are eigenfunctions of the shift operator $\hat{F}$,

$$\hat{F} u_m(r, z) = u_m(r + \lambda_r, z) = e^{i\alpha_m} u_m(r, z). \tag{3.137}$$

Since the shift operator is Hermitian, $\Psi_m(r, z)$ form a complete orthonormal set.

The Floquet modes can be expressed through the eigenfunctions of the unperturbed waveguide (with $\delta c = 0$) $\varphi_m(z)$. Let us discuss this issue. Take functions

$$\psi_m(z) \equiv \Psi_m(0, z) \tag{3.138}$$

and present each of them as a decomposition

$$\psi_m(z) = \sum_q C_{qm} \varphi_q(z). \tag{3.139}$$

To number the Floquet modes we shall associate a parameter μ with each function $\psi_m(z)$, where

$$\mu = \sum_q q \left| C_{qm} \right|^2, \tag{3.140}$$

and sort the Floquet modes in the order of increasing values of μ. Note that eigenfunctions $\psi_m(z)$ with small numbers m describe waves localized at depths z in the vicinity of the minimum of $U(z)$. The greater is the number

m the steeper are the grazing angles of waves forming the corresponding Floquet mode.

Consider a unitary Floquet matrix F with elements

$$F_{mn} = \int dz\ \varphi_m(z)\hat{F}\varphi_n(z), \tag{3.141}$$

where $\hat{F}\varphi_n(z)$ denotes a solution to parabolic equation (1.17) at range $r = \lambda_r$ obtained for an initial condition $u(0,z) = \varphi_n(z)$. Substituting Eq. (3.139) into Eq. (3.137) with $r = 0$, it is easy to see that eigenvalues ξ_m and eigenvectors $\mathbf{X}_m$ of matrix F, satisfying the equation

$$F\mathbf{X}_m = \xi_m\mathbf{X}_m, \tag{3.142}$$

determine the connection between the modes of the unperturbed waveguide and the Floquet ones. Namely, we get

$$\xi_m = e^{i\alpha_m}$$

and components of the m-th eigenvector $\mathbf{X}_m$ are $(C_{1m}, C_{2m}, \ldots)$.

Using the Floquet modes yields a simple expression for cross-sections of the wave field at discrete ranges $n\lambda_r$ in the form of a decomposition

$$u(n\lambda_r, z) = \sum_m g_m(n)\psi_m(z), \tag{3.143}$$

where

$$g_m(n) = e^{in\alpha_m} g_m(0). \tag{3.144}$$

At any intermediate range, a similar decomposition is valid with the same α_m but different eigenfunctions $\psi_m(z)$. Equation (3.143) may be considered as an analogue to Eq. (3.31) for a waveguide with a periodic range dependence.

3.3 Mode Coupling Under Chaotic Conditions

Manifestations of the chaotic ray dynamics in the modal structure of a wave field is an important aspect of the wave chaos problem. In this section, the issue is examined in a waveguide with a periodic range dependence. The dynamics of normal modes in a more realistic environmental model with random perturbations is considered in Chapter 4.

We already know the important role which a nonlinear ray-medium resonance plays in the emergence of ray chaos. At a finite wavelength, this phenomenon has a counterpart called the *mode-medium resonance*, which is a complete analogue of the nonlinear quantum resonance [Reichl (1992); Sagdeev *et al.* (1988); Zaslavsky (1998)]. While overlapping of the ray-medium resonances causes chaotic ray motion, overlapping of the mode-medium resonances increases mode coupling and gives rise to strong variations in mode amplitudes.

However, there is a qualitative difference between the dynamics exhibited by rays and modes. In the presence of a periodic perturbation, part of ray trajectories become chaotic and behave as random processes. Nevertheless, the normal modes formed by these rays evolve in a much more regular way. Due to the Floquet theorem, the solution to the parabolic equation (1.17) is a superposition of terms with simple range dependencies. At a finite wavelength, the chaotic ray motion in a periodic waveguide manifests itself in the structures of the Floquet modes. Below it will be demonstrated that the complicated structure of the phase space, which is a combination of stable islands submerged in a chaotic sea, can be observed in the Wigner distribution functions of the Floquet modes. Similar effects are known in the theory of quantum chaos.

We will see an analogue to another phenomenon from the theory of the quantum chaos — the so-called scarring. This effect consists in increase of the amplitude of the wave field in the vicinity of a periodic ray path.

3.3.1 *Spatial mode-medium resonance*

As it was shown in Sec. 2.3, the phenomenon of ray-medium resonance plays a crucial role in the emergence of ray chaos in periodic waveguides. At a finite wavelength, this phenomenon manifests itself as the mode-medium resonance. The relationship between the ray- and mode-medium resonances was considered in Refs. [Virovlyansky and Zaslavsky (1999, 2000); Virovlyansky (2000)]. We know that the m-th mode is formed by contributions from the rays with action variables I_m satisfying the quantization rule (3.34). In an unperturbed (range-independent) waveguide, all of them have the same cycle length D_m. A periodic perturbation with period λ_r affects mainly the modes with D_m satisfying the resonance condition

$$n_1 D_m = n_2 \lambda_r, \tag{3.145}$$

where n_1 and n_2 are integers. This is easily seen from Eq. (3.82). Using the relation

$$E_{m+\Delta m} - E_m = \frac{2\pi}{k_0\, D(I_m)} \Delta m, \tag{3.146}$$

which follows from Eq. (3.36), we rewrite Eq. (3.82) as

$$\frac{db_m}{dr} = -ik_0 \sum_{\Delta m} V_{m,m+\Delta m} b_{m+\Delta m} e^{i2\pi \Delta m r / D_m}. \tag{3.147}$$

Approximation (3.146) is valid only for not large Δm. Therefore, Eq. (3.147) can be employed only if $|V_{m,m+\Delta m}|$ rapidly decreases with increasing $|\Delta m|$. The matrix element $V_{m,m+\Delta m}$ is a periodic function of range r with the same period λ_r as the perturbation. Therefore, the most significant range variations in the mode amplitudes b_m are expected for modes satisfying the resonance condition (3.145). A similar effect occurs for a quantum-mechanical particle oscillating in a potential well whose parameters are periodic functions of time [Zaslavsky (1984); Reichl (1992)].

Due to the ray-mode duality, each ray-medium resonance is associated with the mode-medium one. Both resonances are determined by practically equivalent conditions (2.37) and (3.145). The number Δm of normal modes, captured into the mode-medium resonance, can be estimated as $\Delta m = k_0 \Delta I$, where ΔI is the width of the ray-medium resonance given by Eq. (2.52).

Normal modes, captured into an isolated resonance, weakly interact with other modes. The overlapping of resonances (it occurs when the amplitude of the perturbation is sufficiently large) leads to the onset of ray chaos. On the other hand, it causes a sharp increase in the number of coupling modes. These modes are formed by chaotic rays, and their amplitudes exhibit irregular range dependencies. However, the energy distribution between the normal modes, even under conditions of ray chaos, manifests regular (coherent) features. Let us illustrate this point with a numerical example.

Take the environmental model used in Sec. 2.6. This waveguide has the canonical background sound speed profile determined by Eq. (1.102) and a periodic perturbation modeling a single internal wave (2.129). The phase portrait of the ray system in this waveguide is shown in Fig. 2.19. Figure 3.2 shows the same portrait in the action-angle variables. This representation is more convenient for studying the ray-mode relations. Using the quantization rule (3.34), we expressed the mode number m, which is

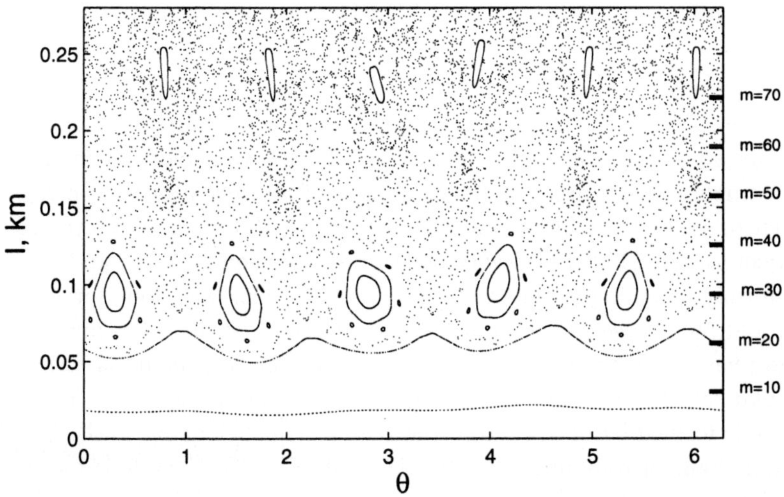

Fig. 3.2 The same phase portrait of the ray system as in Fig. 2.19, but represented in the action-angle variables.

formally considered as a continuous parameter, through the action I as $m = k_0 I - 1/2$ with k_0 corresponding to 75 Hz. Some values of m are indicated on the right vertical axis. The lower part of the phase portrait, corresponding to $I \lesssim 0.05$ ($m < 20$), is formed by flat rays belonging to the big central island in the center of Fig. 2.19. Other islands from Fig. 2.19 are also distinguished.

It was indicated in Sec. 2.6 that for rays propagating without reflection from the surface and the bottom, there are two primary resonances determined by the conditions $5\lambda_r = D(I)$ and $6\lambda_r = D(I)$. For the first of these resonances, we have $I = 0.0884$ km. This resonance is associated with the five-fold island chain. The islands, belonging to this chain in Fig. 3.2, are centered at actions close to the above resonant value. The second resonance corresponds to $I = 0.2503$ km. A six-fold chain in the upper part of the phase portrait in Fig. 3.2 is located in the vicinity of this resonant value. Estimates, made on the basis of formulas (2.61), show that these primary resonances strongly overlap giving rise to ray chaos. The chaotic sea fills most of the phase portrait.

Following Refs. [Smirnov *et al.* (2004)] and [Smirnov *et al.* (2005a)], we will study the evolution of the modal structure with range by solving numerically parabolic equation (1.17) with the initial condition

$$u(0,z) = \varphi_{m_0}(z). \tag{3.148}$$

The solution will be decomposed into a sum of eigenfunctions of the unperturbed waveguide $\varphi_m(z)$. From Sec. 3.2.4.1 we know that in the case of a single mode source the initial actions of all rays are equal to I_{m_0}. In order to see how the interval of actions grows with the range, we traced the trajectories of 120000 rays with starting actions equal to I_{m_0} and angles θ uniformly sampling an interval from 0 to 2π. This was done for $m_0 = 5$, 24, and 58. These mode numbers were selected for the following reasons. The phase portrait in Fig. 3.2 suggests that the 5-th mode is formed by regular rays (they belong to the big central island in Fig. 2.19). The 24-th mode is formed by both chaotic and regular rays belonging to the five-fold island chain surrounding the central island. The 58-th mode is formed by predominantly chaotic rays.

Using averaging over the fan rays, we evaluated range dependencies of the mean action $\bar{I}$ and the standard deviation of action σ for a dense set of distances up to 3000 km. At any particular range r, the interval $(\bar{I(r)} - \sigma(r), \bar{I}(r) + \sigma(r))$ characterizes the spread of action at this range. On the basis of the ray-mode relations discussed in Sec. 3.2, it is natural to expect that most of the modes with relatively large amplitudes will be concentrated within an interval

$$\frac{\bar{I(r)} - \sigma(r)}{k_0} < m < \frac{\bar{I}(r) + \sigma(r)}{k_0} \tag{3.149}$$

In Fig. 3.3 we show curves connecting the borders of such intervals computed for three cases where the 5-th (thin solid lines), 24-th (thick solid lines), and 58-th (dashed lines) are initially excited.

The results presented in Fig. 3.3 are consistent with our expectation. Indeed, the mode 5 is formed by regular rays and, therefore, it is not surprising that the spread of action in this case is very small. In contrast, the rays forming mode 24 start from points belonging to 5 stable islands as well as from the points belonging to the chaotic sea. Correspondingly, our ray-based prediction suggests that this mode should break up into a large group of modes. In the case of the 58-th mode we deal with predominantly chaotic rays, and, therefore, a number of excited modes becomes larger.

The ray based predictions for intervals of modes with relatively large amplitudes are in a reasonable agreement with results obtained by direct solution of the parabolic equation (1.17). In the case when only the

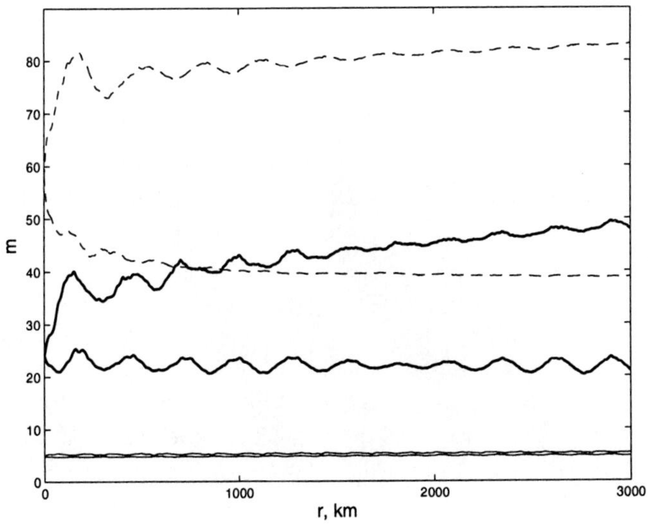

Fig. 3.3 Intervals of modes with comparatively large amplitudes predicted on the basis of Eq. (3.149). Modes with numbers between thin solid, thick solid, and dashed lines should have comparatively large amplitudes at $r > 0$ if at $r = 0$ only the 5-th, 24-th, or 58-th mode, respectively, is excited.

5-th mode is excited at $r = 0$, its amplitude remains practically constant up to 3000 km, the maximum range in our computations. Amplitudes of all other modes are negligible in comparison with $|b_5|$. In Figs. 3.4 and 3.5 we see mode amplitudes as functions of range and mode number when the source at $r = 0$ excites the 24-th and 58-th mode, respectively. In these figures we see two interesting features. (i) The initially excited mode breaks up into a large group of modes already at very short ranges on the order of a few dozens of km (in accord with ray-based predictions shown in Fig. 3.3). (ii) Interference patterns presented in Figs. 3.4 and 3.5 reveal periodic dependencies on the mode number. The periodicity is clearly seen even for modes with numbers $m > 40$ formed by predominantly chaotic rays.

Since the range dependence of perturbation δc in our model is expressed by the factor $\sin(2\pi r/\lambda_r)$ (see Eq. (2.129)) from Eq. (3.147), it follows that the m-th mode interacts most effectively with the modes whose numbers $\tilde{m}$ satisfy, at least approximately, the relation

$$\tilde{m} = m \pm jD/\lambda_r, \tag{3.150}$$

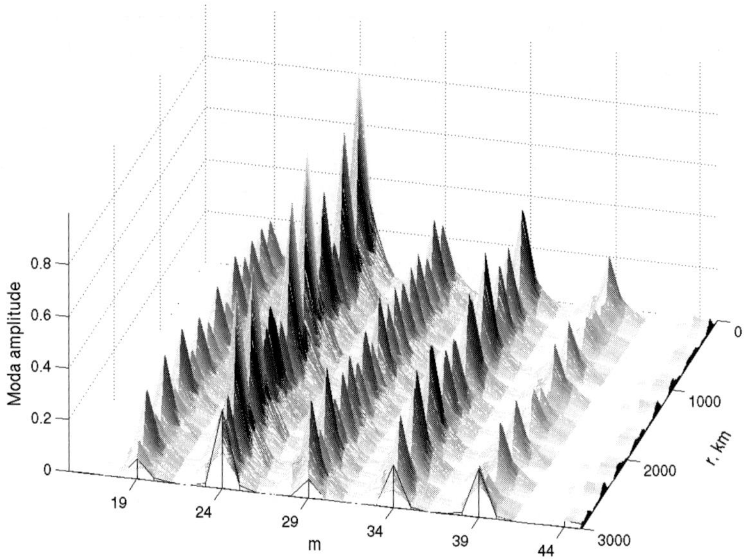

Fig. 3.4 Mode amplitude as a function of range r and mode number m. At $r = 0$, only the 24-th mode is excited [Smirnov *et al.* (2004)].

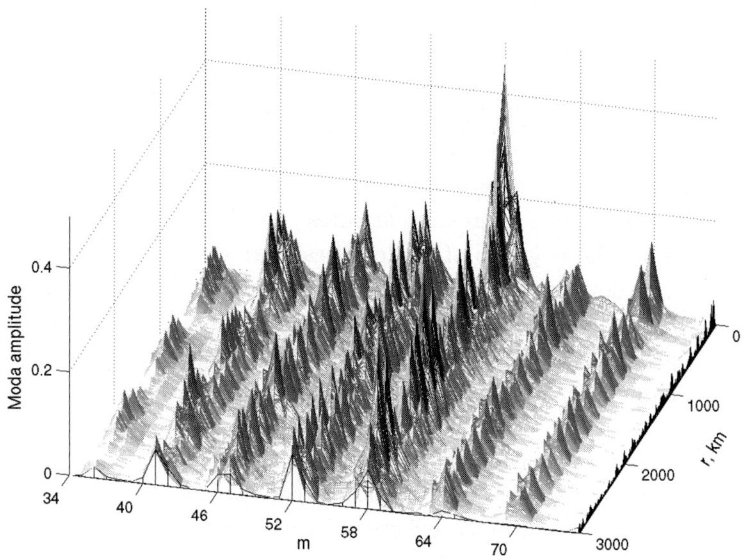

Fig. 3.5 The same as in Fig. 3.4 but at $r = 0$ the 58-th mode is excited [Smirnov *et al.* (2004)].

with j being an integer. For modes with the numbers close to 30, we get $D/\lambda_r \approx 5$. This explains why amplitudes of modes 19, 24, 29, 34, and 39 are most pronounced in Fig. 3.4. In this figure we see also that amplitudes of these modes oscillate with range. Moreover, the amplitude of the 24-th mode almost restores its starting value from time to time.

For modes with the numbers closer to 60, we have $D/\lambda_r \approx 6$. Therefore it is clear why the modes with numbers that differ from 58 by integer multiples of 6 dominate Fig. 3.5. On the other hand, it is rather unexpected that energy distribution over modes formed by predominantly chaotic rays reveal surprisingly regular structure. Even though range dependencies of mode amplitudes are irregular, this effect demonstrates that even under condition of ray chaos the wave field may exhibit coherent properties.

This fact is readily explained from the viewpoint of the Floquet theory. As it was indicated in Sec. 3.2.5, the solution to the parabolic equation (1.17) represents a sum of range-periodic terms. Therefore, in spite of the chaotic ray motion, at a finite wavelength the wave field has pronounced regular features. In fact, Figs. 3.4 and 3.5 illustrate this statement.

In the remainder of this section we will show how the mode-medium resonance and chaotic ray motion manifest themselves in the structures of the Floquet modes at a finite wavelength.

3.3.2 *Mode medium resonance and Floquet modes*

The role of a mode-medium resonance in connection between the Floquet modes and the modes of the unperturbed waveguide was studied in Refs. [Smirnov *et al.* (2004)] and [Smirnov *et al.* (2005a)]. Here we present some results obtained in these works.

Figure 3.6 shows absolute values of C_{qm}, dimensionless coefficients in the decomposition (3.139), computed for the environmental model under consideration. As it was indicated in Sec. 3.2.5, the m-th column of the matrix $\|C_{qm}\|$ defines the m-th Floquet modes. The upper panel of Fig. 3.7 graphs the dependence of parameter μ defined by Eq. (3.140) on m. This dependence is not absolutely monotonic. When numbering the Floquet modes we have slightly broken the monotony in order to make the pattern presented in Fig. 3.6 as regular as possible. As a result, we see that the off-diagonal elements of the matrix $\|C_{qm}\|$ form stripes parallel to the main diagonal in Fig. 3.6. The presence of these stripes reflects the fact that the quantity $|C_{qm}|$ as a function of m has maxima spaced apart by 5 for $q \lesssim 40$. The spacing between maxima changes from 5 to 6 as q approaches 60.

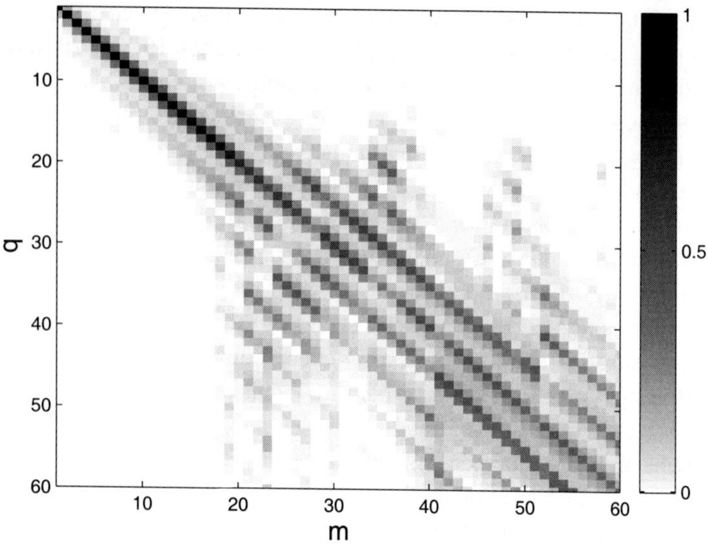

Fig. 3.6 Absolute values of the coefficients C_{qm} determining the decomposition (3.139).

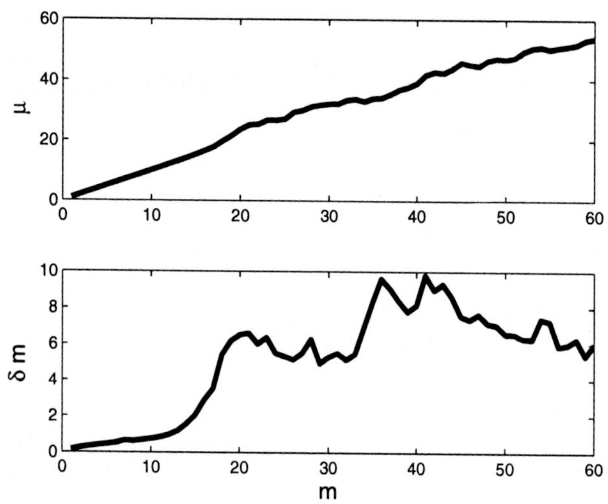

Fig. 3.7 *Upper panel*: Parameter μ (defined by Eq. (3.140)) associated with the m-th Floquet mode. *Lower panel*: The number of unperturbed modes forming the m-th Floquet mode, δm, estimated according to Eq. (3.151).

This periodicity of maxima explains the structures of the patterns seen in Figs. 3.4 and 3.5. Below we shall consider the perturbation theory that relates the appearance of the maxima to the phenomenon of the mode-medium resonance.

The lower panel of Fig. 3.7 presents an estimate for the number of unperturbed modes forming the m-th Floquet one

$$\delta m = \left(\sum_q (\mu - q)^2 \left| C_{mq} \right|^2 \right)^{1/2} . \tag{3.151}$$

This quantity is comparatively small for modes with $m < 20$ formed by regular rays belonging to the central island. For other modes the values of δm are significantly larger. A similar effect was observed in Ref. [Berman and Kolovsky (1992)] and called the delocalization. This term emphasizes that typically the Floquet eigenfunction $\psi_m(z)$ consists of many eigenfunctions $\varphi_m(z)$ corresponding to a smooth unperturbed waveguide. Since each $\varphi_m(z)$ represents a superposition of two quasi-plane (Brillouin) waves, the angular structure of the wave field formed by $\psi_m(z)$ is much more detailed than that of $\varphi_m(z)$.

In order to get insight into the role of the mode-medium resonance in the connection between the Floquet modes and the modes of the unperturbed waveguide, let us estimate the coefficients C_{qm} on the basis of a perturbation theory. Our starting point is the mode coupling equation (3.82). Assume that only one mode is excited at $r = 0$, i.e.,

$$b_m(0) = \delta_{m,n}. \tag{3.152}$$

We shall use a perturbation solution to Eq. (3.82) $b_m(r) = b_m^{(0)}(r) + b_m^{(1)}(r)$, where

$$b_m^{(0)}(r) = \delta_{m,n} e^{-ik_0 E_n r}, \tag{3.153}$$

and

$$b_m^{(1)}(r) = -ik_0 e^{-ik_0 E_m r} \int_0^r dr' \, V_{m,n}(r') e^{ik_0 (E_m - E_n) r'} . \tag{3.154}$$

According to Eq. (3.153) and (3.154), the elements of the Floquet matrix F, determined by Eq. (3.141), can be approximately presented in the form

$$F_{mn} = F_{mn}^{(0)} + F_{mn}^{(1)}, \tag{3.155}$$

where

$$F_{mn}^{(0)} = \delta_{mn} e^{-ik_0 E_n \lambda_r}, \tag{3.156}$$

and

$$F_{mn}^{(1)} = -ik_0 e^{-ik_0 E_m l} \int_0^{\lambda_r} dr\, V_{mn}(r) e^{ik_0 (E_m - E_n) r}. \tag{3.157}$$

Using the standard perturbation theory for matrices [Landau and Lifshitz (1977)] yields the following approximate expressions for the eigenvalues and eigenvectors of F:

$$\xi_m = e^{-ik_0 E_m \lambda_r} \left(1 - ik_0 \int_0^{\lambda_r} dr'\, V_{mm}(r') \right), \tag{3.158}$$

$$C_{mm} = 1, \tag{3.159}$$

and for $m \neq q$

$$C_{qm} = -\frac{k_0 e^{-ik_0 (E_q - E_m) l/2}}{2 \sin \left[k_0 \left(E_q - E_m \right) l/2 \right)]} \int_0^{\lambda_r} dr\, V_{qm}(r) e^{ik_0 (E_q - E_m) r}, \tag{3.160}$$

With our environmental model defined by Eqs. (1.102) and (2.129), we get

$$V(r, z) = \delta c / c_0 = Y(z) \sin \left(2\pi r / \lambda_r \right), \tag{3.161}$$

where $Y(z) = 2\varepsilon z B^{-1} \exp(2z/B)$. Substituting this in Eq. (3.160) yields for $m \neq q$

$$C_{qm} = \frac{ik_0}{2} Y_{qm} \left(\frac{1}{k_0 \left(E_q - E_m \right) + 2\pi/\lambda_r} - \frac{1}{k_0 (E_q - E_m) - 2\pi/\lambda_r} \right), \tag{3.162}$$

where

$$Y_{qm} = \int dz\, \varphi_q(z) Y(z) \varphi_m(z). \tag{3.163}$$

This result can be further simplified if the carrier frequency is so high that the WKB approximation is valid. Making use of Eq. (3.146) yields

$$E_q - E_m = (q - m) \frac{2\pi}{k_0 D_m}. \tag{3.164}$$

Substituting this relation in Eq. (3.162) and taking into account Eq. (3.51), we finally obtain for $m \neq q$

$$C_{qm} = \frac{ik_0 D_m \lambda_r}{8\pi} Y_{|q-m|}(I_m) \left[\frac{1}{D_m + (q-m)\lambda_r} + \frac{1}{D_m - (q-m)\lambda_r} \right], \tag{3.165}$$

where the symbol $Y_\nu(I)$ denotes an amplitude at $\cos(\nu\theta)$ in the expansion (1.67) with $F(z,p) = Y(z)$.

According to this result, the coefficients C_{qm} may be arbitrary large if the cycle length D_m is close to an integer multiple of λ_r, that is, if the condition of the mode-medium resonance is met. For modes of the unperturbed waveguide with the numbers $m \leq 60$, the cycle length D_m monotonically grows with m from 41.8 km for $m = 0$ to 52.4 km for $m = 60$. The only one integer multiple of $\lambda_r = 10$ km in this interval is 50 km which corresponds to the 28-th mode ($D_{28} = 49.9$ km). The applicability of our perturbation theory requires that $|C_{qm}| \leq 1$ for $m \neq q$. Therefore, Eqs. (3.159) and (3.165) cannot describe quantitatively Floquet modes with numbers close to 28. Numeric simulations have shown that the most accurate predictions of Eqs. (3.159) and (3.165) have been obtained for modes with $m < 20$ formed by predominantly regular rays belonging to the central island. Nevertheless, Eq. (3.165) properly predicts positions of main maxima in the dependence of $|C_{qm}|$ on q for practically all modes. This is illustrated in Fig. 3.8. However, it should be noted that while the perturbation theory predicts only 3 peaks for each mode, Figs. 3.6 and 3.8 show that the modes typically have more peaks. In spite of this restriction, the perturbation theory demonstrates a close connection between the phenomenon of the mode-medium resonance and the structure of the Floquet mode.

3.3.3 *Manifestations of stable islands and chaotic sea in structures of the Floquet modes*

Let us demonstrate that the complicated structure of the phase space, seen in the phase portrait presented in Figs. 2.19 and 3.2, reveals itself in the Floquet eigenfunctions $\psi_m(z)$ [Virovlyansky and Zaslavsky (2005)]. In order to link the structures of these eigenfunctions with the classical phase space, we shall consider the Wigner function defined by the relation [Tatarskii (1983)]

$$W_m(p,z) = \frac{k_0}{2\pi} \int d\zeta \psi_m(z - \zeta/2) \psi_m^*(z + \zeta/2)\, e^{ik_0 p \zeta}, \tag{3.166}$$

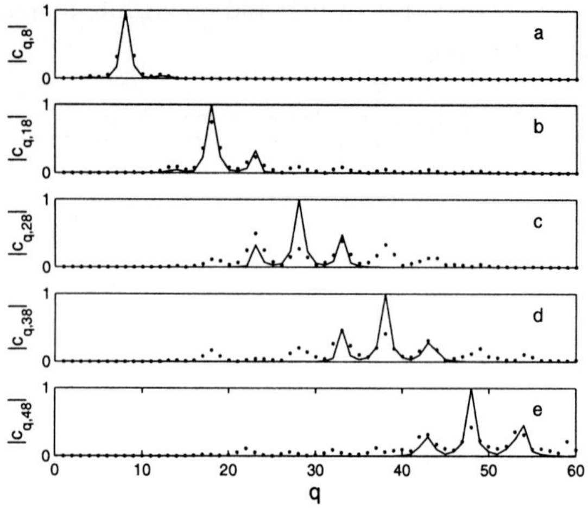

Fig. 3.8 Coefficients $|C_{qm}|$ found numerically (points) and predicted by Eqs. (3.159) and (3.165) (solid lines) as functions of q for $m = 8$ (a), $m = 18$ (b), $m = 28$ (c), $m = 38$ (d), and $m = 48$ (e).

where the asterisk denotes the complex conjugation. Equation (3.166) can be interpreted as an amplitude of a plane wave propagating at the grazing angle $\chi = \arctan p$ with respect to r-axis. Thus, the Wigner function dependence on momentum p in the wave theory characterizes a local angular structure of the wave field. To smooth out small-scale features of this structure and make large-scale ones more noticeable, we shall use a coarse-grained Wigner function [Tatarskii (1983)]

$$w_m(p,z) = \frac{1}{2\pi}\int dz'dp' \exp\left[-\frac{(z-z')^2}{2\Delta_z^2} - \frac{(p-p')^2}{2\Delta_p^2}\right] W_m(p',z'). \quad (3.167)$$

Selecting the smoothing scales satisfying the condition $\Delta_z\Delta_p = 1/2k_0$, we arrive at the Husimi distribution function [Gutzwiller (1990); Reichl (1992); Sundaram and Zaslavsky (1999)]: after some algebra Eq. (3.167) reduces to

$$w_m(p,z) = \left|\frac{1}{\sqrt[4]{2\pi\Delta_z^2}}\int dz'\psi_m^*(z') \exp\left[ik_0p(z'-z) - \frac{(z'-z)^2}{4\Delta_z^2}\right]\right|^2. \quad (3.168)$$

Thus, $w_m(p,z)$ represents a projection of the wave field onto a Gaussian state with a minimum uncertainty (coherent state) centered at a point (p,z).

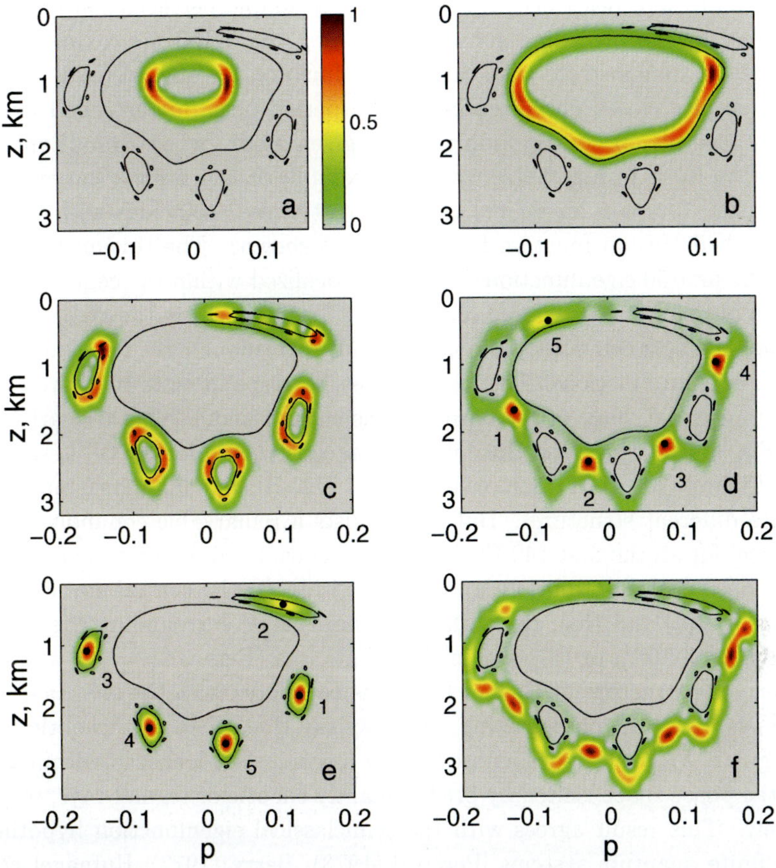

Fig. 3.9 Husimi distributions for 6 Floquet eigenfunctions at a carrier frequency of 200 Hz. The selected eigenfunctions: $\psi_{10}(z)$ (a), $\psi_{40}(z)$ (b), $\psi_{66}(z)$ (c), $\psi_{77}(z)$ (d), $\psi_{88}(z)$ (e), and $\psi_{105}(z)$ (f). [Virovlyansky and Zaslavsky (2005)]

Consider wave propagation at a carrier frequency of 200 Hz. Figure 3.9 shows Husimi functions evaluated for six Floquet modes. These examples have been selected to demonstrate a variety of structures of Floquet modes in our environmental model. At each panel we have outlined borders of stable islands discernible in the central part of the phase portrait 2.19. Although the range variations of the sound speed perturbation δc are most pronounced at the sound channel axis, the first Floquet eigenfunctions $\psi_m(z)$ are most close to their unperturbed counterparts, i.e., to the functions $\varphi_m(z)$ with the same numbers m. Therefore, the low Floquet modes are well described in the scope of a simple perturbation approach derived in the preceding subsection. Using the WKB approximation, it is easy to demonstrate that the Husimi function for any $\varphi_m(z)$ is a fuzzy version of a closed curve representing a period of an unperturbed ray (in the phase plane) corresponding to the given mode (whose Hamiltonian is equal to E_m). In Fig. 3.9(a) we see an example of such a fuzzy curve. Here, a Husimi function for $\psi_{10}(z)$ is plotted. It almost coincides with the unperturbed Husimi function for $\varphi_{10}(z)$ (not shown). The Husimi functions for the first 50 eigenfunctions $\psi_m(z)$ are localized within the central island. Each of them forms a closed fuzzy curve. The larger is m the greater is the difference between $\psi_m(z)$ and $\varphi_m(z)$ and, correspondingly, less similar are their Husimi functions. The fuzzy curves, corresponding to m approaching 50, repeat the shape of the border of the central island. This is clearly seen in Fig. 3.9(b) where the Husimi function constructed for $\psi_{40}(z)$ is shown.

Figures 3.9(b)–(f) demonstrate that the Husimi functions may have quite different structures. But there exists a remarkable common feature typical for all the first 140 Floquet modes considered in our numerical simulations: each distribution is localized within (i) the central stable island (Figs. 3.9(a) and (b)), (ii) within five sub-islands surrounding the central one (Fig. 3.9(e)), or (iii) within the chaotic sea (Figs. 3.9(c), (d), and (f)). Thus, the structure of Husimi functions correlates with the shapes of stable islands. Roughly speaking, one may say that there are "chaotic" and "regular" Floquet modes whose Husimi functions are localized within areas of the phase space visited by predominantly chaotic or regular rays, respectively. This result agrees with the semiclassical eigenfunction hypothesis for finite quantum systems [Percival (1973); Berry (1977); Hufnagel *et al.* (2002)]. According to this hypothesis, in the semiclassical limit, almost all eigenstates are either regular or chaotic, i.e., their phase-space representations are concentrated in regions with either regular or chaotic dynamics. However, in Ref. [Hufnagel *et al.* (2002)] it is shown that in the extended systems "the very concept of regular and chaotic states does not apply in the presence of transporting islands". In such systems all eigenstates may spread over regular islands as well as a chaotic sea of the classical phase space.

Notice that Fig. 3.9(c) presents a Husimi function localized outside the five sub-islands but in the vicinity of their borders. We conjecture that this is a finite-wavelength analogue of the stickiness to the borders of islands-surround-islands, i.e., the presence of such parts of the chaotic ray

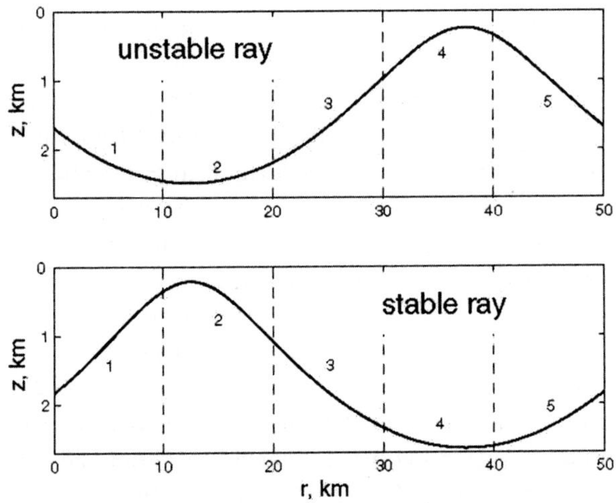

Fig. 3.10 Unstable (upper panel) and stable (lower panel) periodic rays with the period 50 km.

trajectory that correspond to almost regular rotation of rays around the islands [Zaslavsky *et al.* (1997)].

Figures 3.9(d) and 3.9(e) need a more discussion. Each of them has five clearly seen maxima. In Fig. 3.9(d) the maxima are located within the chaotic sea, near the destroyed saddles, while in Fig. 3.9(e) they are located near the elliptic points of the sub-islands. We argue that these maxima are associated with trajectories of the two periodic rays graphed in Fig. 3.10. Both rays have periods equal to $5\lambda_r = 50$ km, while there are no other rays with the period of 50 km. The ray shown in the upper panel of Fig. 3.10 belongs to the chaotic sea. The difference $|\delta z|$ between its coordinate and that of a ray with infinitesimally close starting parameters, on the average, grows as $\exp(\nu r)$ with the Lyapunov exponent $\nu = 1/60$ km^{-1}. In contrast, the ray in the lower panel has a zero Lyapunov exponent, i.e., it is regular. Both rays have been divided into five 10 km segments. The segments connect the consecutive waveguide cross-sections at range points $r = n\lambda_r$ chosen for the stroboscopic observation of both the ray structure (using the Poincaré map, see the respective article in Glossary)) and the Floquet modes (at these ranges, they are presented by $\psi_m(z)$). Small black circles within red and yellow spots (corresponding to local maxima of Husimi function) in Fig. 3.9(d) indicate positions (in the phase plane) of starting points of five segments shown in the upper panel of Fig. 3.10. In agreement with our expectation, these points lie at centers of five maxima of the Husimi function. The number of each segment is assigned to a corresponding circle and is indicated next to it in the figure. Similarly, the five maxima in Fig. 3.9(e) are associated with starting points of segments of the stable ray shown in the lower panel of Fig. 3.10. It is natural that the black circles depicting positions of these points are located in the centers of five stable sub-islands.

Since the ray trajectories in acoustic waveguides do not depend on frequency, it may be expected that the Husimi functions of Floquet modes at different carrier frequencies have similar structures. Results presented in Fig. 3.11 support this expectation. This figure demonstrates that at a carrier frequency of 75 Hz there are Floquet modes whose Husimi functions are very similar to that observed at a frequency of 200 Hz. Correspondingly, our division of mode into "chaotic" and "regular" also remains at 75 Hz: the Husimi functions depicted in Figs. 3.11(a), (b), and (e) are localized within stable islands, while that shown in Figs. 3.11(c), (d), and (f) lie within the chaotic sea. Again, we may expect that Fig. 3.11(e) demonstrates an analogue to the phenomenon of stickiness at a finite wavelength.

3.3.4 *Scarring*

In the theory of quantum chaos it is well-known that the distribution of energy levels and structures of eigenfunctions can be expressed through the specific properties of classical periodic orbits [Gutzwiller (1990); Reichl (1992); Stöckmann (1999)]. In particular, the amplitudes are high for some eigenfunctions in the vicinity of unstable periodic orbits and low elsewhere. This phenomenon, first discovered in a quantum billiard, is called scarring [Heller (1984)]. In the following years it has been established that scarring is a common property for many autonomous chaotic systems [Stöckmann (1999); Bies *et al.* (2001); Takagaki and Ploog (2004)]. It also has been shown that the scarring effect may be observed in quantum maps related to kicked systems [Agam and Brenner (1995)]. The description of scarring is closely related to the problem of construction of eigenfunctions on the basis of purely semiclassical calculations [Bogomolny (1988); Berry (1989); Tomsovic and Heller (1991); Berry and Keating (1992); Tomsovic and Heller (1993); Agam and Fishman (1994)].

Pushing further an analogy between the wave and quantum chaos, we will show that the scarring phenomenon may be observed in a model of the ocean acoustic waveguide with a periodic range-dependence. Our task is to demonstrate that periodic rays determine the structure of Floquet modes

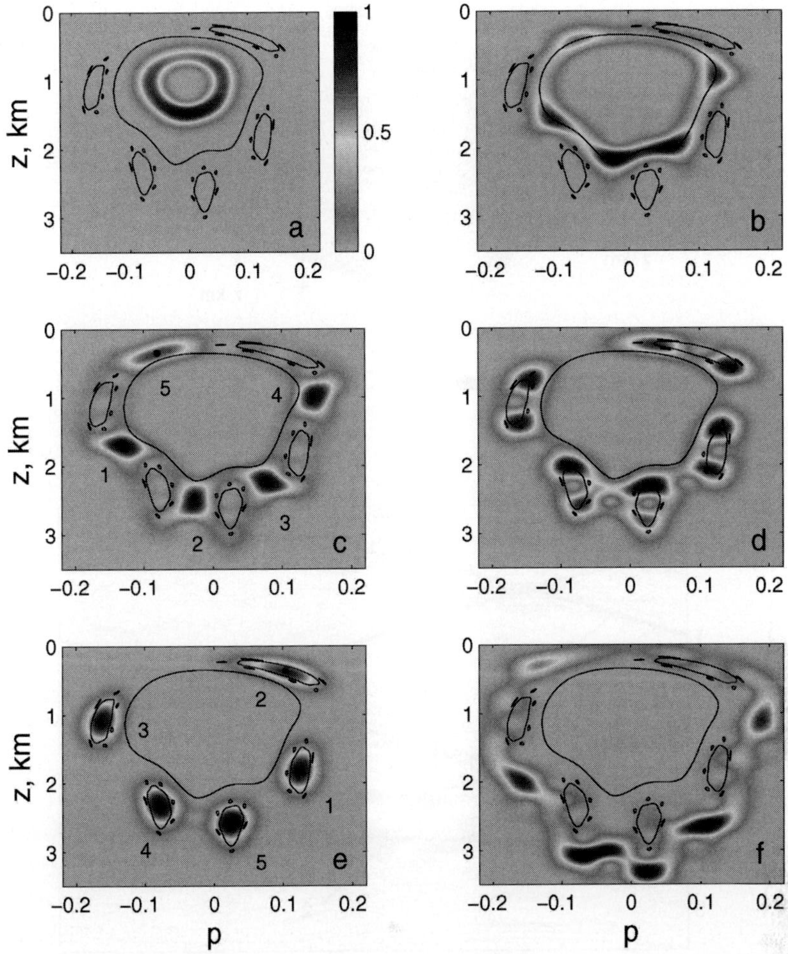

Fig. 3.11 The same as in Fig. 3.9 but at a carrier frequency of 75 Hz. The Husimi distributions are computed for the Floquet eigenfunctions $\psi_5(z)$ (a), $\psi_{18}(z)$ (b), $\psi_{23}(z)$ (c), $\psi_{24}(z)$ (d), $\psi_{29}(z)$ (e), and $\psi_{45}(z)$ (f) [Virovlyansky and Zaslavsky (2005)].

in approximately the same way as periodic orbits determine structures of the wave functions.

Results presented in Figs. 3.9(d) and 3.11(c) suggest that the unstable periodic ray shown in the upper panel of Fig. 3.10 determines essentially the structure of the 77-th Floquet mode at a carrier frequency of 200 Hz and that of the 23-rd Floquet mode at 75 Hz. This conjecture is confirmed

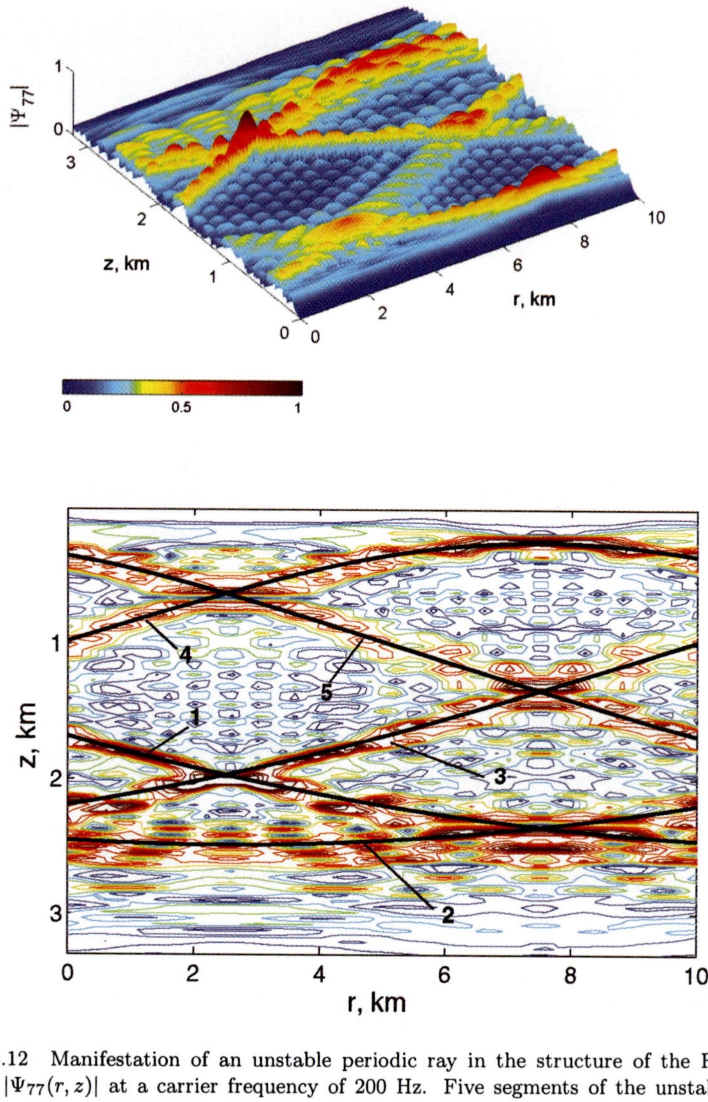

Fig. 3.12 Manifestation of an unstable periodic ray in the structure of the Floquet mode $|\Psi_{77}(r, z)|$ at a carrier frequency of 200 Hz. Five segments of the unstable ray from the upper panel of Fig. 3.10 are superimposed on the contour plot at the bottom [Virovlyansky and Zaslavsky (2005)].

in Figs. 3.12 and 3.13, where we present distributions of intensities of these modes within a range interval $(0, l)$, i.e., over a spatial period of all the Floquet modes. Both functions $|\Psi_{77}(r, z)|$ and $|\Psi_{23}(r, z)|$, as well as similar functions shown in Figs. 3.14–3.16 below, are scaled to have absolute

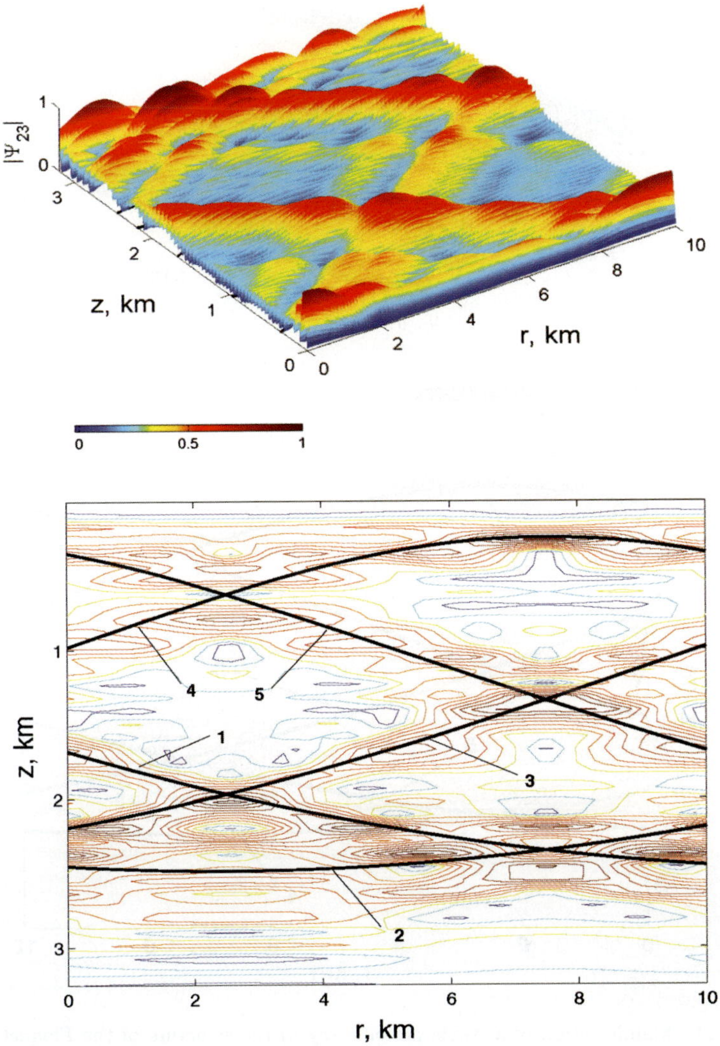

Fig. 3.13 The same as in Fig. 3.12 but for the Floquet mode $|\Psi_{23}(r,z)|$ at a carrier frequency of 75 Hz [Virovlyansky and Zaslavsky (2005)].

maxima equal to one. In Fig. 3.12 we see two representations of the function $|\Psi_{77}(r,z)|$. In the lower panel five segments of the unstable ray from the upper panel of Fig. 3.10 are superimposed on the contour plot representing isolines of the intensity distribution. It is clearly seen that the intensity is high at the segments of the unstable periodic ray and low elsewhere.

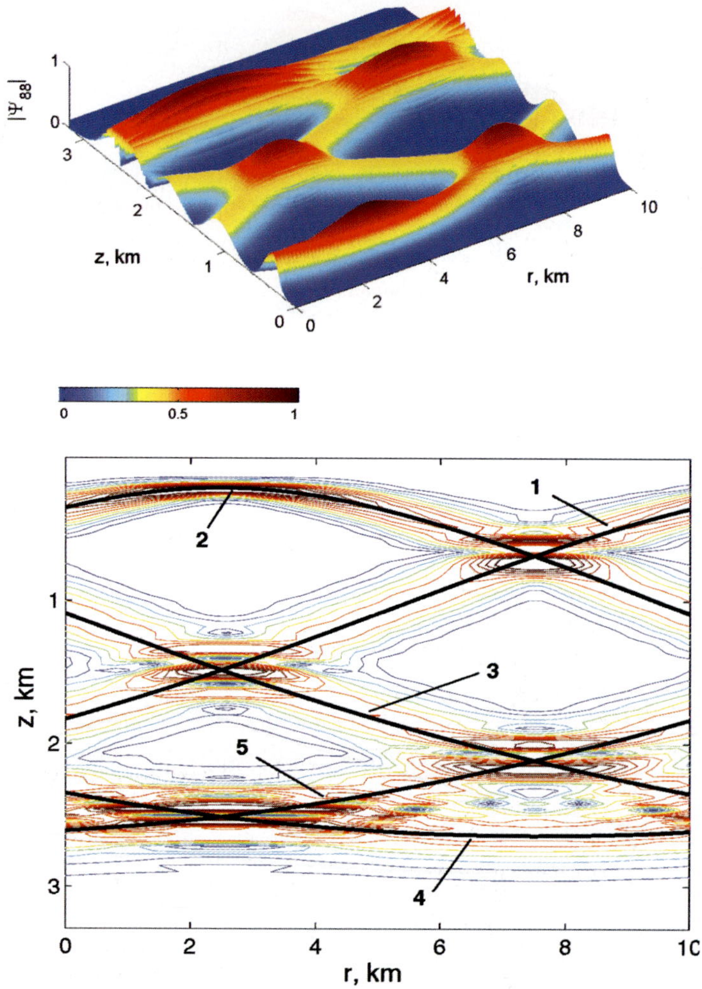

Fig. 3.14 Manifestation of a stable periodic ray in the structure of the Floquet mode $|\Psi_{88}(r,z)|$ at a carrier frequency of 200 Hz. Five segments superimposed on the contour plot correspond to the stable ray from the lower panel of Fig. 3.10 [Virovlyansky and Zaslavsky (2005)].

Figure 3.13 demonstrates the same effect at the carrier frequency 75 Hz. So, the periodic unstable ray path scars the Floquet eigenfunction. This is an analogue to the scarring phenomenon observed in different autonomous quantum systems [Heller (1984); Bies *et al.* (2001)] (see also examples in [Stöckmann (1999)]).

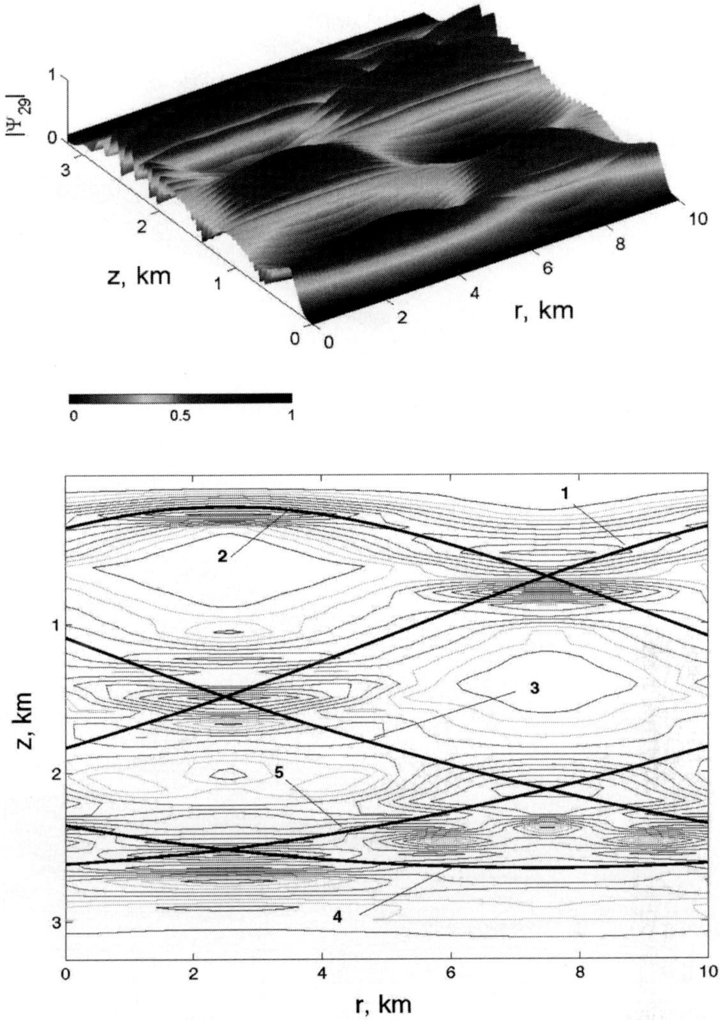

Fig. 3.15 The same as in Fig. 3.12 but for Floquet mode $|\Psi_{29}(r,z)|$ at a carrier frequency of 75 Hz [Virovlyansky and Zaslavsky (2005)].

In Figs. 3.13 and 3.15 it is seen that the 88-th Floquet mode at 200 Hz and the 29-th Floquet mode at 75 Hz localized within the five sub-islands (see Figs. 3.9(e) and 3.11(e)) also have relatively large amplitudes in the vicinity of segments of the stable ray shown in the lower panel of Fig. 3.10. Moreover, practically all the functions $|\Psi_m(r,z)|$ whose Husimi

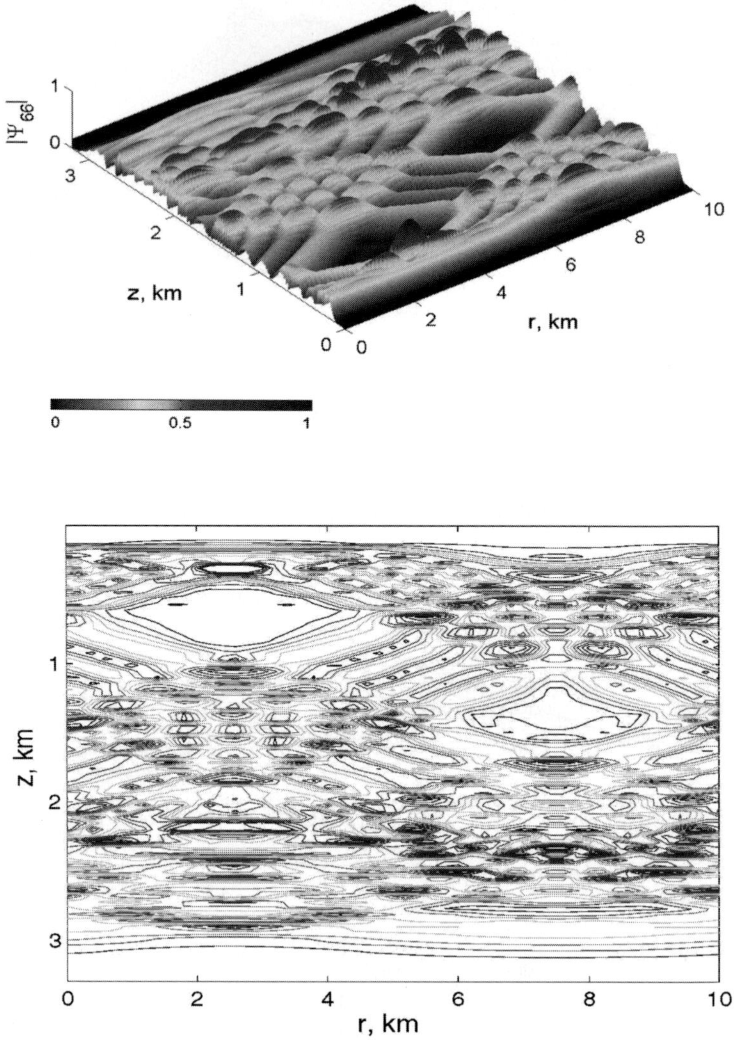

Fig. 3.16 The structure of the Floquet mode $|\Psi_{66}(r, z)|$ at a carrier frequency of 200 Hz [Virovlyansky and Zaslavsky (2005)].

distributions are localized outside the central island display some patterns analogous to those shown in Figs. 3.12–3.15 which, probably, may be associated with other periodic rays. This statement is illustrated in Fig. 3.16 where the 66-th Floquet mode at 200 Hz is shown.

Figures 3.12–3.15 suggest that the Floquet mode associated with one of the periodic rays plotted in Fig. 3.10 may be considered as a superposition of five 10 km pieces of a wave beam propagating along the corresponding ray. We conjecture that scars observed in intensity distributions of other modes, like that shown in Fig. 3.16, also may be associated with segments of some periodic rays. However, this issue needs a more detailed investigation.

3.4 Influence of Fine-Scale Inhomogeneities on Wave Dynamics

3.4.1 *Introductory remarks*

When we study a wavefield pattern using the ray approximation, it is necessary to keep in mind the well-known fact that the ray approximation is just a short-wavelength limit of a wavefield, and its relevance demands a sound wavelength to be small compared with a length scale of ocean variability L, i.e.

$$\lambda_0 \ll L. \tag{3.169}$$

Typically, the deep ocean can be regarded as a stratified media, and its variability in the horizontal direction is by $10^2 \div 10^3$ weaker than in the vertical one. Thus, the horizontal variability satisfies well the above condition, even for low sound frequencies of about 1 Hz. In the case of a vertical variability, things are not so simple. The criterion (3.169) can be reformulated in terms of the transversal (vertical) acoustic wavenumber as

$$k_\perp = k_0 \sin\chi \gg L_z^{-1}, \tag{3.170}$$

where L_z is a length scale of the vertical variability, and χ is a grazing angle of a wave. Depth dependence of a sound speed can be expressed as a sum of two components. The first component slowly varies, resulting in forming an USC, and also satisfies the condition (3.169). The second one corresponds to small-scale variations mainly caused by high-number modes of an internal-wave field. If we deal with low-frequency signals, which are able to propagate over long distances, fulfillment of inequality (3.170) appears to be very questionable. Although the latter component has a significantly smaller amplitude than the former one, their contributions into a vertical gradient of the sound-speed can be comparable, especially in the neighborhood of the channel axis. As it was shown in Sec. 2.5, this circumstance results in strong chaos of near-axial rays. But, according to Eq. (3.170), ray approximation cannot be used for near-axial wavepackets, when $\sin\chi$ is close to zero.

On the other hand, as it was argued by Hegewisch, Cerruti and Tomsovic [Hegewisch *et al.* (2005)], small-scale features of a sound-speed profile should be irrelevant for refraction of waves with low frequencies. In particular, Hegewisch with coauthors showed that removing of features being smaller than some frequency-dependent threshold length scale doesn't alter a wavefield a lot, although these features enhance remarkably ray instability. Thus, we arrive at the intricate conflict of wave and ray descriptions: small-scale sound-speed variations give rise to strong ray chaos, as in the case of a vertical ray-medium resonance, but appear insignificant for wavefield evolution. So, there arise several important questions. What is the extent to which we can trust the ray approximation for low-frequency sound propagation? As rapid vertical oscillations of a sound-speed perturbation are an important source of ray chaos, do chaos-induced phenomena cease completely at low acoustic frequencies? The same should be asked about other features having a ray origin, for example, peaks on Husimi plots of Floquet modes, which are contributed from periodic ray orbits.

The model of a waveguide we use in numerical simulations is the biexponential model

$$U(z) = \frac{b^2}{2}\left(e^{-az} - \eta\right)^2, \tag{3.171}$$

with the sound-speed perturbation given by the formula

$$V(z,r) = \frac{\delta c}{c_0} = \varepsilon Y(z)\sin\frac{2\pi z}{\lambda_z}\sin\frac{2\pi r}{\lambda_r}, \tag{3.172}$$

where the perturbation envelope $Y(z)$ is specified as

$$Y(z) = \frac{z}{B}e^{-2z/B}. \tag{3.173}$$

3.4.2 *Mode coupling in the presence of a fine structure*

Here we shall examine the effect of a fine-scale sound-speed perturbation on a wave motion from the viewpoint of the modal representation. Equation (3.80) can be rewritten as

$$\frac{i}{k_0}\frac{da_m}{dr} = E_m a_m + \sum_m V_{mn} a_n, \tag{3.174}$$

where matrix elements V_{mn} are calculated as

$$V_{mn}(r) = \int \phi_m^*(z)\delta V(z,r)\phi_n(z)\, dz. \tag{3.175}$$

This representation allows us to construct convergent series of the perturbation theory for mode amplitudes, mode functions, and the corresponding eigenvalues E_m, provided that the following condition is satisfied [Landau and Lifshitz (1977)]

$$\frac{|V_{mn}|}{E_m - E_n} \ll 1. \tag{3.176}$$

In the cases where criterion (3.176) is not satisfied, the analysis of the set of equations (3.174) doesn't allow us to obtain quantitative estimates but may be useful for a qualitative description of the spatial structure of a wavefield. In particular, the wavefield expansion in the modes of the range-independent waveguide allows us to estimate the effect of the vertical structure of inhomogeneity on the mode coupling. In the case of the inhomogeneity given by Eq. (3.172), the matrix elements can be represented as the following products:

$$V_{mn} = J_{mn} \sin k_r r, \tag{3.177}$$

where

$$J_{mn} = \frac{\varepsilon}{B} \int \phi_m^*(z)\, Y(z) \sin k_z z\, \phi_n(z)\, dz. \tag{3.178}$$

To simplify the subsequent analysis, we restrict our attention by purely-water modes and use the WKB approximation by representing the mode functions as a sum of Brillouin waves (3.37)

$$\phi_m(z) = Q_m(z) \left(e^{ik_0 g_m(z) + i\pi/4} + e^{-ik_0 g_m(z) - i\pi/4} \right), \tag{3.179}$$

where $g_m(z) = g(I_m, z)$. Function $g(I_m)$ is determined by Eq. (3.39). Substituting (3.179) into (3.178), we obtain

$$J_{mn} = \sum_{l=1}^{8} J_{mn,l}, \quad J_{mn,l} = \int dz\, \Upsilon_{mn}(z) \exp(ik_0 \Psi_{mn,l}), \tag{3.180}$$

where

$$\Upsilon_{mn} = \frac{\varepsilon z}{2iB} \exp\left(-2z/B\right) Q_m Q_n, \tag{3.181}$$

$$\begin{aligned}
\Psi_{mn,1} &= g_m - g_n + \frac{k_z}{k_0} z, & \Psi_{mn,2} &= g_n - g_m - \frac{k_z}{k_0} z, \\
\Psi_{mn,3} &= g_m - g_n - \frac{k_z}{k_0} z, & \Psi_{mn,4} &= g_n - g_m + \frac{k_z}{k_0} z, \\
\Psi_{mn,5} &= g_m + g_n - \frac{k_z}{k_0} z + \frac{\pi}{2}, & \Psi_{mn,6} &= -g_n - g_m + \frac{k_z}{k_0} z - \frac{\pi}{2}, \\
\Psi_{mn,7} &= g_m + g_n + \frac{k_z}{k_0} z + \frac{\pi}{2}, & \Psi_{mn,8} &= -g_m - g_n - \frac{k_z}{k_0} z - \frac{\pi}{2}.
\end{aligned} \tag{3.182}$$

Each of the integrals $J_{mn,l}$ can be evaluated using the stationary phase method (see Appendix D). Taking into account that

$$\frac{\partial g_m}{\partial z} = p_m, \tag{3.183}$$

one finds that the stationary phase conditions reduce to the following two

$$\frac{k_z}{k_0} = p_m(z) + p_n(z), \tag{3.184}$$

and

$$\frac{k_z}{k_0} = |p_m(z) - p_n(z)|. \tag{3.185}$$

The last two integrals, $J_{mn,7}$ and $J_{mn,8}$, vanish because

$$\frac{d\Psi_{mn,7}}{dz} = -\frac{d\Psi_{mn,8}}{dz} = p_m + p_n + \frac{k_z}{k_0} \neq 0 \tag{3.186}$$

for any z. The second derivatives of the phases $\Psi_{mn,l}$ are given by

$$\begin{aligned}
\Psi''_{mn,1} = -\Psi''_{mn,2} = \Psi''_{mn,3} = -\Psi''_{mn,4} &= \frac{dp_m}{dz} - \frac{dp_n}{dz}, \\
\Psi''_{mn,5} = -\Psi''_{mn,6} &= \frac{dp_m}{dz} + \frac{dp_n}{dz}.
\end{aligned} \tag{3.187}$$

Thus, the solution obtained by means of the stationary phase method is

$$\begin{aligned}
J_{mn,l} = \Upsilon_{mn}(z_l) \left| \frac{dp_m}{dz} \mp \frac{dp_n}{dz} \right|^{-1/2} \\
\times \sqrt{\frac{2\pi}{k_0}} \exp\left[ik_0 \Psi_{mn,l}(z_l) + i\,\mathrm{sgn}(\Psi''_{mn,l})\pi/4 \right],
\end{aligned} \tag{3.188}$$

where z_l is a solution of the equation $d\Psi_{mn,l}/dz = 0$, the upper sign corresponds to $1 \leq l \leq 4$ and the lower one corresponds to $l = 5, 6$.

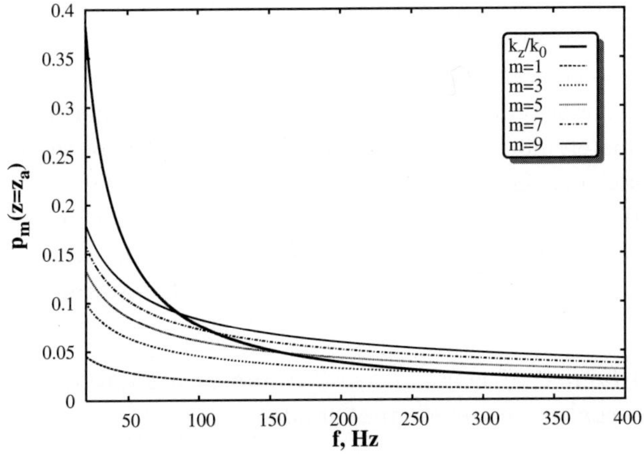

Fig. 3.17 Frequency dependence of the maximal ray momentum of the modal ray. The thick line represents the frequency dependence of the quantity k_z/k_0 [Makarov *et al.* (2008)]. $k_z = 2\pi/200$ m^{-1}.

The functions $|J_{mn,l}|$ will reach its maximal value when z_l is close to the depth of the waveguide axis, z_{a}, where $dp_m/dz = 0$. Taking that into account, we can find the conditions

$$\frac{k_z}{k_0} = p_m(z_{\mathrm{a}}) + p_n(z_{\mathrm{a}}), \tag{3.189}$$

and

$$\frac{k_z}{k_0} = |p_m(z_{\mathrm{a}}) - p_n(z_{\mathrm{a}})|, \tag{3.190}$$

under which vertical oscillations of the sound-speed perturbation enhance the mode coupling the most efficiently. On the other hand, if stationary conditions are not satisfied for any z, then the mode interaction is expected to be weak.

The ray-based theory of mode coupling implies that acoustic energy is transferring from one mode to another due to a variation of ray action. However, it is well-known that ray action is a slowly-varying quantity (except for passages through resonances). It means the dominance of the interaction of neighboring modes. To check whether it holds in the presence of fine-scale depth oscillations of the sound-speed perturbation or not, we calculate the dependence of $p_m(z_a)$ on the acoustic frequency for the 1th, 3th, 5th, and 9th modes and illustrate the results in Fig. 3.17. For comparison, we present the frequency dependence of the quantity k_z/k_0, where

$k_z = 2\pi/200\ \mathrm{m}^{-1}$. From these plots, it follows that, at low frequencies, the term k_z/k_0 dominates, and only higher-order modes or modes with large number difference can interact with each other. The interaction between neighboring lower-order modes arise only when the signal frequency f is of approximately 80 Hz, and the term k_z/k_0 becomes of the same order as the modal ray momenta. As the frequency grows further, conditions (3.184) and (3.185) are satisfied for progressively greater numbers of modes with close numbers. Thus, at low frequencies the ray-based theory fails to describe mode coupling in the presence of fine-scale features.

3.4.3 *Influence of vertical oscillations of a sound-speed perturbation on Floquet modes*

As Poincaré map serves as a convenient tool for exploring phase space patterns of ray motion in range-periodic waveguides, Floquet modes depicted via Husimi plots provide phase space representation of wave dynamics. Here we are going to use Floquet modes for studying changes in wave motion with gradually decreasing depth scale of a sound-speed perturbation. Our goal is to explore differences between actual wavefields at various frequencies and their ray-based predictions and to track how those differences grow with decreasing λ_z. The other parameters are fixed at the following values: $c_0 = 1480$ m/s, $a = 0.5\ \mathrm{km}^{-1}$, $b = 0.557$, $\eta = 0.6065$, $\varepsilon = 0.005$, $B = 1$ km, $\lambda_r = 5$ km.

The Floquet modes can be found numerically using the procedure described in the preceding Section, i.e. by solving repeatedly the parabolic equation with different modes of the unperturbed waveguide as initial conditions, then constructing a Floquet matrix and finding its eigenvectors. The parabolic equation was solved numerically with the idealized boundary conditions

$$u|_{z=0} = 0, \qquad \left.\frac{\partial u}{\partial z}\right|_{z=h} = 0. \tag{3.191}$$

The depth of the ocean bottom h is taken of 4 km. Four acoustic frequencies are considered: 20, 50, 100, and 200 Hz. Floquet modes are numbered according to the principle presented in the preceding Section, i.e. in the order of increasing values of the parameter μ. It allows one to find one-to-one correspondence between a Floquet mode and a corresponding phase-space region. Indeed, the value of μ can be related to a mean value of the action of the respective Floquet mode as

$$\mu \simeq k_0\langle I\rangle. \tag{3.192}$$

Phase-space representation of Floquet modes is obtained by means of the Husimi function which is proven as a suitable tool for this purpose. The parameter Δ_z of the Husimi function, i.e., the depth scale of a Gaussian wavepacket, a wavefield is projected to, is taken of 100 m.

Before we begin to analyze immediately the Floquet modes, some words about general principles of the ray-wave correspondence regarding the phase space should be said. According to Eq. (3.150), the range-periodic perturbation causes resonant transitions of acoustic energy between modes m and $m+m'$, where

$$m' = j\frac{k_r}{\omega}, \tag{3.193}$$

and j is an integer. As we saw in the preceding section, distinct phase space structures, like stable islands or chaotic layers, give rise to the Floquet modes localized within the boundaries of those structures. However, one should remember that a degree of manifestation of any feature of the ray phase space depends on its "mode capacity" δm, i.e. the number of modes of the unperturbed waveguide captured into this feature. The larger δm, the stronger the correspondence between wavepacket dynamics and the ray-based predictions. On the contrary, the limit $\delta m \to 0$ implies suppression of ray-based effects and prevalence of purely wave phenomena. The mode capacity can be expressed as

$$\delta m = k_0 \delta I, \tag{3.194}$$

where δI is a size of the feature in the action space. Therefore, any phase space structure to be reflected in Floquet modes must comprise, at least, several modes coupled to each other according to Eq. (3.193). In the case of ray-medium resonance $m:n$, δI is equal to the resonance width ΔI, and one can introduce the parameter

$$C = \frac{\delta m}{m'} = \frac{nk_0\Delta I}{m}. \tag{3.195}$$

Condition

$$C \gg 1 \tag{3.196}$$

infers that a phase space structure arising from the ray equations should well revealed in Floquet modes. Otherwise,

$$C < 1 \tag{3.197}$$

implies irrelevance of the respective resonance for wave-based phase-space patterns. Obviously, the latter should be expected at low acoustic frequencies, when k_0 is of a small value. Under this regime, some small-size KAM islands should be weakly revealed or cease completely.

As to a vertical resonance, three levels of manifestation, depending on the "strength" of resonance, should be distinguished. "Weak" vertical resonance just reveals itself in broadening the nearest ray-medium resonances, therefore, its relevance for wave dynamics depends on the parameter C introduced by Eq. (3.195). "Moderate" vertical resonance scatters rays and leads to the emergence of a bounded chaotic layer in the phase space. This layer influences wave motion if its width is large enough and comprises many modes of the unperturbed waveguide. Assuming that the width of a layer is approximately equal to the width of the vertical resonance (actually, it can be significantly larger, see subsection 2.5.4), we can quantify also the degree of its manifestation using the parameter C, just replacing ΔI in Eq. (3.195) by the width of the vertical resonance in the action space ΔI_{vert}. The formula linking ΔI_{vert} to Δp given by Eq. (2.95) can be easily derived as

$$\Delta I_{\text{vert}} = \frac{D}{2\pi}\Delta H_{\text{vert}} \simeq \frac{Dp_{\max}}{2\pi}\Delta p, \tag{3.198}$$

where H_{vert} is the resonance width in terms of H, and it is taken into account that $dI/dH = D/2\pi$ and $dH/dp = p$. "Strong" vertical resonance can capture rays like that is illustrated in Fig. 2.12. This process should be reflected in a wavefield as a formation of intensity spots spaced by λ_z in depth. These spots correspond to the states localized within the microchannels emerging due to the perturbation. Such localized states are allowed if the motion within a microchannel satisfies the condition

$$\frac{k_0}{2\pi k_z}\oint \tilde{p}(\Psi,\, \tilde{H}_s)\, d\Psi > 1, \tag{3.199}$$

where $\tilde{p}(\Psi,\, \tilde{H}_s)$ is given by Eq. 2.94, and integration is carried over a separatrix of the resonance domain. Inequality (3.199) means that the area of the resonance domain in the space of variables p and z must be larger than the area of the so-called unit cell volume $1/k_0$ (recall that it is the analogue of the Planck constant in quantum mechanics).

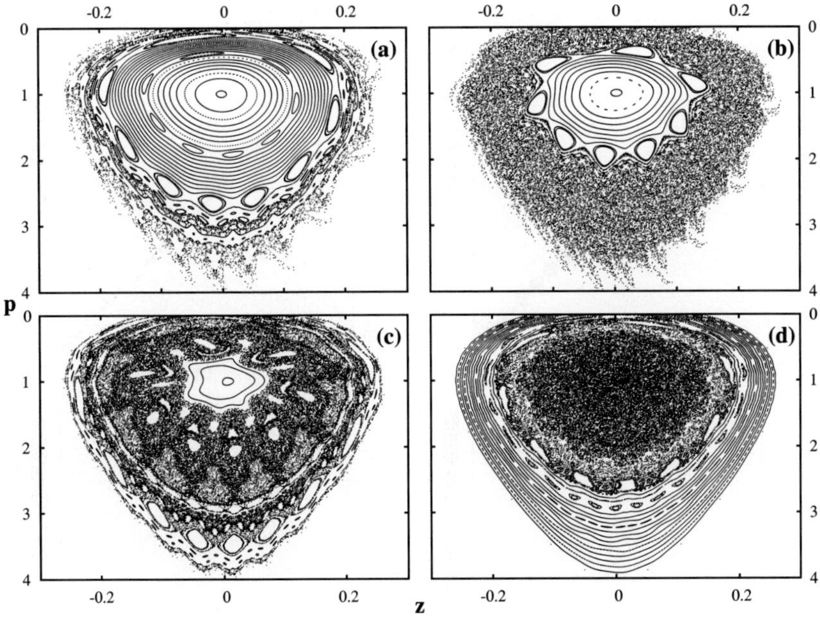

Fig. 3.18 Poincaré maps computed with λ_z = 2000 m (a), λ_z = 1000 m (b), λ_z = 500 m (c), and λ_z = 200 m (d).

We shall consider the cases of $\lambda_z = 2000$ m, $\lambda_z = 1000$ m, $\lambda_z = 500$ m, and $\lambda_z = 200$ m. Below we shall present a brief analysis of each of these cases. In order to facilitate to a reader the comparison of the patterns of Floquet modes with the respective ray-based predictions, we repeat here at once the figure with the respective Poincaré maps (see Fig. 3.18), earlier presented in subsection 2.5.4.

3.4.3.1 $\lambda_z = 2000$ *m*

We begin with considering the sound-speed inhomogeneity with $\lambda_z = 2000$ m. Figure 3.19 represents the matrices of absolute values of C_{qm} calculated with the frequencies of 20 Hz (a), 50 Hz (b), 100 Hz (c), and 200 Hz (d). In the case of $f = 20$ Hz, the elements along the main diagonal have the values close to 1, while other ones are near zero. It means that the perturbation doesn't cause significant mode coupling, and each Floquet mode consists of only one mode of the unperturbed waveguide. All the ray-based predictions fail. As f increases, the situation becomes more complicated. In the case of $f = 50$ Hz, the diagonal elements still dominate,

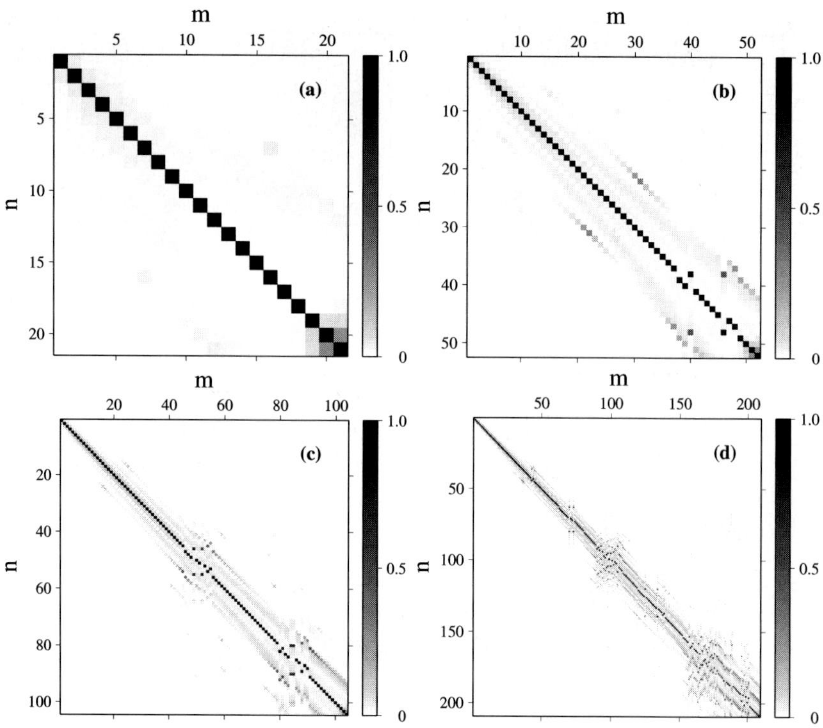

Fig. 3.19 Absolute values of the coefficients C_{qm} determining decomposition of the Floquet modes over modes of the unperturbed waveguide. The vertical wavelength of the sound-speed perturbation λ_z is 2000 m. Acoustic frequency is 20 Hz (a), 50 Hz (b), 100 Hz (c), and 200 Hz (d).

but an additional pair of parallel stripes appears in two blocks of the matrix $\|C_{qm}\|$. The vertical spacings between stripes and the main diagonal obey the relation (3.193). So, the presence of those stripes originates from the mode-medium resonances, 9 : 1 in the center block and 10 : 1 in the right lower block.

The triplets bear smeared patterns that look like "dial-plates" on the Husimi plots of the respective Floquet modes. Examples of such "dial-plates" are shown in Fig. 3.20(a) and (b). The ray-medium resonance 8 : 1 (corresponding to the block in the upper left corner of the matrix) appears too narrow to form the triplet structure with the frequency of 50 Hz, but with larger frequencies, 100 and 200 Hz, a weak triplet occurs (in the upper left corners of the matrices). Besides that, in the cases of higher frequencies the triplets corresponding to the resonances 9 : 1 or 10 : 1 transform

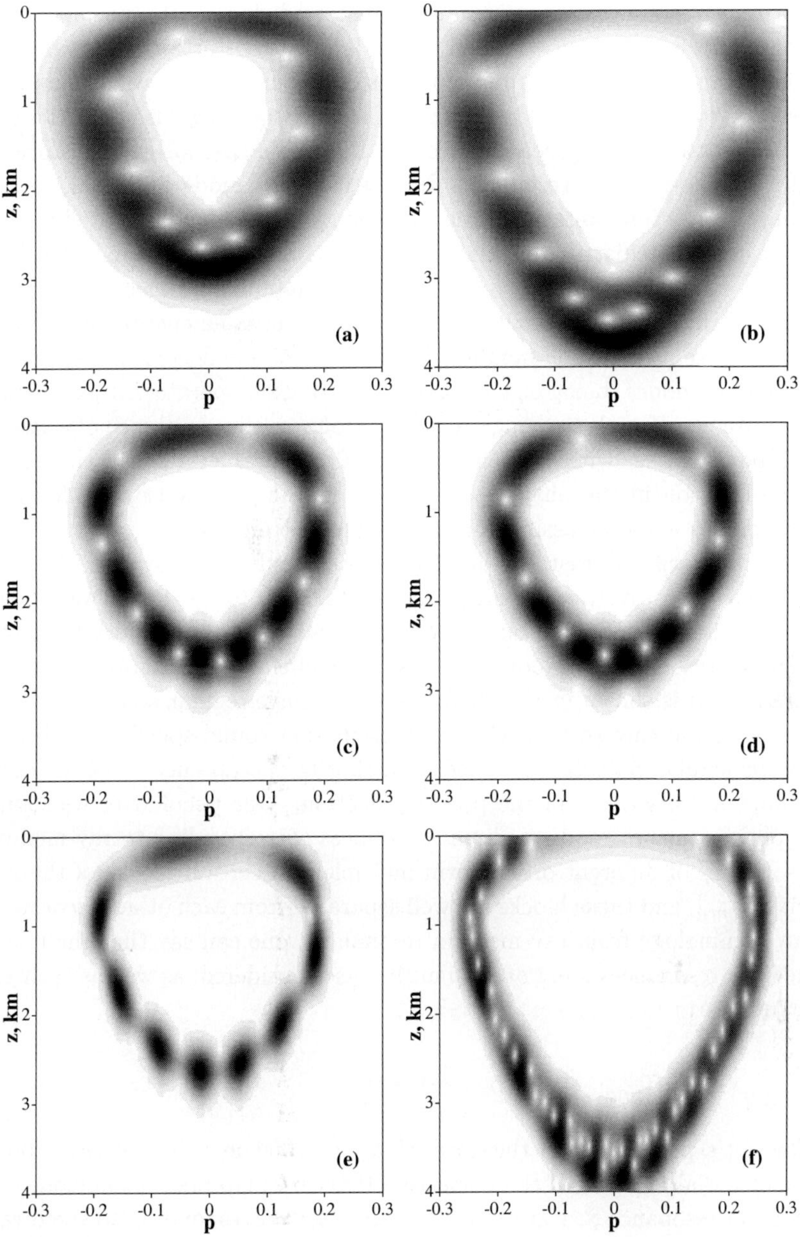

Fig. 3.20 Examples of Floquet modes computed with $\lambda_z = 2000$ m and illustrated via the Husimi plots. (a) $f = 50$ Hz, $m = 31$, (b) $f = 50$ Hz, $m = 46$, (c) $f = 100$ Hz, $m = 49$, (d) $f = 100$ Hz, $m = 52$, (e) $f = 200$ Hz, $m = 101$, (f) $f = 200$ Hz, $m = 189$.

into multiplets, which are somewhat distorted due to the peculiarity of the numbering procedure chosen. Indeed, such a distortion occurs if there exists a compact limited group of modes of an unperturbed waveguide with a strong coupling within the group and a weak coupling with other modes. As was shown in the preceding Section, this is the case for modes captured into a mode-medium resonance. If we excite any mode belonging to this group, then, after some evolution, the energy initially concentrated in that mode will be redistributed among all the modes of the group. The final energy distribution will have some mean mode number. Exciting another mode of this group, we shall arrive at almost the same energy distribution with a close mean mode number. This situation is reflected in a spectrum of Floquet modes, being eigenstates of an evolution operator. In particular, the presence of such a strongly-coupled group implies existence of a family of Floquet modes with very close values of the parameter μ and almost the same location in the phase space. Perhaps, the parameter μ couldn't be regarded as a good classificator of such Floquet modes.

In a sharp contrast to the triplets, the multiplets give rise to well-resolved peaks on the Husimi plots (see Fig. 3.20(c)–(e)). As was shown earlier, these peaks arise on the stable or unstable periodic orbits of ray-medium resonances. Floquet modes, belonging to the chaotic layer, are spread as it is shown in Fig. 3.20(f) but also have regular structure.

Thus, the case of a slowly depth-oscillating sound-speed perturbation reveals similar patterns as those occurring in the absence of depth oscillations and discussed in the previous Section. One peculiarity we should emphasize is that mode-medium resonances corresponding to ray-medium resonances of different orders form multiplets in certain blocks of the matrix $\|C_{qm}\|$, and those blocks are well separated from each other. Borrowing the terminology from ray-medium resonances, one can say that the mode-medium resonances *don't overlap* in the case considered, as well as their ray counterparts.

3.4.3.2 $\lambda_z = 1000$ *m*

Now let's proceed with the case of $\lambda_z = 1000$ m. As we saw above, the essential features of the respective Poincaré map are a prominent ray-medium resonance $8:1$ and a wide chaotic sea corresponding to steep rays and, according to the ray-mode duality, high-number modes. The matrix $\|C_{qm}\|$ is presented in Fig. 3.21. Note that there is no such a salient dominance of the main diagonal as in the previous case, and the mode coupling

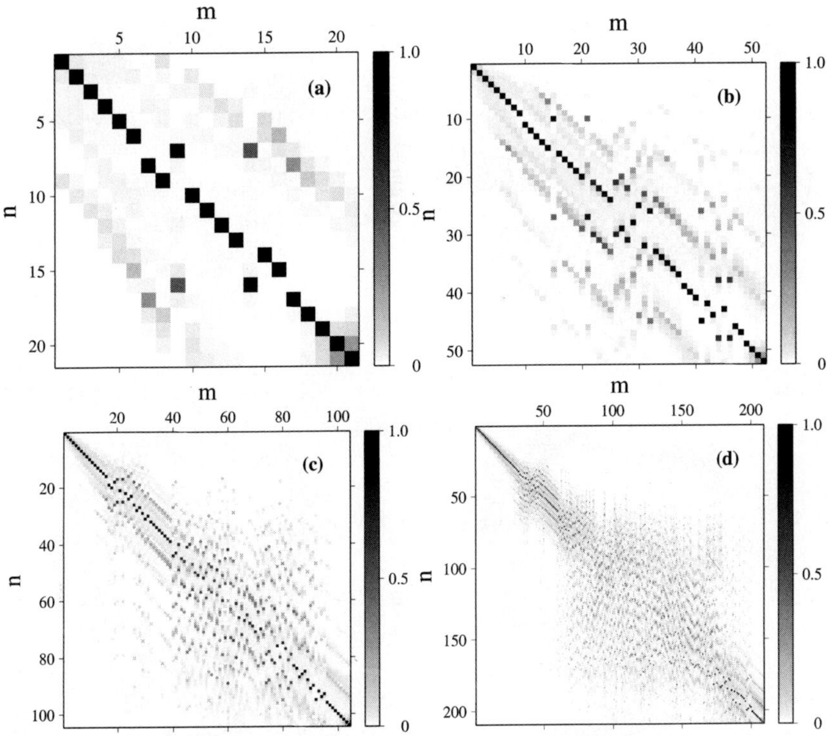

Fig. 3.21 The same as in Fig. 3.19 but for $\lambda_z = 1000$ m.

is significantly stronger. Remarkable triplet pattern appears even with $f = 20$ Hz. Owing to that, some of the Floquet modes computed with $f = 20$ Hz reveal a "dial-plate" pattern of the resonance 9 : 1, as that is depicted in Fig. 3.22(a). The matrix with 50 Hz represents an almost ordered multiplet pattern, a part of which transforms into a large fuzzy block with higher frequencies. There arise the modes like the one presented in Fig. 3.22(b). It consists of distinct peaks of the resonance 8 : 1 and expands into the phase space domain corresponding to the chaotic sea. On the ray level, the ray-medium resonance 8 : 1 is isolated from the chaotic sea by invariant curves which are impenetrable for rays. Presence of such Floquet modes is a manifestation of the phenomenon *dynamical tunneling* [Tomsovic and Ullmo (1994); Mouchet *et al.* (2006)] known in the theory of quantum chaos. It means penetration of wave functions through classical phase space barriers. Another feature we should stress is the

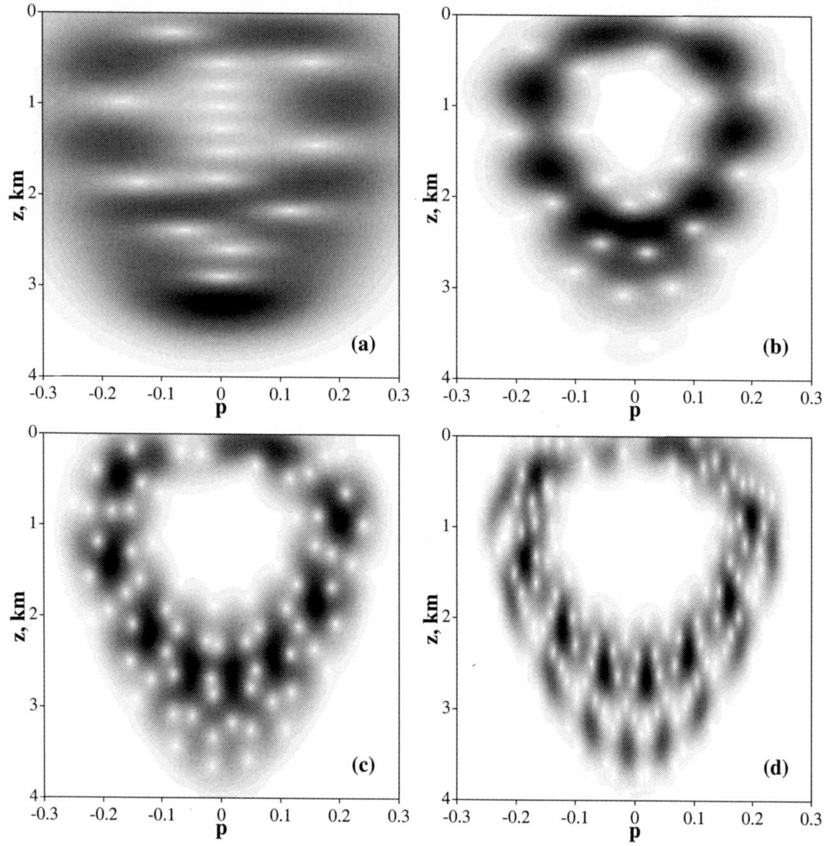

Fig. 3.22 Examples of Floquet modes computed with $\lambda_z = 1000$ m and illustrated via the Husimi plots. (a) $f = 20$ Hz, $m = 14$, (b) $f = 50$ Hz, $m = 20$, (c) $f = 100$ Hz, $m = 53$, (d) $f = 200$ Hz, $m = 109$.

presence of identifiable chain of peaks corresponding to the resonance 9 : 1 (see Fig. 3.22(c)), which is destroyed on the ray level, but its periodic ray orbits do exist and give the contribution into the scarred Floquet modes. Figure 3.22(d) illustrates the Floquet mode involving the peaks of resonances 9 : 1 and 10 : 1, simultaneously.

Thus, decreasing twice the vertical length scale leads to the amplification of mode coupling. However, the phenomena observed are still well-consistent with the ray-based pattern, except for the tunneling modes. It is not surprising, because the depth oscillations are still very slow, and short-wavelength approximation still works well.

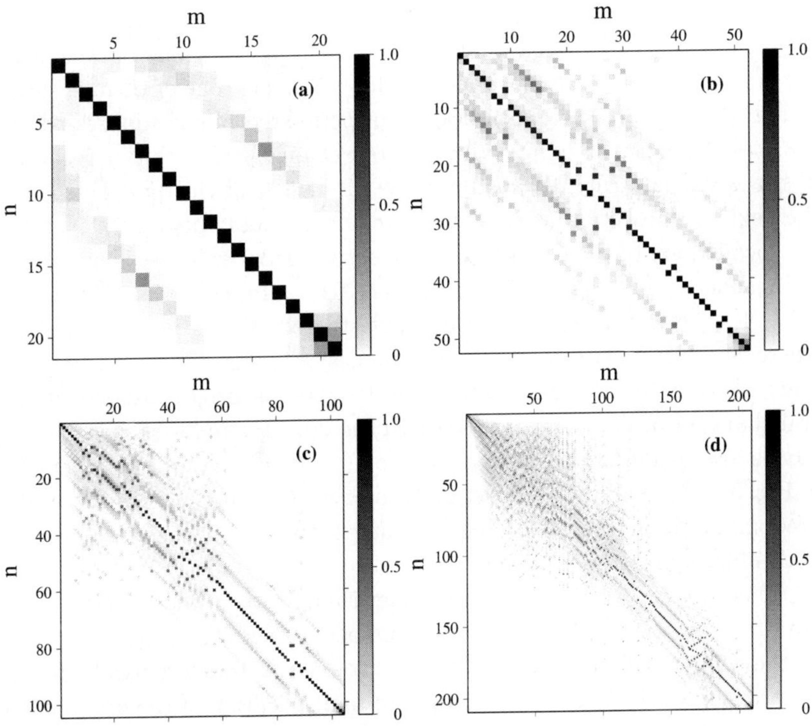

Fig. 3.23 The same as in Fig. 3.19 and 3.21 but for $\lambda_z = 500$ m.

3.4.3.3 $\lambda_z = 500$ *m*

A sound-speed perturbation with $\lambda_z = 500$ m induces two separate nested chaotic layers in the phase space (see Fig. 3.18(c)). The wide inner layer is induced by a vertical resonance and contains several chains of islands within. The outer one is significantly smaller. Even with $f = 20$ Hz, the matrix $\|C_{qm}\|$ has an tridiagonal form. With higher frequencies, the matrix has multiplets being drawn along the upper half of the main diagonal, and Floquet modes reveal well-resolved ordered peaks, as well as extended patterns covering the inner chaotic layer. Notably, the outer chaotic layer appears too narrow and is not reflected significantly in the matrix. We focus our attention at the case with $f = 200$ Hz. The ray-medium resonances $8:1$ and $9:1$ are merged into the chaotic sea, but give rise to ordered chains of peaks, attributed to scarring described in subsection 3.3.4. Although the resonance 8 : 1 has three chains of stable islands due to the perturbation-induced degeneracy, only one chain of eight peaks is present in the corresponding Floquet modes, one of them is shown in Fig. 3.24(a). So, the multiplication of periodic orbits is suppressed and the pre-bifurcation state involving only one chain of eight orbits is recovered in the wave pattern! We postpone the discussion of this phenomenon to the next subsubsection when it will reveal itself more apparently. Notably, the localization of the Floquet modes near its periodic orbits is weaker for the resonance 8 : 1 than for the resonance 9 : 1 (compare Figs. 3.24(a) and (b)), despite the latter one hasn't undestroyed islands in the ray limit. There are also a family of Floquet modes covering almost entirely the inner chaotic layer. They expose spectacular patterns consisted of several chains of peaks, being well-ordered though. Examples of such Floquet modes are demonstrated in Figs. 3.24(c)–(e). The Floquet mode depicted in Fig. 3.24(f) exhibits the effect of dynamical tunneling. It is concentrated predominantly at the stable region, but expands simultaneously over the both inner and outer chaotic layers.

Thus, we can see that features inconsistent with ray-based description tend to grow with decreasing the perturbation wavelength λ_z. In addition to the dynamical tunneling, here we encountered the nontrivial effect of suppression of the bifurcation leading to multiplication of the orbits of the resonance 8 : 1.

3.4.3.4 $\lambda_z = 200$ *m*

In the case of $\lambda_z = 200$ m, the wide chaotic sea induced by the vertical resonance occupies the central domain of the phase space corresponding to near-axial rays. It should lead to a multiplet pattern in the upper left corner of the matrix $||C_{qm}||$ which corresponds to the Floquet modes propagating near the waveguide axis. This expectation is fulfilled with the frequencies in the range 50–100 Hz, but fails with 20 Hz (see Fig. 3.25), when the main diagonal is dominant, i.e. modes are decoupled.

At the frequencies of 100 and 200 Hz, the chaotic sea gives rise to widespread Floquet modes like that shown in Figs. 3.26(a) and (b). These modes are responsible for rapid broadening of a wavepacket until all the chaotic sea will be covered. This regime is well-consistent with the ray limit which anticipates rapid spreading and entanglement of a Lagrangian manifold belonging to the chaotic sea, as it is demonstrated in Fig. 2.16. With lower frequencies this effect is suppressed or completely disappears.

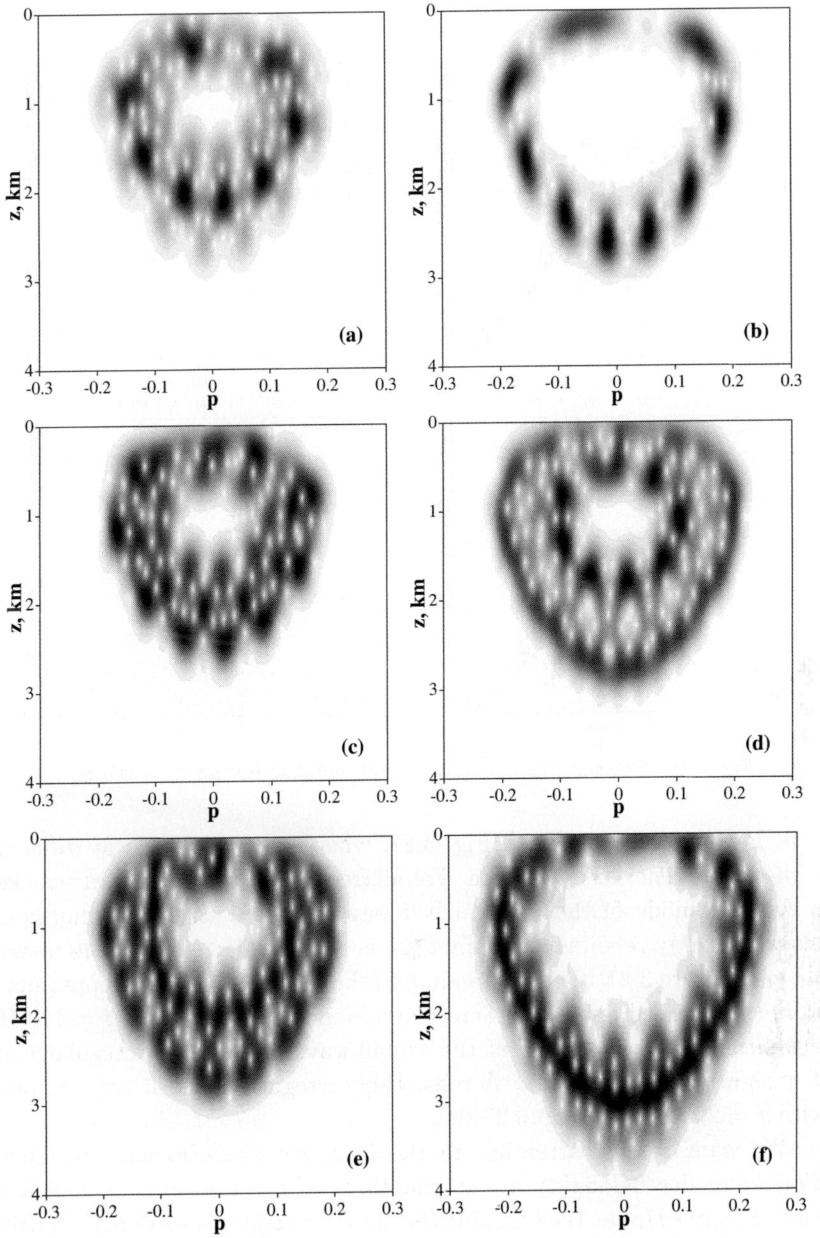

Fig. 3.24 Examples of Floquet modes computed with $\lambda_z = 500$ m and illustrated via the Husimi plots. (a) $f = 20$ Hz, $m = 16$, (b) $f = 200$ Hz, $m = 57$, (c) $f = 200$ Hz, $m = 96$, (d) $f = 200$ Hz, $m = 65$, (e) $f = 200$ Hz, $m = 75$, (f) $f = 200$ Hz, $m = 138$.

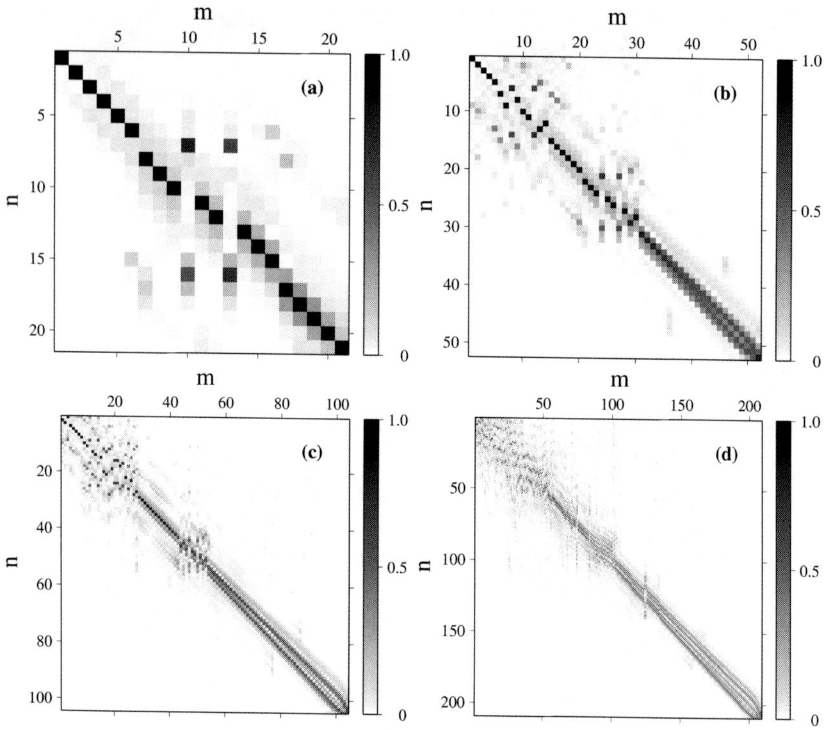

Fig. 3.25 The same as in Fig. 3.19, 3.21 and 3.23 but for $\lambda_z = 200$ m.

It is demonstrably shown in Fig. 3.27, where wave evolution at different acoustic frequencies is presented. For all frequencies, the initial wavepacket is the first mode of the unperturbed waveguide. With 20 Hz almost all acoustic energy remains in the first mode as it is inferred from the matrix $||C_{qm}||$. Figure 3.27(b) corresponding to the case of $f = 50$ Hz represents a more complicated but well-ordered interference pattern. With $f = 100$ Hz weak irregularity emerges, but the overall wavefield still looks regularly, as it is shown in Fig. 3.27(c). A remarkably irregular pattern appears only with $f = 200$ Hz (see Fig. 3.27(d)).

We want to pay attention to the following phenomenon. Many of Floquet modes belonging to the chaotic sea have the form as shown in Figs. 3.26(c)–(f), i.e. they exhibit the chain of eight well-ordered intensity peaks. The peaks with 50 Hz look as the aforementioned "dial-plates". The chains are placed in the area where one should expect the ray-medium resonance $8:1$, therefore, they should be associated with its periodic ray

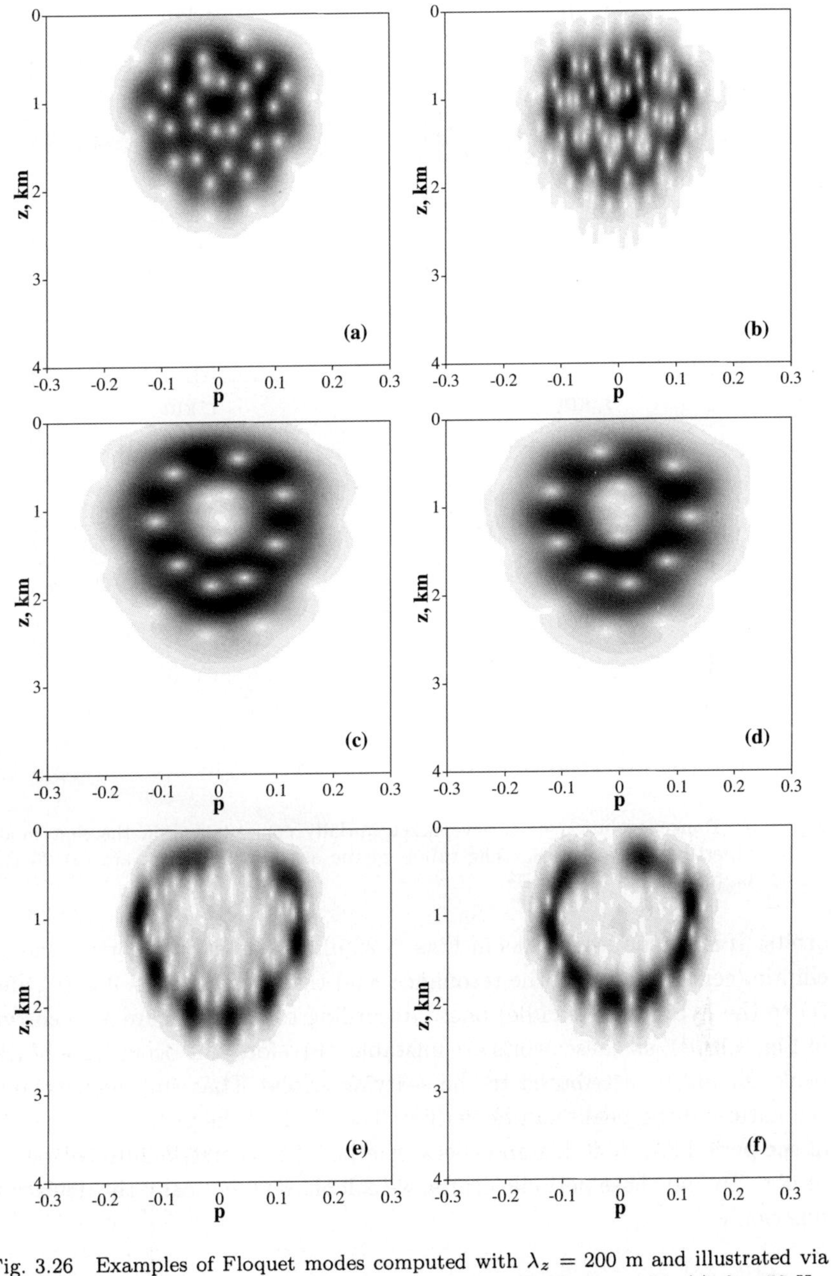

Fig. 3.26 Examples of Floquet modes computed with $\lambda_z = 200$ m and illustrated via the Husimi plots. (a) $f = 100$ Hz, $m = 11$, (b) $f = 200$ Hz, $m = 24$, (c) $f = 50$ Hz, $m = 12$, (d) $f = 50$ Hz, $m = 8$, (e) $f = 200$ Hz, $m = 57$, (f) $f = 200$ Hz, $m = 47$.

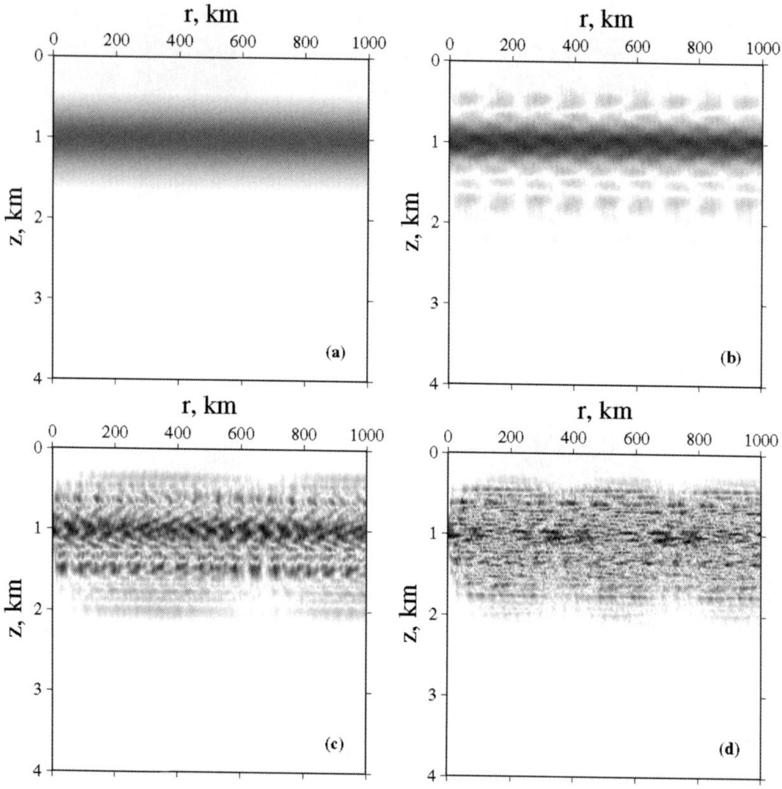

Fig. 3.27 Range evolution of a wavepacket initially concentrated in the first mode of the unperturbed waveguide. The values of the acoustic frequency are (a) 20 Hz, (b) 50 Hz, (c) 100 Hz, (d) 200 Hz.

orbits. If it is true, the peaks in Figs. 3.26(c) and (e) should correspond to elliptic (center) orbits of the resonance, and the peaks in Figs. 3.26(d) and (f) to the hyperbolic (saddle) ones. According to the Poincaré map shown in Fig. 3.18(d), all these orbits are unstable, therefore, the occurrence of the peaks should be attributed to the scarring effect. That suggestion about the nature of the peaks can be verified if we find all the periodic ray orbits of the period $8\lambda_r = 40$ km and check whether they correspond to the peaks or not. To find those periodic orbits, we calculate numerically the deviation function as

$$d(p_0, z_0) = \sqrt{[p(p_0, z_0; r = 8\lambda_r) - p_0]^2 + [z(p_0, z_0; r = 8\lambda_r) - z_0]^2}, \quad (3.200)$$

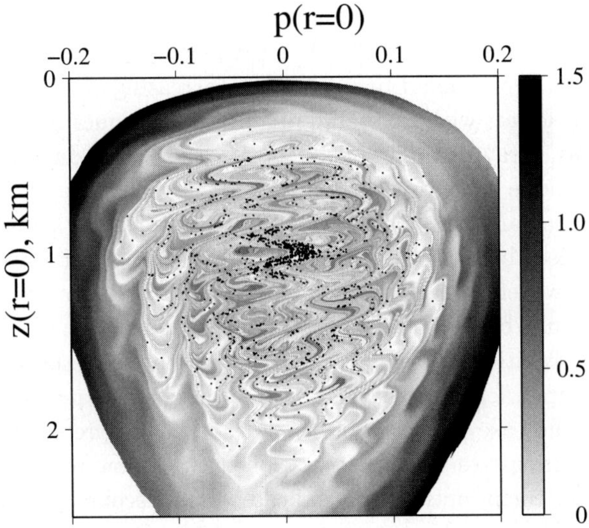

Fig. 3.28 Deviation function d and locations of ray periodic orbits of the period $D = 8\lambda_r$ in the phase space. Values of d are indicated by color modulation. The periodic orbits are marked by black points. The vertical length scale of the perturbation is taken to be $\lambda_z = 200$ m.

where $p_0 = p(r = 0)$ and $z_0 = z(r = 0)$. At periodic orbits, we have $d = 0$. The result of the computation, presented in Fig. 3.28, is somewhat surprising. There are a plenty of periodic orbits, and their distribution is strongly disordered, in marked contrast with the regular patterns of Floquet modes. Taking into account that a contribution of a periodic orbit depends on its stability [Gutzwiller (1990)], the correspondence is partially recovered: the most stable periodic orbits form clusters placed near the peaks of the Floquet modes (not shown).

The origin of proliferation of periodic orbits of the period $8\lambda_r$ becomes clear if we remind ourselves that a vertical resonance produces chaos of near-axial rays by means of the scenario described in subsection 2.5.3. It relies upon the multiplications of ray periodic orbits due to bifurcations. As a consequence of a cascade of these bifurcations, locations of the periodic orbits form the disordered pattern depicted in Fig. 3.28. However, the wavefield ignores these bifurcations. The reason is likely to be concerned with a weak sensitivity of wave refraction at low frequencies to small-scale sound-speed variations that cause the bifurcations. Therefore, the patterns of the Floquet modes look like before the orbit proliferation and reveal the peaks relating to the orbits that are actually destroyed. This effect was referred to by Kon'kov *et al.* (2007) as the recovery of periodic orbits.

To summarize, it should be noted that decreasing of λ_z enhances mode coupling that agrees with ray-based predictions. Floquet modes reflect the main features of ray dynamics, such as ray-medium resonances or chaotic layers if the mode capacity of these features is sufficient. The strong conflict between ray and wave descriptions emerges when the sound speed perturbation causes bifurcations which appear irrelevant for wave dynamics. Therefore, a wavefield recovers the pattern destroyed by the bifurcation. It is also should be noted that the Floquet modes with $f = 20$ Hz are very weakly sensitive to peculiarities of ray dynamics, especially to those induced by rapid depth oscillations of the sound-speed perturbation. It is apparent in the case of $\lambda_z = 200$ m. Although the presence of the wide chaotic sea anticipates strong interaction between low modes, the matrix of the Floquet coefficients $\|C_{qm}\|$ indicates their decoupling.

3.5 Summary

We discussed the concept of wave chaos in underwater waveguide propagation, its connection to the phenomenon of quantum chaos, and general aspects of the ray-wave correspondence.

We introduced the modal representation of the wave field in an unperturbed (range-independent) waveguide and discussed its generalization to the case of a perturbed (range-dependent) environment. One of our objective in this chapter was the investigation of the relationships between the rays and modes. We derived a ray-based approach which represents an analogue of the geometrical optics for modes. In the scope of this approach, the mode amplitude is expressed through solutions of the ray equations. This approach allows one to use the results obtained in the study of ray dynamics to predict the range variations of the mode amplitudes. It was shown that the ray-medium resonance has its analogue for modes, called the mode-medium resonance. This phenomenon plays a significant role in the mode coupling and leads to a strong and irregular range dependencies of mode amplitudes under conditions of ray chaos.

It is well known that the Floquet theory gives an important insight into the structure of the wave field in a range-periodic waveguides. This theory explains why the distribution of energy over the normal modes has regular features even under conditions of ray chaos. We demonstrated that the structure of the phase space observed in the phase portrait of the ray system clearly manifests itself in the Wigner (Husimi) functions of the Floquet modes. From the analysis of Floquet modes it was also shown that in a range-periodic waveguide there exists a "scarring" effect similar to that observed in autonomous quantum systems. It turns out that the segments of an unstable periodic ray trajectory may be distinguished in the spatial distribution of the wave field intensity at a finite wavelength.

The problem of the ray-wave correspondence in the presence of strong fine-scale depth oscillations of a waveguide inhomogeneity was studied. At the frequencies of the order of 100–200 Hz, the wave-field structure correlates, in general, with ray-based predictions. However, manifestations of ray chaos in the wave field turns out to be weaker than the ray theory predicts. In particular, the effect of proliferation of periodic ray orbits due to vertical resonance is suppressed by wave corrections.

Chapter 4

Chaotic Phenomena in Random Environment

4.1 Ray Chaos in a Random Medium

So far, we considered ray and wave chaos in acoustic waveguides with a periodic range dependence. In the present chapter we proceed to the analysis of a more realistic environmental model with a random perturbation $\delta c(r, z)$. Due to this perturbation, the ray dynamics at long ranges becomes stochastic and, therefore, it should be treated by statistical methods. Traditional approaches for describing stochastic rays are associated with ideas of the study of wave propagation in random media (WPRM) [Rytov *et al.* (1978); Flatté *et al.* (1979)].

The theory of WPRM and the theory of ray chaos investigate the ray structure from different and complementary viewpoints. Let us formulate the difference in statements of problems in these two approaches [Virovlyansky *et al.* (2007)].

(i) The description of WPRM is based on the notion of statistical ensemble. The latter consists of infinitely many realizations of the waveguide specified by different $\delta c(r, z)$. Statistical characteristics of a ray with the given starting parameters z_0 and χ_0 – initial depth and grazing angle, respectively, – are determined by averaging over rays with the same initial parameters in all realizations.

(ii) The theory of ray chaos deals with a **deterministic** medium. In our case the latter is specified by a **single** realization of random perturbation. At ranges $r \gg \nu^{-1}$, where ν is a typical Lyapunov exponent, initially close ray paths become practically independent and the averaging over their starting parameters can be considered as the statistical averaging.

4.1.1 *Statistical description of chaotic ray dynamics*

Statistical description of chaotic rays is based on a property of the chaotic ray dynamics called *mixing* [Sagdeev *et al.* (1988); Lichtenberg and Lieberman (1992)]. Take a bundle of rays starting from a small area of the phase space $\mathcal{R}$ centered at a point (p_0, z_0). At ranges $r \gg \nu^{-1}$, points depicting the ray paths in the phase plane are scattered over much larger area $\mathcal{R}'$. Consider a small portion of this area $\Delta\mathcal{R}'$ centered at an observation point (p, z). A fraction of trajectories arriving at $\Delta\mathcal{R}'$ may be treated as an expectancy of hitting this area. In order to quantify this statement, introduce a function

$$P_{pz|p_0z_0}(p, z, r|p_0, z_0)$$
$$= \frac{1}{S_{\mathcal{R}}} \int\!\!\int_{\mathcal{R}} dp_0' dz_0' \; \delta(z - z(r, p_0', z_0', 0)\delta(p - p(r, p_0', z_0', 0), \tag{4.1}$$

where $S_{\mathcal{R}}$ is the square of $\mathcal{R}$. Function $P_{pz|p_0z_0}$ may be interpreted as a conditional probability density function (PDF) of ray coordinates in the phase plane. Then an integral over the area $\Delta\mathcal{R}'$

$$P_{\Delta\mathcal{R}'} = \int_{\Delta\mathcal{R}'} dpdz \; P_{pz|p_0z_0}(p, z, r - r_0|p_0, z_0) \tag{4.2}$$

determines the probability that a ray path starting from (p_0, z_0) arrives at $\Delta\mathcal{R}'$. The mixing phenomenon manifests itself in the fact that at long ranges $P_{\Delta\mathcal{R}'}$ weakly depends on the size and shape of $\mathcal{R}$. When r tends to infinity and sizes of $\mathcal{R}$ tend to zero, the dependence of $P_{\Delta\mathcal{R}'}$ on the shape of $\mathcal{R}$ should vanish.

But at finite ranges r, we cannot take too small $\mathcal{R}$'s and consider probabilities of hitting too small $\Delta\mathcal{R}'$'s. The sizes of $\mathcal{R}$ along the p- and z-axes, Δp_0 and Δz_0, respectively, should satisfy the conditions

$$\Delta p_0 \ll \Delta p, \; \Delta z_0 \ll \Delta z, \tag{4.3}$$

where Δp and Δz are characteristic sizes of the area $\mathcal{R}'$ available for rays at observation range r. On the other hand, at least one of the conditions

$$\Delta p_0 \gg \Delta p \; e^{-\beta(r-r_0)}, \quad \Delta z_0 \gg \Delta z \; e^{-\beta(r-r_0)}, \tag{4.4}$$

should also be satisfied. These inequalities imply that at range r there are many ray paths (escaped from $\mathcal{R}$) which can be considered as independent.

In what follows, we will consider the random field $\delta c(r,z)$ statistically homogeneous along the r axis. Its correlation scale along this coordinate denote by l_r. Assuming that

$$r - r_0 \gg l_r, \tag{4.5}$$

we will present the PDF as $P_{pz|p_0 z_0}(p, z, r - r_0|p_0, z_0)$, that is, as a function of the difference $r - r_0$.

The area $\mathcal{R}$ may be not only two-dimensional but one dimensional, as well. For example, it may be a small interval of the p axis. Then Eq. (4.1) reduces to

$$\begin{aligned} &P_{pz|p_0 z_0}(p, z, r - r_0|p_0, z_0) \\ &\quad = \frac{1}{\Delta p_0} \int\limits_{p_0 - \Delta p_0/2}^{p_0 + \Delta p_0/2} dp_0' \delta(z - z(r, p_0', z_0, r_0)\delta(p - p(r, p_0', z_0, r_0). \end{aligned} \tag{4.6}$$

The product $P_{pz|p_0 z_0}(p, z, r - r_0|p_0, z_0)dpdz$ gives a fraction of rays (out of all rays escaped from $\mathcal{R}$) whose coordinates in the phase plane at range r belong to a small area with the square $dpdz$ centered at point (p, z). This definition makes sense only if the conditions (4.3) and (4.4) are met and the function $P_{pz|p_0 z_0}(p, z, r - r_0|p_0, z_0)$, smoothed over p and z, weakly depends on the shape and size of the area $\mathcal{R}$. The requirement of smoothing is essential. Two PDFs, determined using one-dimensional and two-dimensional areas $\mathcal{R}$ and centered at the same point (p_0, z_0), cannot coincide without smoothing because one of them is singular (for one-dimensional $\mathcal{R}$) while another one is regular. Thus, the PDFs, determined by Eqs. (4.1) and (4.6), may be used for averaging only sufficiently smooth functions of p and z.

In a similar way we define a conditional PDF $P_{I\theta|I_0\theta_0}(I, \theta, r - r_0|I_0, \theta_0)$ for the action-angle variables. In particular, an analogue of Eq. (4.6) has the form

$$\begin{aligned} &P_{I\theta|I_0\theta_0}(I, \theta, r - r_0|I_0, \theta_0) \\ &\quad = \frac{1}{\Delta p_0} \int\limits_{p_0 - \Delta p_0/2}^{p_0 + \Delta p_0/2} dp_0' \delta(I - I(r, p_0', z_0, r_0)\delta(\theta - (r, p_0', z_0, r_0). \end{aligned} \tag{4.7}$$

The connection between the PDFs $P_{pz|p_0 z_0}$ and $P_{I\theta|I_0\theta_0}$ is readily established using the standard formulas of the probability theory. Equations (1.48) and

(1.49) formally determine a nonlinear change of variables. The Liouville theorem (1.51) simplifies the application of the standard relations and we arrive at

$$P_{I\theta|I_0\theta_0}(I,\theta,\Delta r|I_0,\theta_0) = P_{pz|p_0z_0}(p(I,\theta),z(I,\theta),\Delta r|p(I_0,\theta_0),z(I_0,\theta_0)) \tag{4.8}$$

and

$$P_{pz|p_0z_0}(p,z,\Delta r|p_0,z_0) = P_{I\theta|I_0\theta_0}(I(p,z),\theta(p,z),\Delta r|I(p_0,z_0),\theta(p_0,z_0)), \tag{4.9}$$

where $\Delta r = r - r_0$.

The PDF introduced by Eq. (4.1) can be used for any environmental model in which rays exhibit chaotic behavior. But the application of this approach in a waveguide representing a realization of a random medium may have the following specific feature. At long ranges, initially close ray paths diverge so significantly that they are spaced apart from each other by intervals exceeding correlation scales of the medium. Then the rays travel through practically independent inhomogeneities and behave as if they propagate in different realizations of the medium. Therefore, it is natural to expect that the averaging over initial conditions may give results close to those obtained by the ensemble averaging. It means that the PDF defined by Eq. (4.1) may depend comparatively weakly on a particular realization of the waveguide. Numerical simulations presented below support this expectation.

4.1.2 *Environmental model*

In all numerical simulations presented in this chapter we use an environmental model with an unperturbed profile $c_0(z)$ (borrowed from Ref. [Colosi and Flatté (1996)]) shown in the left panel of Fig. 4.1. It is determined by the same relation as the Munk profile (1.102)

$$\bar{c}(z) = c_0\left[1 + \varepsilon\left(e^{-2(z-z_\mathrm{a})/w_a} + 2(z-z_\mathrm{a})/w_a - 1\right)\right], \tag{4.10}$$

but with different values of parameters: $c_0 = 1.47833$ km/s, $\varepsilon = 0.00238$, $w_a = 0.485$ km, and $z_\mathrm{a} = 0.7$ km. Figure 4.2 shows the action variable as a function of χ, the ray grazing angle at the sound channel axis z_a. The plot is constructed for rays propagating without striking the surface and the bottom. The latter is set at a depth of 5 km. The cycle length D grows with χ and varies from 30.5 km at $\chi = 0$ to 64 km at $\chi = 14°$. Correspondingly,

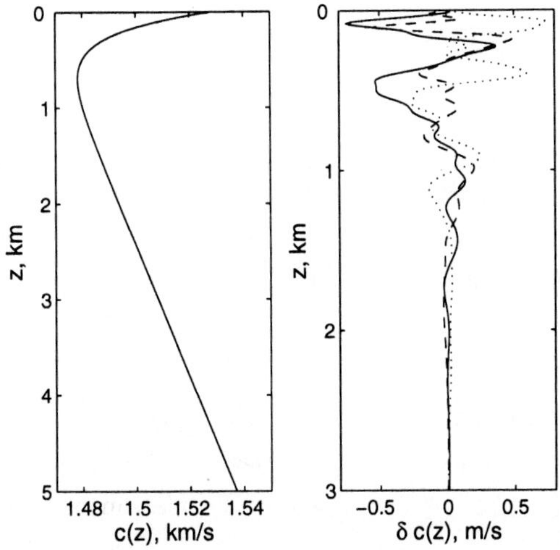

Fig. 4.1 Unperturbed sound speed profile $\bar{c}(z)$ (left panel) and perturbation δc in vertical sections of the waveguide at three different ranges (right panel).

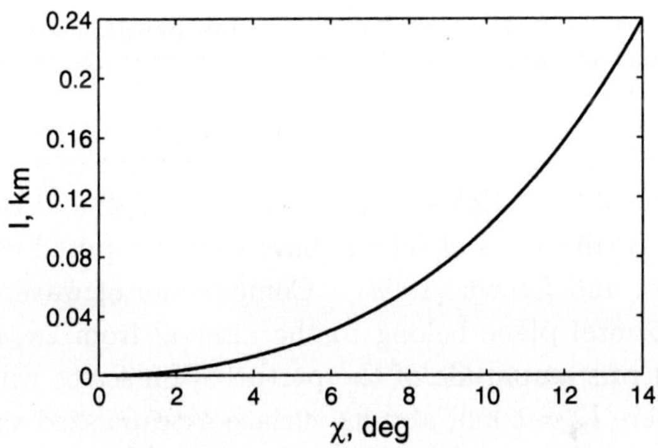

Fig. 4.2 Action I versus χ, the ray grazing angle at the sound channel axis.

the spatial frequency of the ray path oscillations $\omega = 2\pi/D$ decreases with χ. Figure 4.3 presents ω and its derivative with respect to I, denoted by ω_I', as functions of χ.

It is assumed that the weak perturbation $\delta c(r, z)$ superimposed on the unperturbed profile is caused by random internal waves with statistics determined by the empirical Garrett-Munk spectrum [Flatté *et al.* (1979)]. To generate realizations of a random field $\delta c(r, z)$ we apply a numerical

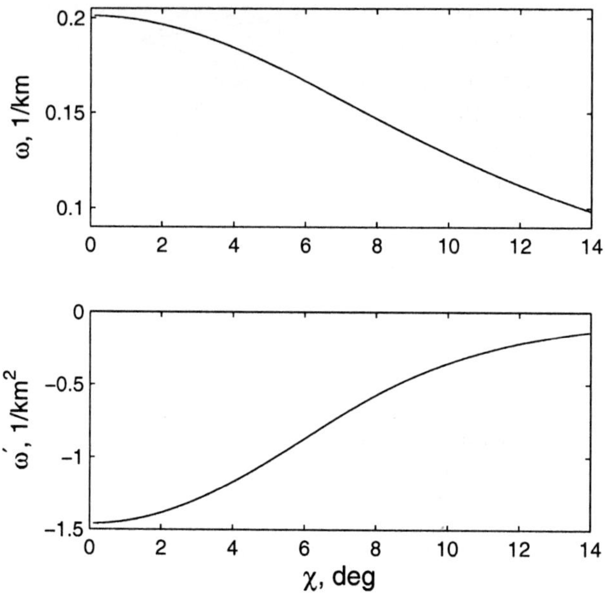

Fig. 4.3 Angular frequency of ray oscillations ω (upper panel) and its derivative with respect to the action variable $\omega'_I \equiv d\omega/dI$ (lower panel) versus the ray grazing angle χ at the sound channel axis.

technique developed by J. Colosi and M. Brown [Colosi and Brown (1998)] (see Sec. 1.5.1). Realizations of $\delta c(r, z)$ have been computed using Eq. (19) from Ref. [Colosi and Brown (1998)]. Components of wave number vectors in the horizontal plane belong to the interval from $2\pi/100$ km^{-1} to $2\pi/4$ km^{-1}. An rms amplitude of the perturbation scales with depth like $\exp(3z/2l_z)$, where $l_z = 1$ km, and its surface-extrapolated value is about 0.5 m/s. Depth dependencies of δc at three different ranges are shown in the right panel of Fig. 4.1.

4.1.3 *Fokker-Planck equation for action*

Numerical simulations show that the angle variable rapidly randomizes and already at relatively short ranges (hundreds of kilometers) the wrapped θ becomes uniformly distributed on $(0, 2\pi)$. Then we get

$$P_{I\theta|I_0\theta_0}(I, \theta, r - r_0|I_0, \theta_0) = \frac{1}{2\pi} P_{I|I_0}(I, r - r_0|I_0) \tag{4.11}$$

where the conditional PDF $P_{I|I_0}(I, r - r_0|I_0)$ describes the diffusion of action in the phase space. Thus, the problem reduces to the description of diffusion of action. A similar approach is commonly used in the theory of dynamical chaos [Sagdeev *et al.* (1988); Lichtenberg and Lieberman (1992)].

Note that according to Eqs. (4.4) and (4.5), the function $P_{I|I_0}(I, r - r_0|I_0)$ (as $P_{pz|p_0z_0}$ and $P_{I\theta|I_0\theta_0}$) is determined only for sufficiently large values of $r - r_0$. Let us redefine this function as

$$P_{I|I_0}(I, r - r_0|I_0) = \frac{1}{2\pi R}\int_0^R dr' \int_0^{2\pi} d\theta_0\, \delta(I - I(r + r', I_0, \theta_0, r_0 + r')), \quad (4.12)$$

where $R \gg \max(\nu^{-1}, l_r)$. This extension of definition does not change $P_{I|I_0}(I, r - r_0|I_0)$ for large $r - r_0$. At the same time, this function is now determined for an arbitrary $r - r_0$. Statistical moments of $I - I_0$ are defined as

$$\begin{aligned}\langle (I - I_0)^m \rangle &= \int dI (I - I_0)^m P_{I|I_0}(I, r - r_0|I_0) \\ &= \frac{1}{2\pi R}\int_0^R dr' \int_0^{2\pi} d\theta_0 (I(r + r', I_0, \theta_0, r_0 + r') - I_0)^m. \quad (4.13)\end{aligned}$$

A characteristic scale of range variation of the action variable denote l_I. It is assumed that, due to the weakness of perturbation, this scale is large compared with the longitudinal correlation scale l_r, i.e.,

$$l_r \ll l_I. \quad (4.14)$$

If this condition is met the right hand side of Eq. (1.74) may be formally considered as a delta-correlated random function. Then $I(r)$ is a Markov process whose PDF $P_{I|I_0}(I, r - r_0|I_0)$ obeys the Fokker-Planck equation [Tikhonov and Mironov (1977); Gardiner (1985)]

$$\frac{\partial P}{\partial r} = -\frac{\partial}{\partial I}(AP) + \frac{1}{2}\frac{\partial^2}{\partial I^2}(BP). \quad (4.15)$$

Here and in the remaining part of this section, for short, we omit the subscripts at P. The coefficients A and B are determined as

$$A(I) = \lim_{\Delta r \longrightarrow 0} \frac{1}{\Delta r}\int \Delta I\, P(I + \Delta I, \Delta r|I)\, d\Delta I \quad (4.16)$$

and

$$B(I) = \lim_{\Delta r \longrightarrow 0} \frac{1}{\Delta r} \int \Delta I^2 \, P(I + \Delta I, \Delta r | I) \, d\Delta I. \tag{4.17}$$

There exists a simple dependence between A and B which can be easily established using the principle of detailed balance [Gardiner (1985)]. The latter imposes a restriction on the function $P(I + \Delta I, \Delta r | I)$ caused by the symmetry of initial ray equations. In our case this is the symmetry of the Hamilton equations with respect to range reversal, a direct analogue of time reversal in mechanics. The restriction is expressed by the relation

$$P(I + \Delta I, \Delta r | I) = P(I, \Delta r | I + \Delta I). \tag{4.18}$$

Since $P(I + \Delta I, 0 | I) = \delta(\Delta I)$, it is clear that for small Δr the conditional PDF $P(I + \Delta I, \Delta r | I)$, as a function of ΔI, is negligible everywhere except a small neighborhood of the point $\Delta I = 0$. On the other hand, $P(I + \Delta I, \Delta r | I)$ is a smooth function of I and on the basis of Eq. (4.18) we can apply an expansion

$$P(I + \Delta I, \Delta r | I) = P(I - \Delta I, \Delta r | I) + \Delta I \frac{\partial}{\partial I} P(I - \Delta I, \Delta r | I) + O(\Delta I^2).$$

Neglecting terms $O(\Delta I^2)$, we find

$$\int \Delta I \, P(I + \Delta I, \Delta r | I) \, d\Delta I = \int \Delta I \, P(I - \Delta I, \Delta r | I) \, d\Delta I + \frac{\partial}{\partial I} \int \Delta I^2 \, P(I - \Delta I, \Delta r | I) \, d\Delta I.$$

This gives the desired connection between A and B:

$$A = \frac{1}{2} \frac{\partial B}{\partial I}. \tag{4.19}$$

Using this formula, the Fokker-Planck equation (4.15) can be presented in the form

$$\frac{\partial P}{\partial r} = \frac{\partial}{\partial I} B(I) \frac{\partial P}{\partial I}. \tag{4.20}$$

A derivation of Eq. (4.20) without use of the principle of detailed balance is given in Ref. [Virovlyansky (2001)].

4.1.4 *Evaluation of diffusivity B*

The diffusivity B can be evaluated numerically in the following way. Take an arbitrary range point r and select range $r_0 < r$ such that

$$l_r \ll r - r_0 \ll l_I. \tag{4.21}$$

According to Eq. (4.14), the range r_0, satisfying the above condition, always exists. The action and (unwrapped) angle variables at the point r_0 denote by I_0 and θ_0, respectively. Similarly, the values of these variables at point r denote by I and θ. According to the left inequality in Eq. (4.21) and Eq. (1.74) we can exploit an approximate relation

$$\Delta I = I - I_0 = -\int_{r_0}^{r} dr'\, V_\theta(I_0, \theta_0, r'), \tag{4.22}$$

where

$$V_\theta(I_0, \theta_0, r) = \left.\frac{\partial V(I_0, \theta, r)}{\partial \theta}\right|_{\theta=\theta_0+\omega(I_0)(r-r_0)}. \tag{4.23}$$

Substitute this in Eq. (4.13) with $m = 2$. Taking into account the right inequality in Eq. (4.21) yields

$$\langle \Delta I^2 \rangle = 2(r - r_0) \int_0^{r-r_0} d\rho\, K(I, \rho), \tag{4.24}$$

where

$$K(I, \rho) = \frac{1}{2\pi R} \int_0^R dr \int_0^{2\pi} d\theta_0\; V_\theta(I, \theta_0, r + \rho) V_\theta(I, \theta_0, r). \tag{4.25}$$

On the one hand, the left part of condition (4.21) allows one to tend the interval of integration in Eq. (4.24) to infinity. On the other hand, from the right part of Eq. (4.21) it follows that the difference $r - r_0$ is small as compared to the range variation scale of I. Therefore, Eq. (4.24) can be substituted in Eq. (4.17) with $(r - r_0)$ replaced by Δr. Since $K(I, \rho) = K(I, -\rho)$, we get

$$B(I) = \int_{-\infty}^{\infty} d\rho\, K(I, \rho). \tag{4.26}$$

The function $B(I)$ can be evaluated using a simple numerical ray tracing. From Eq. (1.19) it follows that

$$\frac{\partial V}{\partial \theta} = \frac{1}{c_0}\frac{\partial \delta c}{\partial z}\frac{\partial z(I,\theta)}{\partial \theta}, \tag{4.27}$$

where $z(I,\theta)$ is a function present in the second of Eqs. (1.49) determining the canonical transformation. Note that in a range-independent waveguide we get

$$\frac{1}{\omega(I)}\frac{dz}{dr} = \frac{p}{\omega(I)} = \frac{\partial z}{\partial \theta}. \tag{4.28}$$

Consider an **unperturbed** ray path with action variable I. The range dependencies of its momentum and coordinate are given by functions $p(r)$ and $z(r)$, respectively. These functions can be easily computed using a standard ray code. The derivative (4.27) at point $(r, z(r))$ in the perturbed waveguide according to Eqs. (4.27) and (4.28) is

$$\frac{\partial V}{\partial \theta} = \frac{\delta c_z(r)}{c_0}\frac{p(r)}{\omega}, \tag{4.29}$$

where

$$\delta c_z(r) = \left.\frac{\partial \delta c(r,z)}{\partial z}\right|_{z=z(r)}. \tag{4.30}$$

An estimate of $K(I,\rho)$ evaluated by integrating over the trajectory of the unperturbed ray path is given by an expression

$$\hat{K}(I,\rho) = \frac{1}{c_0^2 R\omega^2(I)}\int_0^R dr\ \delta c_z(r+\rho)p(r+\rho)\delta c_z(r)p(r). \tag{4.31}$$

The range R should be much greater than both ν^{-1} and l_r. A more accurate estimate of $K(I,\rho)$ can be obtained by averaging over integrals (4.31) computed along a group of unperturbed paths with equal actions I but different initial values of the angle variable. This averaging corresponds to the integration over θ_0 in Eq. (4.25).

Figure 4.4 presents the function $B(I)$ computed using the above procedure for a single realization of the sound speed field. The spatial averaging was performed over range interval $R = 3000$ km. 20 unperturbed trajectories with θ_0 equally spaced within interval $(0, 2\pi)$ were computed for each value of I. Similar calculations demonstrated that the numerical estimate of $B(I)$ weakly depends on a particular realization of the random perturbation $\delta c(r,z)$.

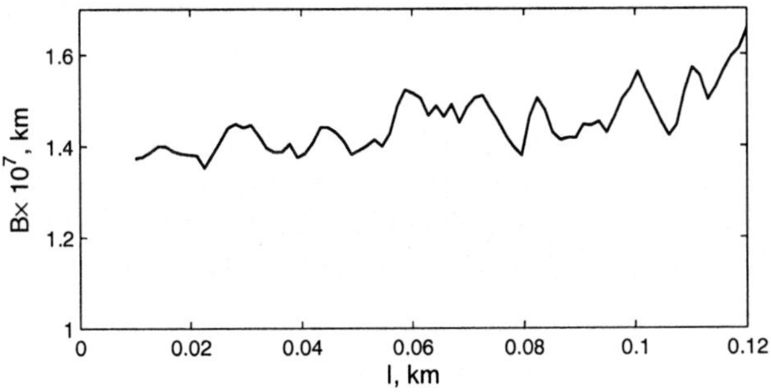

Fig. 4.4 Diffusivity B as a function of the action variable I.

4.1.5 *Wiener process approximation*

According to the data presented in Fig. 4.4, the diffusivity B in our environmental model weakly depends on I. Below, we will approximate it by the constant

$$B = 1.4 \times 10^{-7} \text{ km}. \tag{4.32}$$

This value is in approximate agreement with the estimates of B reported in Refs. [Virovlyansky (2001, 2005b); Rypina and Brown (2007); Udovidchenkov and Brown (2008)]. The Fokker-Planck equation with constant B is readily solved analytically [Rytov *et al.* (1978); Gardiner (1985)]. In this approximation, the range dependence of action is presented as

$$I(r) = I_0 + x(r), \tag{4.33}$$

where $I_0 \equiv I(0)$ and $x(r)$ is the so-called Wiener process [Gardiner (1985)]. The latter is defined by a stochastic Langevin equation

$$\frac{dx}{dr} = \xi(r), \quad x(0) = 0, \tag{4.34}$$

where $\xi(r)$ is a zero mean white noise with a correlation function

$$\langle \xi(r)\xi(r') \rangle = B\delta(r - r'). \tag{4.35}$$

Note that Eq. (4.34) represents an idealized form of the Hamilton equation (1.74). The PDF of the Wiener process is well-known [Gardiner (1985)]:

$$P_x(x) = \frac{1}{\sqrt{2\pi Br}} \exp\left(-\frac{x^2}{2Br}\right). \tag{4.36}$$

There is a subtlety to this result. According to Eqs. (4.33) and (4.36), action I may take on both positive and negative values. But the action is nonnegative by definition. In order to overcome this contradiction, we introduce a reflecting boundary for the Wiener process at $x = -I_0$. This is an additional assumption whose applicability is justified by numerical simulation. From the viewpoint of Fokker-Planck equation, the presence of the boundary is accounted for by the condition [Gardiner (1985)]

$$\left.\frac{\partial P_{I|I_0}}{\partial I}\right|_{I=0} = 0. \tag{4.37}$$

A solution to the Fokker-Planck equation (4.20) with an initial condition $P_{I|I_0}(I, 0|I_0) = \delta(I - I_0)$ and boundary condition (4.37) is [Gardiner (1985)]

$$P_{I|I_0}(I, r|I_0) = \frac{1}{\sqrt{2\pi Br}}\left[\exp\left(-\frac{(I-I_0)^2}{2Br}\right) + \exp\left(-\frac{(I+I_0)^2}{2Br}\right)\right]. \tag{4.38}$$

This gives explicit expressions for the mean deviation of action from its starting value,

$$\langle I - I_0\rangle = +\sqrt{\frac{2Br}{\pi}}\exp\left(-\frac{{I_0}^2}{2Br}\right) - I_0\,\mathrm{erfc}\left(\frac{I_0}{\sqrt{2Br}}\right), \tag{4.39}$$

and the dispersion of this deviation,

$$\langle (I - I_0)^2\rangle = Br - 2\sqrt{\frac{2Br}{\pi}} I_0 \exp\left(-\frac{I_0}{2Br}\right) + 2I_0^2\mathrm{erfc}\left(\frac{I_0}{\sqrt{2Br}}\right), \tag{4.40}$$

where

$$\mathrm{erfc}(u) = \frac{2}{\sqrt{\pi}}\int_u^\infty e^{-v^2}\,dv \tag{4.41}$$

is a complementary error function [Abramovitz and Stegun (1964)].

For steep rays with I_0 satisfying condition

$$I_0 > (Br)^{1/2}, \tag{4.42}$$

the second exponential on the right of Eq. (4.38) may be neglected. Then Eq. (4.38) is approximately replaced by

$$P_{I|I_0}(I, r|I_0) = \frac{1}{\sqrt{2\pi Br}}\exp\left(-\frac{(I-I_0)^2}{2Br}\right). \tag{4.43}$$

Most part of such rays have not reflected off the boundary $I = 0$. Then $\langle I \rangle = I_0$ and we get

$$\sigma_I \equiv \langle (I - I_0)^2 \rangle^{1/2} = (Br)^{1/2}. \tag{4.44}$$

In numerical examples presented below, we shall study statistics of chaotic rays at $r = 3000$ km. At this range, $(Br)^{1/2} = 0.02$ km which is approximately equal to I_0 of an unperturbed ray whose grazing angle at the sound channel axis χ_a is close to $\pm 5°$. In this case, the condition (4.42) can be presented in the form

$$|\chi_a| > 5°. \tag{4.45}$$

Turn our attention to the derivation of a stochastic Langevin equation for the angle variable θ. We will consider an "unwrapped" θ monotonically increasing with range (see comment after Eq. (1.49)). Let us present it in the form

$$\theta = \theta_0 + \omega(I_0)r + y(r), \tag{4.46}$$

where $y(r)$ is a difference between angle variables of perturbed and unperturbed paths with equal starting parameters $I(0) = I_0$ and $\theta(0) = \theta_0$.

Using an approximation $\omega(I) = \omega(I_0) + \omega_I'(I_0)x$ and neglecting the last term on the right of Eq. (1.75) yields

$$\frac{dy}{dr} = \omega_I'(I_0)x. \tag{4.47}$$

The applicability condition of this approximation is considered in Ref. [Virovlyansky (2001)]. Equation (4.47) is an idealized form of the Hamilton equation for the angle variable (1.75). According to Eq. (4.47) we have

$$y(r) = \omega_I'(I_0) \int_0^r x(r')dr', \tag{4.48}$$

that is, the random function $y(r)$ is proportional to an integral of the Wiener process. For steep rays satisfying condition (4.42), we have $\langle y \rangle = 0$ and

$$\sigma_\theta \equiv \langle y^2 \rangle^{1/2} = |\omega_I'(I_0)|\,(B/3)^{1/2}\, r^{3/2}. \tag{4.49}$$

Statistical approach for description of chaotic rays, based on Langevin equations (4.34) and (4.47), we shall call the Wiener process approximation.

At short ranges where the ray action is still close to its starting value, the deviation of the ray path from its unperturbed position is determined by the random increment of the angle variable y. According to Eq. (4.47) the rms value of y is proportional to the derivative $\omega_I'(I_0)$ which is defined by the unperturbed sound speed profile $\bar{c}(z)$. This means that the parameter $\omega_I'(I_0)$ to a significant extent determines the sensitivity of a ray path to the sound speed fluctuations. This conclusion is consistent with results obtained in Refs. [Beron-Vera and Brown (2003, 2004); Brown *et al.* (2005); Rypina and Brown (2007); Udovidchenkov and Brown (2008)], where the authors argue that the sensitivity of the ray path is controlled by the stability parameter of the background sound speed profile

$$\alpha = \frac{I\omega_I'(I)}{\omega(I)}. \tag{4.50}$$

Numerical results presented in the above papers show that ray instability increases with increasing the magnitude of α.

Using standard relations of the probability theory, it is easy to show that for steep rays satisfying condition (4.45) the joint probability density function (PDF) of x and y is

$$P_{xy|I_0}(x,y,r|I_0) = \frac{\sqrt{3}}{\pi B r^2 \left|\omega_I'(I_0)\right|} \exp\left[-\frac{x^2}{2Br} - \frac{6}{Br^3}\left(\frac{y}{\omega_I'(I_0)} - \frac{xr}{2}\right)^2\right]. \tag{4.51}$$

This function does not depend on θ_0. Substituting $I - I_0$ and $\theta - \theta_0 - \omega(I_0)r$ for x and y, respectively, yields the PDF $P_{I\theta|I_0\theta_0}(I,\theta,r|I_0,\theta_0)$ for the unwrapped angle variable. Recall that the explicit expression for $P_{I\theta|I_0\theta_0}(I,\theta,r|I_0,\theta_0)$ with θ considered the wrapped angle variable, is given by Eqs. (4.11) and (4.43).

Figure 4.5 compares predictions of Eqs. (4.44) and (4.49) with results of a direct numerical ray tracing. This is done for a ray with the starting action $I_0 = 0.06$ km. In the unperturbed waveguide it crosses the sound channel axis z_a at the grazing angle $\chi_a = 8°$. The latter satisfies condition (4.45) which determines the applicability of Eqs. (4.44) and (4.49). The asterisks show σ_I given by Eq. (4.44) (upper panel) and σ_θ given by Eq. (4.49) (lower panel) as functions of range r. Solid curves present the same functions computed numerically for four particular realizations of the random perturbation δc. For each realization, 100 ray paths with equal starting actions I_0 and different starting angles θ_0 uniformly distributed

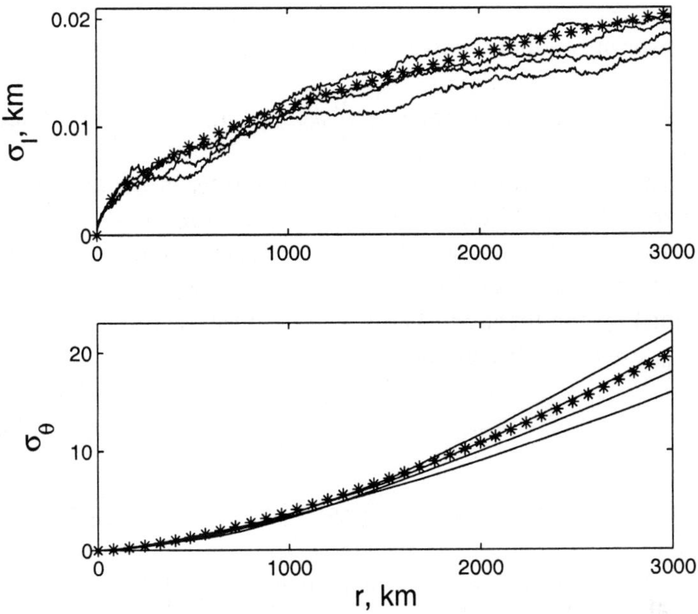

Fig. 4.5 Standard deviation of action (upper panel) and angle (lower panel) variables for a steep chaotic ray with starting action $I_0 = 0.06$ km. Asterisks: analytical estimates (4.44) (upper panel) and (4.49) (lower panel). Thin solid curves: results of numerical ray tracing in four realizations of the random medium. Each curve has been computed by tracing 100 ray paths with equal I_0 and different θ_0 uniformly distributed in the interval $(0, 2\pi)$.

in the interval $(0, 2\pi)$ were traced. Each solid curve shows the rms spread of actions I (upper panel) or angles θ (lower panel) for the 100 ray paths traced in one of the realizations of our random medium. It is seen that, consistent with our expectation, results of numerical ray tracing weakly depend on particular realizations of perturbation and are in a reasonable agreement with analytical estimates (4.44) and (4.49).

Analytical evaluation of statistical moments of y in the presence of reflecting boundary at $x = -I_0$ is tedious. But this problem can be easily solved numerically. To this end, we approximately replace the continuous random function $x(r)$ by a sequence of random variables x_m associated with discrete range points $r_m = m\varepsilon$, where $m = 0, \dots N$, and $\varepsilon = r/N$. Values of x_m are evaluated using a recurrence formula

$$x_{m+1} = x_m + \xi_m, \quad x_0 = 0, \tag{4.52}$$

where ξ_n are zero mean Gaussian random variables (produced by a standard generator of random numbers) with the correlation function

$$\langle \xi_{m_1} \xi_{m_2} \rangle = \frac{B}{\varepsilon} \delta_{m_1 m_2}. \tag{4.53}$$

The presence of a reflecting boundary is accounted for in the following way. At each step the value of x_{m+1} is compared with $-I_0$. If x_{m+1} turns out to be less than $-I_0$, it is replaced by $-2I_0 - x_{m+1}$. Simple numerical tests confirm that at $N \gg 1$ the PDF of the random variable $I = I_0 + x_N$ is well described by the function (4.38). Values of the random function $y(r)$ at points r_m are computed by the formula

$$y_m = \omega_I'(I_0) \sum_{q=1}^{m} x_q \varepsilon. \tag{4.54}$$

In what follows, we shall use Eqs. (4.52)–(4.54) to model realizations of $x(r)$ and $y(r)$ and evaluate their PDFs and statistical moments at a range of 3000 km. These calculations will be done with $\varepsilon = 10$ km.

Statistical approach considered in the present subsection is based on approximating the Hamilton equations for the action and angle variables by stochastic Langevin equations 4.34 and 4.47. In Ref. [Brown and Viechnicki (1998)] a similar approach was applied to the Hamilton equations for the momentum and position variables.

4.1.6 *Probability density functions of ray parameters*

To test the validity of the Wiener process approximation, we compare in this section its predictions with results obtained by direct numerical ray tracing.

4.1.6.1 *Rays starting from a small area in the phase space*

Our analytic description is based on Eq. (4.38) that predicts a distribution of actions for rays starting from a small area $\mathcal{R}$ in the phase plane. Results of numerical ray tracing carried out to verify this formula are presented in Figs. 4.6(a)–(f). We have selected three areas $\mathcal{R}$ shown in plots a, c, and e by closed rectangles. The rectangle in plot (a) defines rays with initial depths close to the sound channel axis z_a and launch angles centered at $-3°$. 10 000 rays starting from points (p_0, z_0) uniformly filling this rectangle have been traced up to $r = 3000$ km. Randomly scattered points in plot (a) depict

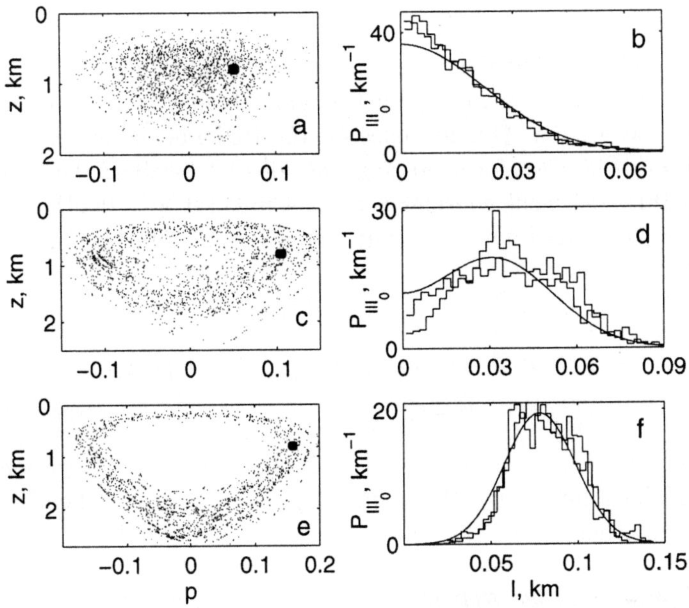

Fig. 4.6 Distributions of rays escaping a small area $\mathcal{R}$ in the phase plane. Three selected areas $\mathcal{R}$ are shown in plots a, c, and e by small closed rectangles. Randomly scattered points in each of these plots depict coordinates of rays (starting from a corresponding rectangle) at a range of 3000 km. Right column in the plots present distributions of action I at 3000 km. Stairstep graphs show normalized histograms computed by direct ray tracing for two different realizations of $\delta c(r, z)$. Smooth curves show PDF predicted by Eq. (4.38). Starting depths of rays escaping all three rectangles are close to the sound channel axis. Launch angles are close to $-3°$ (*a* and *b*), $-6°$ (*c* and *d*), and $-9°$ (*e* and *f*) [Virovlyansky *et al.* (2007)].

the distribution of these rays in the phase plane (p, z) at 3000 km. The distribution of actions of these rays is presented by one of two normalized histograms (stairstep graphs) in plot (b). Another histogram in plot (b) presents an analogous distribution computed for a different realization of $\delta c(r, z)$. The normalized histograms are compared to a PDF predicted by Eq. (4.38) (smooth curve). Similar results for rays with launch angles centered at $-6°$ and $-9°$ are presented in plots (c–d) and (e–f), respectively.

In accord with our conjecture made at the end of Sec. 4.1.1 the normalized histograms obtained for different $\delta c(r, z)$ look alike and they are in a reasonable agreement with estimates following from Eq. (4.38). Similar results have been observed for rays traced using a few other realizations of $\delta c(r, z)$ (not shown).

4.1.6.2 *Averaging over ray starting parameters*

The main idea behind the statistical approach introduced in Sec. 4.1.1 is that the averaging over ray starting parameters may be interpreted as the statistical averaging. Our formulas for the PDFs of the action and angle variables can be used for averaging over all rays escaping the source. To explain this statement consider a point source set at a depth z_s. Initial momenta p_0 of rays leaving the source lie within an interval $(-p_{\max}, p_{\max})$. Consider the quantity

$$\overline{F(I,\theta)} \equiv \int_{-p_{\max}}^{p_{\max}} dp_0\ F(I(r,p_0,z_s),\theta(r,p_0,z_s)), \tag{4.55}$$

where $F(I,\theta)$ is a given function. Divide the interval of integration into a set of small subintervals with end points p_j, $j = 1,\ldots,N$. Then Eq. (4.55) can be rewritten as

$$\overline{F(I,\theta)} = \int dI'd\theta'\ F(I',\theta') \times \sum_{j=1}^{N-1} \int_{p_j}^{p_{j+1}} dp_0\ \delta\left(I' - I(r,p_0,z_s)\right) \delta\left(\theta' - \theta(r,p_0,z_s)\right). \tag{4.56}$$

But each integral over p_0 in this sum is a particular case of Eq. (4.7). Therefore

$$\frac{1}{p_{j+1}-p_j} \int_{p_j}^{p_{j+1}} dp_0\ \delta(I' - I(r,p_0,z_s))\delta\left(\theta' - \theta(r,p_0,z_s)\right) = P_{xy|I_0}(I' - I_0, \theta' - \theta_0 - \omega(I_0)r, r|I_0), \tag{4.57}$$

where I_0 and θ_0 are the starting action and angle of a ray with $p_0 = (p_j+p_{j+1})/2$. Substituting this into Eq. (4.56) and replacing the summation over j by integration over p_0 (with $p_{j+1} - p_j$ replaced by dp_0) yields

$$\overline{F(I,\theta)} = \int_{-p_{\max}}^{p_{\max}} dp_0 \int dxdy \times F\left(I_0 + x, \theta_0 + \omega\left(I_0\right) r + y\right) P_{xy|I_0}(x,y,r|I_0). \tag{4.58}$$

If function F does not depend on θ, Eq. (4.58) reduces to

$$\overline{F(I)} = \int_{-p_{\max}}^{p_{\max}} dp_0 \int dI\ F(I) P_{I|I_0}(I, r|I_0). \tag{4.59}$$

The above expressions for $\overline{F(I,\theta)}$ and $\overline{F(I)}$, as well as the expressions for both PDFs, $P_{xy|I_0}(x, y, r|I_0)$ and $P_I(I, r|I_0)$, do not depend on a particular realization of perturbation δc.

4.1.6.3 *Rays escaping a point source*

Take a point source set at depth z_s and evaluate the distributions of ray parameters at a range of observation $r \gg \nu^{-1}$. Assume that the quantity $\eta(p_0)dp_0$ determines a fraction of rays with starting momenta from an interval $(p_0, p_0 + dp_0)$. The function $\eta(p_0)$ is determined by the radiation pattern of a particular source. Formally considering $\eta(p_0)$ as a PDF of the initial momentum p_0 satisfying the normalization condition

$$\int dp_0\ \eta(p_0) = 1, \tag{4.60}$$

we define the PDF of the action variable I as

$$P_I(I, r) = \int dp_0\ \eta(p_0) \delta(I - I(r, p_0, z_s)).$$

This integral has the form of Eq. (4.55) with the function F independent of θ. Making use of formula (4.59), we get

$$P_I(I, r) = \int dp_0\ \eta(p_0) P_{I|I_0}(I, r|I(p_0, z_s)), \tag{4.61}$$

where $I(p, z)$ is the function present in the right hand side of the first of Eqs. (1.48). At long ranges, the wrapped angle variable θ is independent on I and uniformly distributed over the interval $(0, 2\pi)$. According to Eq. (4.11), the joint PDF of I and θ is

$$P_{I\theta}(I, \theta, r) = \frac{1}{2\pi} P_I(I, r). \tag{4.62}$$

Formally considering the canonical transformation (1.48) as a nonlinear change of variables and exploiting the Liouville theorem (1.51), we find a joint PDF of p and z (cf. Eq. (4.9)):

$$P_{pz}(p, z, r) = \frac{1}{2\pi} P_I(I(p, z), r). \tag{4.63}$$

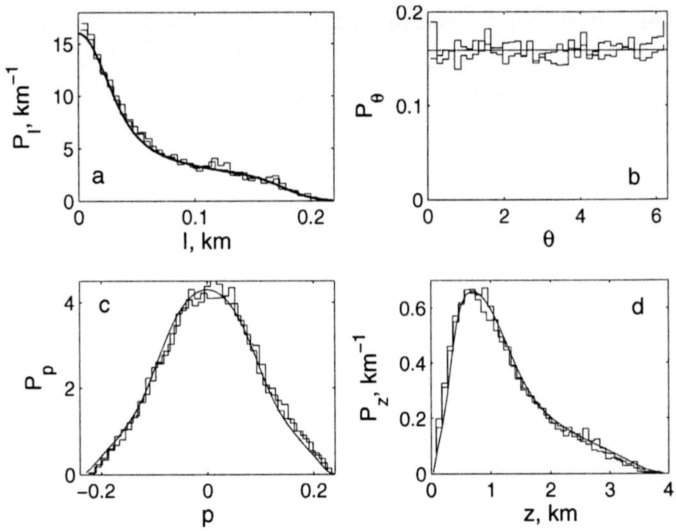

Fig. 4.7 Distributions of ray parameters at $r = 3000$ km. Smooth curves show PDFs of I (a), θ (b), p (c), and z (d) evaluated analytically in the Wiener process approximation for rays escaping a point source set at the depth $z = 0.7$ km. Stairstep graphs present normalized histograms obtained by numerical ray tracing in waveguides with two different realizations of perturbation $\delta c(r, z)$ [Virovlyansky *et al.* (2007)].

Equations (4.61) and (4.63) yield a PDF of the momentum p

$$P_p(p,r) = \frac{1}{2\pi} \int dp_0 dz \; \eta(p_0) P_{I|I_0}(I(p,z), r | I(p_0, z_s)), \tag{4.64}$$

and depth z

$$P_z(z,r) = \frac{1}{2\pi} \int dp_0 dp \; \eta(p_0) P_{I|I_0}(I(p,z), r | I(p_0, z_s)). \tag{4.65}$$

To check the above relations we have traced 50000 rays escaping a source set at the depth $z_s = z_a$ with initial momenta uniformly distributed in an interval $(-p_{max}, p_{max})$, with $p_{max} = 0.22$ corresponding to a maximum launch angle $\chi_{max} = 12.5°$. In this case $\eta(p_0) = (2p_{max})^{-1}$. Smooth lines in Fig. 4.7 show PDFs of I (a), θ (b), p (c), and z (d) at a range of 3000 km obtained in the Wiener process approximation, that is, evaluated by formulas (4.61), (4.64), and (4.65). For the wrapped angle variable θ the uniform distribution is expected. The smooth curves are compared to stairstep graphs representing distributions (normalized histograms) of the same quantities obtained by direct numerical ray tracing for two different realizations of $\delta c(r, z)$. It is clear that predictions made in the Wiener process approximation are in a good agreement with results of the ray tracing. In these figures we see again that distributions of ray parameters weakly depend on a particular realization of the perturbation δc.

4.1.7 *Coarse-grained Wigner function*

Let us show how the above results may be applied for description of the wave field at a finite wavelength. This will be done using the Wigner function which gives a link between the wave field and the phase space of the Hamilton (ray) equations.

4.1.7.1 *Ray-based description of the Wigner function*

The definition of the Wigner function is given by Eq. (3.166). We rewrite it in the form

$$W(r,p,z) = \frac{k_0}{2\pi} \int d\zeta\, u(r, z - \zeta/2)\, u^*(r, z + \zeta/2)\, e^{ik_0 p\zeta}. \tag{4.66}$$

In quantum mechanics this function is often referred to as a quasiprobability while in the wave propagation theory it represents a local spectrum of the wave field in the neighborhood of the point (r, z) (see comment after Eq. (3.166)). The Wigner function dependence on momentum p in the wave theory characterizes a local angular structure of the wave field. Naturally, this interpretation makes sense only for a small enough wavelength $2\pi/k_0$. Function $W(r,p,z)$ is real and provides the field intensity when the momentum p is integrated out:

$$|u(r,z)|^2 = \int_{-\infty}^{\infty} dp\, W(r,p,z). \tag{4.67}$$

The Wigner function satisfies the so-called quantum Liouville equation which is readily derived from Eq. (1.17) [Tatarskii (1983)] and can be written in the form

$$\frac{\partial W}{\partial r} + p\frac{\partial W}{\partial z} - \frac{\partial U}{\partial z}\frac{\partial W}{\partial p} = \sum_{n=1}^{\infty} \frac{(-1)^n}{(2k_0)^{2n}(2n+1)!} \frac{\partial^{2n+1} U(r,z)}{\partial z^{2n+1}} \frac{\partial^{2n+1} W(r,p,z)}{\partial p^{2n+1}}. \tag{4.68}$$

All the terms with coefficients depending on k are put together in the right-hand side. Neglecting these terms, one arrives at the classical Liouville equation

$$\frac{\partial W}{\partial r} + p\frac{\partial W}{\partial z} + \frac{\partial U}{\partial z}\frac{\partial W}{\partial p} = 0, \tag{4.69}$$

whose solution is easily expressed through parameters of the ray trajectories and the Wigner function at $r = 0$ [Tatarskii (1983)]:

$$W(r,p,z) = W(0, p_0(r,p,z), z_0(r,p,z)). \tag{4.70}$$

Here the functions $p_0(r,p,z)$ and $z_0(r,p,z)$ express initial coordinates of the ray at $r = 0$ in the phase plane through its coordinates at a given range r.

In order to illustrate the use of Eq. (4.70) consider a point source. Substituting Eq. (1.20) into Eq. (4.66) yields

$$W(0,p,z) = \frac{k_0}{2\pi}\delta(z - z_s). \tag{4.71}$$

Making use of Eq. (4.70), we find that at $r \geq 0$

$$W(r,p,z) = \frac{k_0}{2\pi}\delta(z_s - z_0(r,p,z)) = \frac{k_0}{2\pi}\sum_j \frac{\delta(p-p_j)}{\left|\frac{\partial z_0(r,p,z)}{\partial p}\right|_{p=p_j}}, \tag{4.72}$$

where the sum goes over all the eigenrays arriving at the observation point (r,z). The subscript j numbers the eigenrays. Using Eqs. (1.38), we can rewrite the above equation as

$$W(r,p,z) = \frac{k_0}{2\pi}\sum_j \frac{\delta(p-p_j)}{\left|\frac{\partial z(r,p_0,z_s)}{\partial p_0}\right|_{p_0=p_{0,j}}}. \tag{4.73}$$

Integrating out the momentum p (see Eq. (4.67)) we present the wave field intensity as an incoherent sum of contributions from the eigenrays:

$$|v(r,z)|^2 = \frac{k_0}{2\pi}\sum_j \frac{1}{\left|\frac{\partial z(r,p_0,z_s)}{\partial p_0}\right|_{p_0=p_{0,j}}}. \tag{4.74}$$

4.1.7.2 *Evaluation of the coarse-grained Wigner function in the Wiener process approximation*

Consider the smoothed Wigner function [Tatarskii (1983)] defined as

$$\overline{W}(r,p,z) = \int dp'dz'\,Q(p-p',z-z')\,W(r,p',z'), \tag{4.75}$$

where $Q(p,z)$ is a weighting function that provides the smoothing over some area of the phase plane. For example, the weighting function can be taken in the form

$$Q(p,z) = \frac{1}{2\pi}e^{-\frac{p^2}{2\Delta_p^2}-\frac{z^2}{2\Delta_z^2}} \tag{4.76}$$

as it is done in Eq. (3.167). If smoothing scales Δ_p and Δ_z satisfy the condition $\Delta_p\Delta_z \geq 1/(2k_0)$, then $\overline{W}(r,p,z)$ becomes nonnegative [Tatarskii (1983)]. If $\Delta_p\Delta_z = 1/(2k_0)$, $\overline{W}(r,p,z)$ represents the Husimi distribution function which can be presented in the form of Eq. (3.168).

Substituting Eq. (4.70) into Eq. (4.75) and changing the variables of integration from final values of momentum and coordinate to their starting values (the Liouville theorem allows one to replace $dp'dz'$ with dp_0dz_0), we find

$$\overline{W}(r,p,z) = \int dz_0\,dp_0\,Q(p-p(r,p_0,z_0),z-z(r,p_0,z_0))W(0,p_0,z_0). \tag{4.77}$$

This equation expresses the smoothed Wigner function as an integral over the ray paths. We know from Sec. 4.1.1 that the integration over the ray starting parameters may be replaced by the statistical averaging. Let us illustrate the application of this approach with an example of a point source. Substituting Eq. (4.71) into Eq. (4.77) yields

$$\overline{W}(r,p,z) = \frac{k_0}{2\pi}\int dp_0\,Q(p-p(r,p_0,z_s),z-z(r,p_0,z_s)). \tag{4.78}$$

Representing the weighting function in Eq. (4.77) as

$$Q(p-p(r,p_0,z_s),z-z(r,p_0,z_s)) = \int dp'dz'\;Q(p-p',z-z')\delta(p'-p(r,p_0,z_s)\delta(z'-z(r,p_0,z_s) \tag{4.79}$$

and making use of Eq. (4.1), we arrive at

$$\overline{W}(r,p,z) = \frac{k_0}{2\pi}\int dp_0dp'dz'\;Q(p-p',z-z')P_{pz|p_0z_0}(p',z',r|p_0,z_s). \tag{4.80}$$

According to Eqs. (4.9) and (4.11), we get

$$P_{pz|p_0z_0}(p', z', r|p_0, z_s) = \frac{1}{2\pi} P_{I|I_0}(I(p', z'), r|I(p_0, z_s)). \tag{4.81}$$

Then Eq. (4.80) translates to

$$\overline{W}(r, p, z) = \frac{k}{(2\pi)^2} \int dp_0 dp' dz' \; Q(p - p', z - z') P_{I|I_0}(I(p', z'), r|I(p_0, z_s)). \tag{4.82}$$

Thus, the smoothed Wigner function is expressed through the three known functions: the weighting function $Q(p, z)$, the PDF $P_{I|I_0}(I(p', z'), r|I(p_0, z_0))$ expressed (in the Wiener process approximation) by Eq. (4.38) or (4.43), and function $I(p, z)$ determining the variable I in an unperturbed waveguide as a function of p and z.

If the weighting function does not depend on p, that is $Q = Q(z)$, Eq. (4.75) can be simplified using Eq. (4.67). Then we get $\bar{W}(r, p, z) = J(r, z)$, where

$$J(r, z) = \int dz \; Q(z - z') \left| u(r, z) \right|^2 \tag{4.83}$$

is a smoothed intensity of the wave field. Assuming that the ray paths escaping the point source have starting momenta p_0 within the interval $-p_{\max} < 0 < p_{\max}$, we find from Eq. (4.82)

$$J(r, z) = \frac{p_{\max} k_0}{\pi} \int_{-p_{\max}}^{p_{\max}} dz \; Q(z - z') P_z(r, z'), \tag{4.84}$$

where $P_z(z, r)$ is the PDF of the ray coordinate z (4.65) with $\eta(p_0) = (2p_{\max})^{-1}$ (as in the numerical example considered in Sec. 4.1.6.3). Taking

$$Q(z) = \frac{1}{\sqrt{2\pi}\Delta_z} \exp\left(-\frac{z^2}{2\Delta_z^2}\right),$$

we get [Virovlyansky (2005b)]

$$J(r, z) = \frac{2k_0 p_{\max}}{(2\pi)^{3/2} \Delta_z} \int dz' e^{-\frac{(z-z')^2}{2\Delta_z^2}} P_z(z', r), \tag{4.85}$$

Figure 4.8 compares the prediction of Eq. (4.85) (dashed line) with results obtained by numerical solving the parabolic equation for four realizations of the random perturbation at frequency 75 Hz. The wave field is excited by a point source set at the sound channel axis z_a. The smoothing scale is $\Delta_z = 0.4$ km. It is seen that Eq. (4.85) gives a rough estimate of the smoothed intensity.

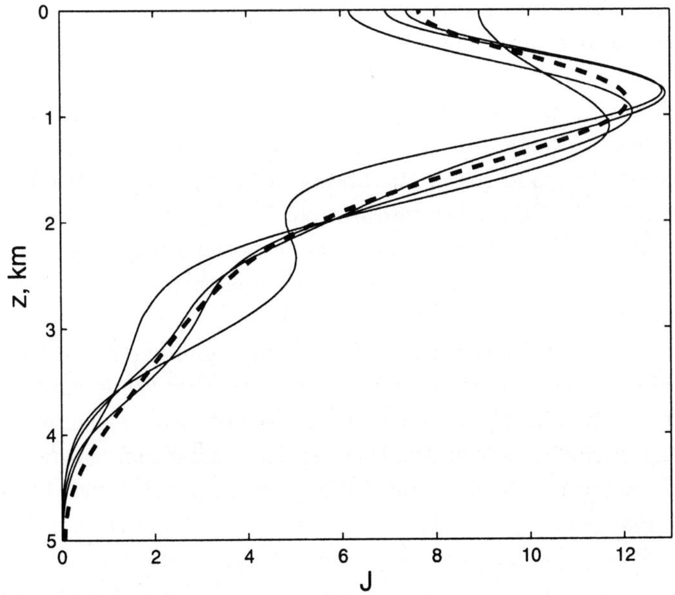

Fig. 4.8 Smoothed intensity J at 3000 km versus depth z. The carrier frequency is $f = 75$ Hz. Dashed line: prediction of Eq. (4.85). Thin solid lines: results of numerical solving the parabolic equation for 4 realizations of the random perturbation.

4.1.8 *Specific Poincaré map for a randomly-inhomogeneous waveguide*

Comparing Chaps. 2 and 4 of the present book, you can see that the methods and approaches to study range-periodic and random waveguides are very different. Ray dynamics in a typical range-periodic waveguide is characterized by mixed phase space with coexisting regions of stable and unstable ray motion. There are islands of regular motion merged into a chaotic sea, islands around islands, stochastic layers confined between invariant tori, cantory nearby the boundaries of islands, and so on. The motion is predictable in some regions but permits a statistical description only in those ones where extremal sensitivity to initial conditions takes place. The predictability horizon of ray motion is proportional to the inverse maximal Lyapunov exponent.

Indeed, the notions of the ray instability and predictability horizon can be applied to ray propagation in a random media as well [Wolfson and Tappert (2000)]. Statistical approaches invoked to study ray motion are usually based on the assumption of fully-developed chaos. This assumption looks quite natural, because the randomness of an inhomogeneity anticipates the absence of any impenetrable stable islands in the phase space. However, as it was recognized by Wolfson and Tomsovic (2001), some bunches of rays in a random environment remain stable by Lyapunov on ranges much beyond the predictability horizon, albeit their fraction decays exponentially. It implies intermittent-like dynamics with chaotic oscillations interrupted by regular ones. It is especially true for guided sound propagation due to resonant interaction of unperturbed ray motion with a random perturbation [Makarov *et al.* (2004)].

In this section we treat a model of a randomly-perturbed waveguide from a purely deterministic viewpoint, and show that there can be regions of stability in the underlying phase space. As opposed to islands of stability in range-periodic waveguides, these stable regions can survive only for finite but comparatively long range. They can be studied with the help of a specific map which plays the same role as the Poincaré map in range-periodic waveguides. The map is intended to find numerically clusters of trajectories with close initial conditions which remain stable by Lyapunov with a given realization of a random sound-speed perturbation.

Ray motion in a waveguide with a weak random sound-speed perturbation caused by internal waves can be described by the following Hamiltonian

$$H = \frac{p^2}{2} + U(z) + \varepsilon Y(z)\xi(z,\, r). \tag{4.86}$$

Here $\varepsilon \ll 1$, $Y(z)$ is a slowly-varying envelope function, and $\xi(z,\, r)$ is a random term normalized as

$$\langle \xi(z,r) \rangle = 0, \quad \langle \xi^2(z,r) \rangle = \frac{1}{2}. \tag{4.87}$$

Ray trajectories obey the equations

$$\frac{dq}{dr} = p, \quad \frac{dp}{dr} = -\frac{dU}{dz} - \varepsilon\left(\frac{dY}{dz}\xi(z,\, r) + Y\frac{d\xi}{dz}\right). \tag{4.88}$$

We formulate our objective as to find some method allowing for analysis of ray equations (4.88) in terms of a deterministic theory. It should facilitate one to make up the gap between statistical and deterministic approaches. To solve this problem, we shall go the reversal way and assume that $\xi(z,\, r)$ is a *deterministic noise-like* function having a broad Fourier-spectrum.

The notion of stability means a weak sensitivity of a solution to small changes in initial conditions that is quantified by the Lyapunov exponent.

As we know from the KAM theory, most of the invariant tori of integrable Hamiltonian systems, where the motion is stable, are preserved under a small perturbation. In spite of each single realization of a random perturbation $\xi(z, r)$ can be treated as a deterministic function than a stochastic one, infinite number of frequencies in the spectrum of $\xi(r)$ leads to densely distributed ray-medium resonances in the phase space and lacking of invariant tori. In the limit $r \to \infty$, no regions in the phase space remain separated under a noise-like sound-speed perturbation.

However, in practice, we always deal with a *finite* range interval $[r_0 : r_0 + R]$. By this reason we can recover "deterministic" methods to explore the sets of stable trajectories satisfying to the condition of the finite-range invariance: *if any set in the phase space at $r = r_0$ transforms to itself at $r = r_0 + R$ without mixing, then it corresponds to an ensemble of ray trajectories which are stable by Lyapunov within the interval $[r_0 : r_0 + R]$.* In order to find such stable sets for an arbitrary spectrum of perturbation, we propose the following map [Makarov and Uleysky (2006); Makarov *et al.* (2006)]

$$p_{i+1} = p(r = r_0 + R \,|\, p_i, z_i, r_0), \quad z_{i+1} = z(r = r_0 + R \,|\, p_i, z_i, r_0), \quad (4.89)$$

where $p(r = r_0 + R \,|\, p_i, z_i, r_0)$ and $z(r = r_0 + R \,|\, p_i, z_i, r_0)$ are the solutions of Eqs. (4.88) with initial conditions $p(r_0) = p_i$, $z(r_0) = z_i$. Those values, p_{i+1} and z_{i+1}, are chosen as new initial conditions at $r = r_0$ for computing the next solution of (4.89), p_{i+2} and z_{i+2}, at the moment $r = r_0 + R$. Since we treat $\xi(z, r)$ as a deterministic function, all steps of the computation are performed with one and the same piece of $\xi(z, r)$, corresponding to $r_0 \leq r \leq r_0 + R$. If we introduce a shifted range variable $r' = r - r_0$, in fact, this map is equivalent to a Poincaré map for the ray equations ensuing from the Hamiltonian

$$H = \frac{p^2}{2} + U(z) + \varepsilon Y(z)\tilde{\xi}(z, r'), \quad (4.90)$$

where $\tilde{\xi}(z, r')$ is a periodic function of r' consisting of identical pieces of $\xi(z, r')$ of the same length R

$$\tilde{\xi}(r' + nR) = \xi(r'), \quad 0 \leq r' \leq R, \quad n = 0, 1, \ldots. \quad (4.91)$$

In this way we replace our original random sound-speed perturbation by a periodic one. It should be emphasized that the validity of this replacement in restricted by the range interval $[r_0 : r_0 + R]$. By analogy with the usual Poincaré map, the key property of the map (4.89) is the following: *each* *point of a continuous closed ray trajectory of the map (4.89) corresponds to a starting point (at $r = r_0$) of the solution of Eqs. (4.88) which remains stable by Lyapunov till the range $r = r_0 + R$.* The inverse statement is not, in general, true. It will be shown below that the map (4.89) provides sufficient but not necessary criterion of stability. Note that the initial range r_0 can be chosen arbitrarily, and the map (4.89) can be used for studying *transient ray dynamics*. In other words, this map is designed to identify regimes of finite-range stability which may occur at any ranges.

Actually, we have only a little information about the sound-speed inhomogeneity of an underwater sound channel. However, a knowledge about its spatial spectrum can help us to predict some generic features of the map (4.89) which are repeating from one realization to another.

Since the Hamiltonian (4.90) is a range-periodic function, topological properties of trajectories of the map (4.89) can be described in the framework of the theory of ray-medium resonance discussed in Section 2.3. To simplify our analysis, let us assume that $\tilde{\xi}(z,r)$ can be written in the factorized form

$$\tilde{\xi}(z,r) = \zeta(z)\varrho(r). \tag{4.92}$$

Function $\varrho(r)$ can be expanded into the Fourier series

$$\varrho(r) = \frac{1}{2}\sum_{m=-\infty}^{\infty} \varrho_m\, e^{ink_r r}, \tag{4.93}$$

where $k_r = 2\pi/R$. Assume that $Y(z)$ is a sum of vertical Garrett-Munk modes (see subsection 1.5.1)

$$\zeta(z) = \frac{1}{2}\sum_{m'=-\infty}^{\infty} \zeta_{m'}\, e^{im'\pi\chi(z)}, \tag{4.94}$$

where $\chi(z) = e^{-z/B} - e^{-h/B}$, and B is a thermocline depth. Transform the ray variables to the canonical action-angle variables

$$I = \frac{1}{2\pi}\oint p\,dz, \quad \vartheta = \frac{\partial}{\partial I}\int\limits_{z_0}^{z} p\,dz. \tag{4.95}$$

Expressing z and p as the functions of I and ϑ, the Hamiltonian (4.90) reads

$$H = H_0(I) + \varepsilon Y(I,\vartheta)\zeta(I,\vartheta)\varrho(r). \tag{4.96}$$

In the absence of vertical oscillations, i.e., at $\zeta(z) = 1$, then ray motion governed by the Hamiltonian (4.96) can be studied in the framework of the theory of ray-medium resonance (see Section 2.3). Then one expands the range-dependent term in the Hamiltonian (4.96) into the double Fourier series, i.e.

$$\varepsilon Y(I, \vartheta)\varrho(r) = \frac{\varepsilon}{2}\sum_{l,m=0}^{\infty} Y_l(I)\varrho_m e^{i(l\vartheta - mk_r)} + \text{c. c.} \tag{4.97}$$

The ray equations look as

$$\begin{aligned} \frac{dI}{dr} &= -\frac{i\varepsilon}{2}\sum_{l,m=0}^{\infty} l|Y_l\varrho_m|e^{i\Psi_{l,m}} + \text{c. c.}, \\ \frac{d\vartheta}{dr} &= \omega + \frac{\varepsilon}{2}\sum_{l,m}\frac{\partial|Y_l|}{\partial I}|\varrho_m|e^{i\Psi_{l,m}} + \text{c. c.}, \end{aligned} \tag{4.98}$$

where we expressed complex Fourier-amplitudes as $Y_l = |Y_l|e^{i\phi_l'}$ and $\varrho_m = |\varrho_m|e^{i\phi_m''}$, and introduced the phase $\Psi_{l,m}$

$$\Psi_{l,m} = l\vartheta - mk_r + \phi_l' - \phi_m''. \tag{4.99}$$

The stationary phase condition $d\Psi_{l,m}/dr = 0$ implies the resonances of the map (4.89)

$$lD(I = I_{\text{res}}) = mR, \tag{4.100}$$

where $D(I) = 2\pi/\omega(I)$ is a ray cycle length. As long as we deal with a stochastic sound-speed perturbation with a broad spectrum, an infinite number of resonances $kl : km$ $(k = 1, 2, \ldots, \infty)$ corresponds to each resonant action. If $\zeta(z)$ varies smoothly with depth, then one can suggest that the product $|Y_{k*l}\varrho_{k*m}|$ decreases rapidly with increasing k. Owing to that, the resonances $kl : km$ with $k > 1$ are weak as compared with the resonance $l : m$. Hence, if $R > D$, only the superior resonant term with $l = 1$ should be taken into account in Eqs. (4.98). Elimination of resonances with $l > 1$ allows one to describe the motion in the vicinity of I_{res} in the same approximation, as we earlier done in Section 2.3. Retaining only the dominant resonant terms, we can rewrite Eqs. (4.98) in the form:

$$\begin{aligned} \frac{dI}{dr} &= \varepsilon|Y_1\varrho_m|\sin\Psi, \\ \frac{d\vartheta}{dr} &= \omega - mk_r + \frac{\partial|Y_1|}{\partial I}|\varrho_m|\cos\Psi. \end{aligned} \tag{4.101}$$

Here we omitted the subscript at Ψ for the brevity. Coupled equations (4.101) ensue from the universal Hamiltonian of nonlinear resonance (see Section 2.3) which now looks as

$$\tilde{H} = \frac{1}{2}|\omega_I'|(\Delta I)^2 + \Omega_0^2 \cos\Psi, \tag{4.102}$$

where $\Delta I = I - I_{\rm res}$, and Ω_0 is given by

$$\Omega_0 = \sqrt{\varepsilon|\omega_I' Y_1 \varrho_m|}. \tag{4.103}$$

The formula for the width of resonance in the space of spatial frequency of ray oscillations is

$$\Delta\omega = \omega_I'\Delta I = 2\sqrt{\varepsilon|\omega_I' Y_1 \varrho_m|}. \tag{4.104}$$

If the criterion of Chirikov (2.61)

$$K_{\rm c} = \frac{\Delta\omega}{\delta\omega} \geq 1 \tag{4.105}$$

holds, neighbouring resonances overlap in the phase space and a global chaotic layer emerges. The distance between resonances $1:m$ and $1:(m+1)$ is given by

$$\delta\omega = \frac{2\pi}{R}, \tag{4.106}$$

i.e. resonances become closer and closer to each other with increasing range. The range beyond which the overlapping appears can be found from the equation

$$R_{\rm c} - \frac{\pi}{\sqrt{\varepsilon}|\omega_I' Y_1 \varrho_m(R_{\rm c})|} = 0. \tag{4.107}$$

The horizon $R_{\rm c}$ can be treated as the lower border of the onset of global chaos, because stable regions in the phase space can persist at $R > R_{\rm c}$ as stable islands. It should be mentioned that fast decay of correlations of $\xi(r)$ doesn't infer, in general, smallness of $R_{\rm c}$. Indeed, fast decay of correlations implies that significant amount of energy is concentrated in harmonics with large horizontal wavenumbers. These harmonics correspond to weak high-order resonances with small widths $\Delta\omega$.

This situation can be changed in the presence of depth oscillations of the function $\xi(z, r)$, when vertical resonances must be taken into account.

The condition of vertical resonance can be derived from the representation of $\xi(z,r)$ by means of Eqs. (4.92),(4.94) and (4.93) as

$$\xi(z,r) = \frac{1}{4}\sum_{n,n'=0}^{\infty}\left(\varrho_n\zeta'_n \exp i\psi^+_{n,n'} + \varrho_n\zeta'^*_n \exp i\psi^-_{n,n'}\right) + \text{c. c.}, \qquad (4.108)$$

where

$$\psi^{\pm}_{n,n'} = nk_r \pm n'\pi\chi(z). \qquad (4.109)$$

Ray paths under vertical resonance obey the formula

$$\frac{d\psi_{n,n'}}{dr} = nk_r \pm \frac{n'\pi p}{B}e^{-z/B} = 0. \qquad (4.110)$$

The zone of the strongest influence of vertical resonance on rays can be approximately found from condition

$$\frac{d^2\psi_{n,n'}}{dr^2} = 0. \qquad (4.111)$$

Taking into account Eq. (2.81) and setting

$$\frac{dp}{dr} \simeq -\frac{dU}{dz}, \qquad (4.112)$$

it yields

$$\frac{p^2(z=z_{\text{res}},\,H)}{B} + \frac{dU}{dz}(z=z_{\text{res}}) = 0, \qquad (4.113)$$

where $p(z,\,H) = \sqrt{2[H-U(z)]}$. Equations (4.110) and (4.113) determine the phase space regions most affected by vertical resonance. As it was shown in subsection 2.5.3, vertical resonance is accompanied by significant enhancing of nearest ray-medium resonances, independently of their orders. According to Eq. (4.100), the phase space locations of dominant resonances of map (4.89) are specified by the value of R. On the other hand, condition (4.110) yields a set of vertical resonances corresponding to different harmonics of $\varrho(r)$. Therefore, it is very probable that resonance domains of map (4.89) overlap with vertical resonances. However, it should be reminded that any resonance domain comprises an infinite number of resonances $k{*}l : k{*}m$. According to subsection 2.5.3, vertical resonance enhances effect of those harmonics of $\varrho(r)$ which satisfy conditions (4.100) and (4.110) simultaneously. One cannot be assured that any of these harmonics necessarily corresponds to the resonance with $k=1$. Therefore, the condition of fast decreasing of $|Y_{k*l}\varrho_{k*m}|$ with decreasing k can be violated, and there can be more than one principal component in the expansion (4.98). Then the Hamiltonian includes terms with $k > 1$ and looks as

$$\tilde{H} = \frac{1}{2}|\omega_I'|(\Delta I)^2 + \sum_k \Omega_k^2 \cos k\Psi, \tag{4.114}$$

where

$$\Omega_k = \sqrt{\varepsilon|\omega_I' Y_k \varrho_{k*m}|}. \tag{4.115}$$

A phase portrait corresponding to the Hamiltonian (4.114) can be very complicated as compared with that of the Hamiltonian (4.102). In particular, terms with $k > 1$ can create multistability inside a resonance domain. It means that there occur additional elliptic and hyperbolic fixed points, and a resonance domain becomes divided into subdomains with different ray dynamics.

We exemplify the map (4.89) using the biexponential sound-speed profile

$$U(z) = \frac{\Delta c(z)}{c_0} = \frac{b^2}{2}(e^{-az} - \eta)^2. \tag{4.116}$$

The range-dependent factor $\varrho(r)$ is modelled as a sum of a large (of the order of one thousand) number of harmonics with random phases and the horizontal wave numbers k_{hor} distributed in the range from $k_{\text{min}} = 2\pi/100$ km^{-1} to $k_{\text{max}} = 2\pi$ km^{-1}. The spectral density of the perturbation decays as k_{hor}^{-2}. Let the vertical oscillations contain a single mode only

$$\zeta(z) = \zeta_1 \sin \pi\chi(z), \tag{4.117}$$

and the other parameters are as follows: $\varepsilon = 0.0025$, $V(z) = (2z/B)\exp(-2z/B)$, and $B = 1$ km. We start with a number of initial values of the momentum-position variables in different regions of the phase space (z_0, p_0) and compute their values at the distance $R = 100$ km (z_1, p_1) from a sound source with the respective equations of ray motion with a chosen realization of the noisy-like horizontal internal-wave perturbation and the single vertical mode (4.117). Then we compute the ray variables with the same realization of noise at the same distance $R = 100$ km, find (z_2, p_2), and so on. The results of computing of the Poincaré map with two different realizations of a noisy-like perturbation in the coordinates $I_x = (I/I_{\text{s}})\cos\vartheta$ and $I_y = (I/I_{\text{s}})\sin\vartheta$ are shown in Fig. 4.9. Here I_{s} is the most accessible value of the action for rays propagating without reflections off the bottom. The formulae for the action-angle variables are presented in subsection A.4. The map reveals numerous island-like regions of stability in the phase space

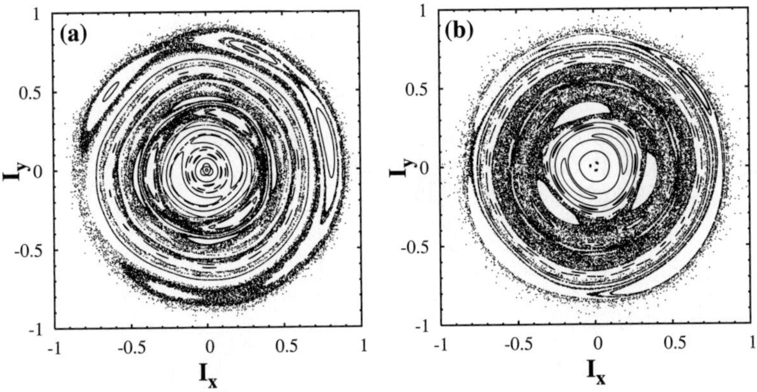

Fig. 4.9 The map (4.89) with $R = 100$ km revealing stable bundles of sound rays surviving in a model underwater waveguide under a noisy-like internal-wave perturbation with a single vertical mode (4.117). The polar action-angle variables $I_x = (I/I_s)\cos\theta$ and $I_y = (I/I_s)\sin\theta$ are in units of the separatrix action I_s. Fragments (a) and (b) correspond to two different realizations of noise at the same other conditions [Makarov *et al.* (2006)].

corresponding to the resonance condition $mD(I) \simeq lR$, where $D(I)$ is a cycle length of a ray with a given action I. The values of the island's areas and of the respective action are close with different realizations, but the island's form and their angle coordinates may be different.

Let us consider now a more complicated model with five vertical modes of internal waves

$$\zeta(z) = \sum_{n=1}^{5} \zeta_n \sin(n\pi\chi(z) + \phi_n), \tag{4.118}$$

where $\zeta_n \propto n^{-2}$, and ϕ_n are random phases. The respective maps (4.89) are shown in Fig. 4.10 with $R = 100$ km and with two different realizations of the noise-like perturbation. It follows from comparing Figs. 4.9 and 4.10 that complication in the vertical perturbation affects, mainly, the most steep rays corresponding to the outer resonance islands. The most outer islands in Fig. 4.10 are separated from the other domains of stability by a layer of escaping rays (by escaping rays we mean the rays reaching the lossy bottom of the waveguide during their propagation). Comparing Figs. 4.10(a) and (b), we see that under different realizations of noise the outer region of stability is of approximately the same area and topology and changes its angular position only. It reflects that global properties of a phase space pattern produced by the map (4.89) are repeating for *typical* realization of a noisy-like function $\xi(z, r)$ with a given spectrum.

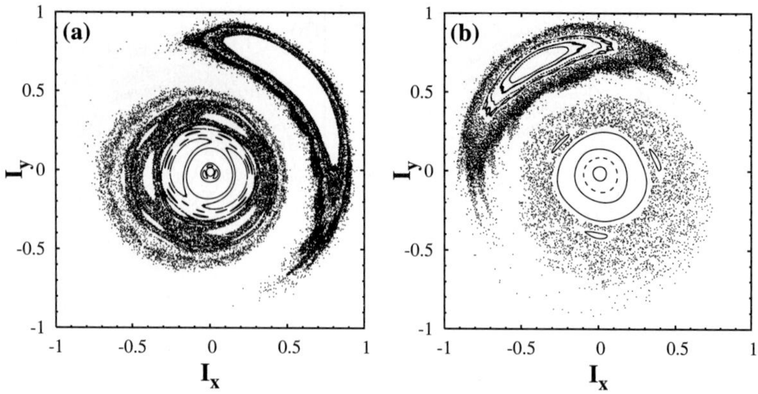

Fig. 4.10 The same as in Fig. 4.9 but with five vertical modes (4.118), $R = 100$ km. Fragments (a) and (b) correspond to two different realizations of noise at the same other conditions [Makarov *et al.* (2006)].

With increasing R, the area of islands decreases but distinct islands of stability remain visible in the phase space at a rather long distance $R = 500$ km (Fig. 4.11) even with five vertical modes of internal waves to be included in the model.

Thus, we described the technique allowing one to study ray motion in randomly-inhomogeneous waveguides within a deterministic approach. As a range-dependent sound-speed perturbation is a stationary stochastic process, one can conclude that large islands, like those shown in Figs. 4.9 and 4.10, are present at almost any range, albeit each island exists only

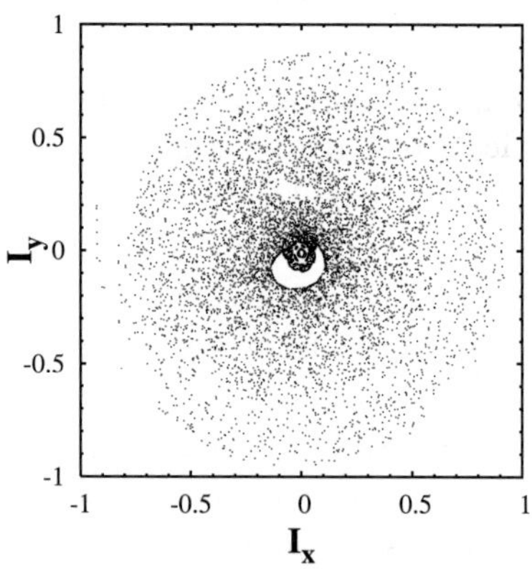

Fig. 4.11 The same as in Fig. 4.10 but with $R = 500$ km [Makarov *et al.* (2006)].

within a limited range interval. It suggests a dynamical mechanism of intermittency in underwater sound channels, associated with resonances between unperturbed ray motion and a random sound-speed perturbation.

4.2 Travel Times of Chaotic Rays

In Sections 1.4 and 3.4 we examined the properties of ray travel times in a range-independent waveguide and in a waveguide with periodic range-dependence. It was found that the travel times of both regular and chaotic rays to a significant extent are determined by the identifiers (topologies) of their paths in the sense that at a given observation range, rays with the same identifier have close travel times. In the present section it is shown that this property remains valid in a more realistic environmental model with the sound speed fluctuations induced by random internal waves. As in the case of a waveguide with periodic range-dependence, arrivals associated with the same identifier form a comparatively compact fuzzy segment of the timefront located in the vicinity of an unperturbed segment corresponding to the same identifier.

Here we obtain approximate analytical relations for the difference between travel times of perturbed and unperturbed rays. Combining these relations with the stochastic ray theory derived in Sec. 4.1 yields analytical estimates for the spread and bias of fuzzy segments in the presence of random internal waves.

Even in the presence of perturbation giving rise to ray chaos, segments of the timefront formed by early arriving steep rays reveal a remarkable stability. The early portion of the timefront still "remembers" its structure in the unperturbed waveguide. This property is well-known and it has been observed in both numerical simulations and field experiments [Simmen *et al.* (1997); Brown and Viechnicki (1998); Worcester *et al.* (1999); Colosi *et al.* (1999)].

4.2.1 *Timefront in a randomly inhomogeneous waveguide*

Figure 4.12 presents timefronts in the perturbed (upper panel) and unperturbed (lower panel) waveguides at the range 3000 km constructed by tracing numerically a fan of 120 000 rays escaping a point source set at the depth $z_s = 0.7$ km. Starting momenta of the fan rays uniformly fill an interval corresponding to launch angles $\pm 12.5°$. The structure of the

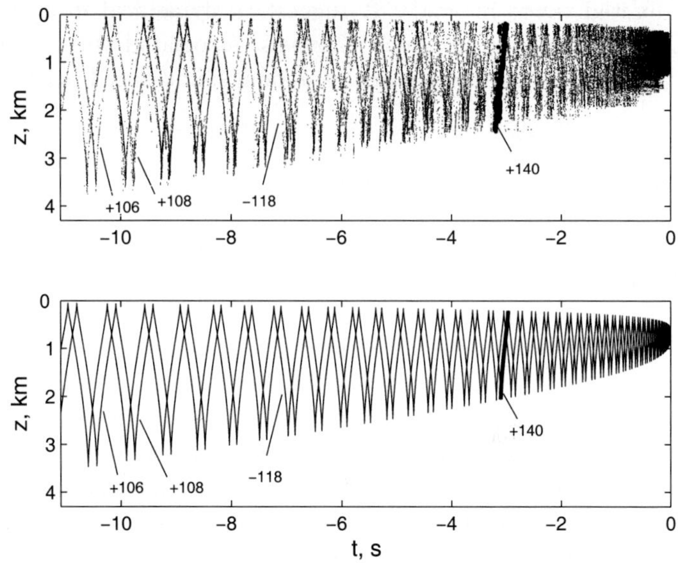

Fig. 4.12 Timefronts in the perturbed (upper panel) and unperturbed (lower panel) waveguide: depth versus ray travel time at 3000 km for a point source set at $z_s = 0.7$ km. Identifiers of rays forming some particular segments are indicated next to the corresponding segments. In the upper panel, arrivals with identifier +140 are marked by thick points. In the lower panel, arrivals with this identifier are depicted by a thick solid line.

perturbed timefront resembles what we have seen in Sec. 2.6.2 (Figs. 2.21, 2.22, and 2.25) when studying a waveguide with a periodic perturbation.

Travel times of chaotic rays in a random environment have the same fundamental property as in a waveguide with periodic range dependence: rays with equal identifiers have close travel times and form compact fuzzy segments in the time-depth plane. Moreover, each fuzzy segment once again occupies an area of the time-depth plane close to the location of its unperturbed counterpart corresponding to the same identifier. This is seen in Fig. 4.12 where identifiers are indicated next to some fuzzy and unperturbed segments.

Note two distinctions between the timefront presented in the upper panel of Fig. 4.12 and the timefronts in the waveguide with periodic range-dependence shown in Figs. 2.21 and 2.25. In the presence of a random perturbation, there are no regular rays and all the segments have some spread. Therefore, in the upper panel of Fig. 4.12 we do not see regular segments present in the late parts of the timefronts in Figs. 2.21 and 2.25.

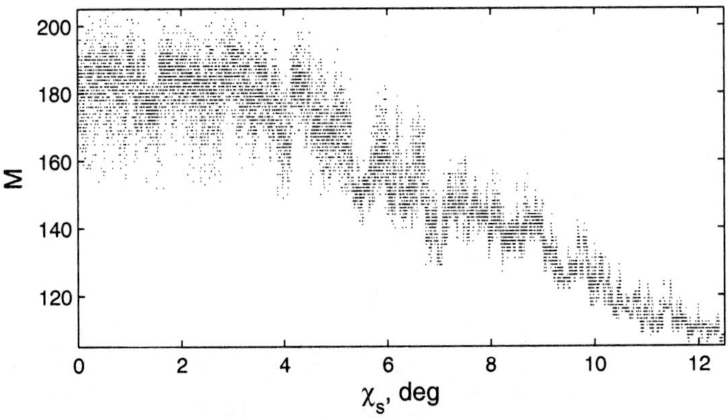

Fig. 4.13 Number of ray turning points M at 3000 km against launch angle χ_s. Rays escape a point source set at the depth $z_s = 0.7$ km.

Another distinction is that in a realistic environmental model the steep rays are less chaotic than the flat ones [Simmen *et al.* (1997); Beron-Vera *et al.* (2003)]. The segments formed by early arriving steep rays are significantly less widened and biased than the segments formed by near axial (flat) rays with greater travel times. It may be said that the early part of the timefront remains relatively stable in the presence of the internal-wave-induced perturbation. It is noteworthy that this stability is combined with high sensitivity of chaotic paths to small variations of their starting parameters. Figure 4.13 shows how the number of ray turning points, M, at the 3000 km range depends on the launch angle, χ_s. It is clearly seen that rays with very close starting parameters may have quite different numbers of turning points, that is, the rays may follow quite different trajectories. This is true not only for near axial rays, but for the less chaotic steep rays, as well.

When internal waves are absent the ray arrivals are spread in a smooth and predictable way. If the launch angle χ_s of an unperturbed ray varies monotonically, then a point depicting this ray on the timefront moves monotonically along a corresponding broken curve from one segment to the neighboring one. In the presence of internal waves, this point "jumps" in the time-depth plane. However, this chaotic behavior obeys a nontrivial deterministic law: at early times the point remains in the vicinity of the unperturbed timefront. It jumps from one segment to another (not necessary to an adjacent one) and back but do not fill some area in the time-depth plane as it might be expected.

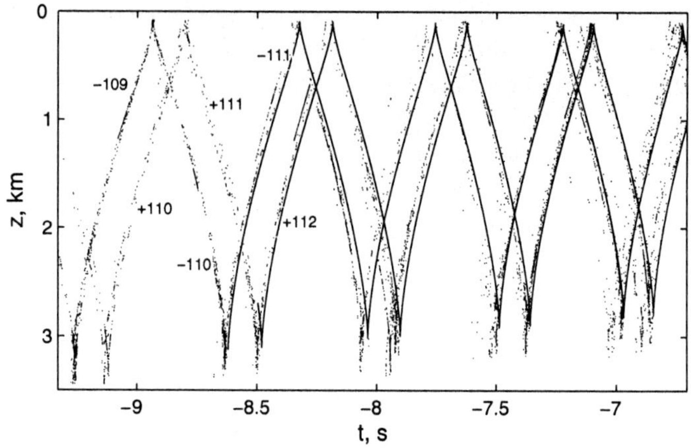

Fig. 4.14 Parts of the unperturbed (solid broken lines) and perturbed (points) timefronts at 3000 km formed by rays with launch angles $|\chi_s| > 11.5°$.

Figure 4.14 presents in the same plot early parts of the unperturbed (solid line) and perturbed (points) timefronts. In this figure only arrivals of ray with launch angles $\chi_0 > 11.5°$ are shown. Note that six earliest perturbed (fuzzy) segments with identifiers -109, ± 111, and $+112$ have no unperturbed counterparts. The point is that the rays with these identifiers in the unperturbed waveguide have launch angles exceeding 11.5°. In the presence of perturbation rays with the above identifiers appear because the perturbation leads to a widening of the interval of ray grazing angles. It is remarkable that the segments, appeared due to scattering at random inhomogeneities, are quite regular and coincide with the segments missed in the broken lines depicting the unperturbed timefront.

Figure 4.14 demonstrates another general property of the perturbed timefront: the presence of perturbation causes not only a spread of segments but their *bias* in the direction of early times, as well. Below this phenomenon will be studied in detail.

4.2.2 *Travel time differences*

Our analysis of travel times is based on a surprisingly simple formula for the difference between eikonal (travel times) of rays with close initial actions. This formula will be derived in the present subsection. It generalizes analytical relations for the difference between travel times of unperturbed rays given by Eqs. (1.110) and (1.114).

In what follows, we will evaluate the difference in eikonals of two rays. To distinguish between similar parameters of these rays, parameters of one of them will be marked with overbars. For example, starting and final points of one trajectory will be denoted by (p_0, z_0) and (p_1, z_1), respectively, while for another trajectory we will have $(\bar{p}_0, \bar{z}_0)$ and $(\bar{p}_1, \bar{z}_1)$. The symbol Δ will be used to denote the difference between any characteristic of one ray and its counterpart for another ray, e.g., $\Delta I = I - \bar{I}$, $\Delta S = S - \bar{S}$, $\Delta z_0 = z_0 - \bar{z}_0$, and so on.

We assume that both rays under consideration have large numbers of cycles $N, \bar{N} \gg 1$, but $\Delta N = N - \bar{N} = O(1)$. This ensures the closeness of mean cycle lengths for our rays and suggests that their mean actions (averaged over the paths), I and $\bar{I}$, should also be close. The difference ΔI is considered as a small parameter. More precisely, it is assumed that

$$\frac{1}{\omega}\left|\frac{d\omega}{dI}\Delta I\right| \ll 1, \tag{4.119}$$

that is, $|\Delta I|$ is small compared to a characteristic scale of function $\omega(I)$.

4.2.2.1 *Two rays in a range-independent waveguide*

Let us begin with a pair of rays in the unperturbed waveguide with $\delta c = 0$. Then $V = 0$ and Eqs. (1.76)–(1.78) yield

$$S = F(I)\,r - G(z_0, I) + G(z_1, I) + I\,(\theta_0 - \theta_1), \tag{4.120}$$

where

$$F(I) = I\,\omega(I) - H_0(I). \tag{4.121}$$

In the unperturbed waveguide $I_0 = I_1$ and we have dropped the subscript at I. It follows from Eq. (1.53) that

$$F' = I\omega_I', \quad F'' = \omega_I' + I\omega''. \tag{4.122}$$

Since ΔI is small, in the expansion

$$F(I) - F(\bar{I}) = \bar{I}\omega_I'(\bar{I})\,\Delta I + O(\Delta I^2) \tag{4.123}$$

we will retain only the first term. According to the second of Eqs. (1.47), we get

$$G(z_{0,1}, I) = G(z_{0,1}, \bar{I}) + \theta_{0,1}\Delta I + O(\Delta I^2). \tag{4.124}$$

Making use of Eqs. (4.123) and (4.124) gives

$$c_0\Delta t = G(z_1,\bar{I}) - G(\bar{z}_1,\bar{I}) - G(z_0,\bar{I}) + G(\bar{z}_0,\bar{I}) + \bar{I}\omega_I'(\bar{I})\Delta I r + \bar{I}(\Delta\theta_0 - \Delta\theta_1). \quad (4.125)$$

In the same approximation, it follows from Eqs. (1.50) and (1.55) that

$$\Delta\theta = \omega_I'(\bar{I})\, r\Delta I = 2\pi\Delta N - (\Delta\theta_0 - \Delta\theta_1). \quad (4.126)$$

Note that according to Eqs. (1.27) and (1.83) we have

$$\Delta S = c_0\Delta t. \quad (4.127)$$

Substituting Eq. (4.126) in Eq. (4.125), one finds

$$c_0\Delta t = 2\pi\bar{I}\Delta N + G(z_1,\bar{I}) - G(\bar{z}_1,\bar{I}) - G(z_0,\bar{I}) + G(\bar{z}_0,\bar{I}). \quad (4.128)$$

When deriving this approximate relation, we have dropped terms proportional to r , and it may seem that Eqs. (4.125) and (4.126) are valid only at short enough ranges. However, it is not so. The terms proportional to r, neglected in Eqs. (4.125) and (4.126), are, on the other hand, proportional to ΔI^2 (see Eq. (4.123)). According to Eq. (4.126), the difference ΔI can be estimated as

$$\Delta I \simeq \frac{2\pi\Delta N}{\omega_I' r}. \quad (4.129)$$

This estimate shows that if ΔN is kept constant then the neglected terms do not increase, but decrease with increasing range.

Formula (4.128) simplifies if both rays under consideration start from two closely spaced depths with starting angles of the same sign, their arrival depths at range r are also close, and arrival angles have the same sign. Then the identifiers of these rays are $\pm(M + 2\Delta N)$ and $\pm M$. If these conditions are met, making use of the first of Eqs. (1.47) transforms Eq. (4.128) to

$$c_0\Delta t = 2\pi\Delta N\,\bar{I} - \bar{p}_0\Delta z_0 + \bar{p}_1\Delta z_1. \quad (4.130)$$

This equation takes the simplest form when $\Delta z_0 = \Delta z_1 = 0$, that is when we compare two eigenrays connecting a point source and a point receiver, and these eigenrays have identifiers $\pm(M+2)$ and $\pm M$ (the ray with action I has an extra cycle). Then Eq. (4.130) reduces to Eq. (1.110).

There is another important particular case of Eq. (4.128). If both rays escape the same depth point ($\Delta z_0 = 0$) and their paths have the same number of minima ($\Delta N = 0$), we have

$$c_0 \Delta t = \frac{1}{c_0} \left[G(z_1, \bar{I}) - G(\bar{z}_1, \bar{I}) \right]. \tag{4.131}$$

This equation gives an approximate quantitative description of the unperturbed timefront segment. If $\bar{z}_1$ is the lower turning point of the ray path, Eq. (4.131) reduces to Eq. (1.114).

4.2.2.2 *Ray in a range-independent waveguide and ray in a range-dependent waveguide*

To find out how the presence of a small range-dependent component of the sound speed $\delta c(r, z)$ affects travel times, we compare a pair of rays one of which – "unperturbed" ray – propagates in a range-independent waveguide (its parameters will be marked with overbar) and another one – "perturbed" ray – propagates in a range-dependent waveguide. Let us think that the perturbation V is proportional to a small dimensionless parameter ε. It is assumed that the initial actions of the rays are close. Then in accord with Eq. (1.74) at $r > 0$, we get

$$\Delta I = I - \bar{I} = O(\varepsilon). \tag{4.132}$$

We do not impose other restrictions on starting and final parameters of rays except the requirements $N, \bar{N} \gg 1$ and $\Delta N = O(1)$. The rays can start from quite different depths z_0 and $\bar{z}_0$ and arrive at different depths z_1 and $\bar{z}_1$.

Using Eqs. (1.47), we obtain an expression for the difference between the components S_a determined by Eq. (1.77) for the eikonals of our rays (cf. Eq. (4.125))

$$S_a - \bar{S}_a = - \left[G(z_0, \bar{I}) - G(\bar{z}_0, \bar{I}) \right] + \left[G(z_1, \bar{I}) - G(\bar{z}_1, \bar{I}) \right] + \bar{I} \left(\Delta\theta_0 - \Delta\theta_1 \right) + O(\varepsilon^2). \tag{4.133}$$

The difference between the components S_b (see Eq. (1.78)) accumulates with range. In order to estimate it, we will retain terms $O(\varepsilon^2)$. Making use of the relations $d\theta = (\omega(I) + \partial V / \partial I)\, dr$ and $d\bar{\theta} = \omega(\bar{I}) dr$, which follows from Eq. (1.75), yields

$$S_b - \bar{S}_b = \int_0^r \left(\bar{I}\omega_I'(\bar{I})\Delta I + \frac{1}{2}\left(\omega_I'(\bar{I}) + \bar{I}\omega''(\bar{I})\right)\Delta I^2 \right) dr$$

$$+ \int_0^r \left(V'\bar{I} - V + V'\Delta I \right) dr + O(\varepsilon^3). \quad (4.134)$$

Below we will take into account that

$$V(I,\theta,r) - \Delta I V'(I,\theta,r) = V(\bar{I},\theta,r) + O(\varepsilon^3). \quad (4.135)$$

Up to terms $O(\varepsilon^2)$, the difference in angle variables of our rays can be estimated as

$$\Delta\theta = 2\pi\Delta N - \Delta\theta_0 + \Delta\theta_1 = \int_0^r \left(\omega_I'(\bar{I})\Delta I + \frac{1}{2}\omega''(\bar{I})\Delta I^2 + V' \right) dr. \quad (4.136)$$

The first equality follows from Eq. (1.50). Substituting Eq. (4.136) into Eq. (4.134), summing up Eqs. (4.134) and (4.133), and taking into account Eq. (4.135), we get

$$c_0\Delta t = S - \bar{S} = -\left[G(z_0,\bar{I}) - G(\bar{z}_0,\bar{I})\right] + \left[G(z_1,\bar{I}) - G(\bar{z}_1,\bar{I})\right]$$

$$+ 2\pi\Delta N\bar{I} + \frac{\omega_I'(\bar{I})}{2} \int_0^r \Delta I^2 dr - \int_0^r V(\bar{I},\theta,r)dr. \quad (4.137)$$

When deriving this formula, we have expanded $\omega(I) - \omega(\bar{I})$ in powers of ΔI and retained only the two first terms of the expansion. This may be done only if $|\Delta I|$ is small compared to the characteristic scale of the function $\omega(I)$, that is if the condition (4.119) is satisfied. In Ref. [Virovlyansky (2001)] it is shown that in a typical deep water environment with the internal-wave-induced perturbation this condition is met even at megameter ranges.

As in the case of Eq. (4.128), formula (4.137) slightly simplifies if the rays have identifiers $\pm(M + 2\Delta N)$ and $\pm M$, start from close points in the phase space, and arrive at also close (but, generally, other) points. Then the quantities $z_{0,1}$ are close to $\bar{z}_{0,1}$, and $p_{0,1}$ are close to $\bar{p}_{0,1}$. In accord with the first of Eqs. (1.47), formula (4.137) can be approximately rewritten as

$$c_0 \Delta t = -\bar{p}_0 \Delta z_0 + \bar{p}_1 \Delta z_1 + 2\pi \Delta N \bar{I} + \frac{\omega_I'(\bar{I})}{2} \int_0^r \Delta I^2 dr - \int_0^r V(\bar{I}, \theta, r) dr. \tag{4.138}$$

In this form the expression for Δt was derived in Ref. [Virovlyansky (2003)] (see also Refs. [Virovlyansky (2001)] and [Virovlyansky (2003a)]). In Ref. [Virovlyansky (2003)] the applicability of our estimate at multimegameter ranges was confirmed by comparison with results of a numerical ray tracing. Note that the last integral in Eqs. (4.137) and (4.138) is evaluated along a path which is not a solution of the ray equations: its action variable in all range points is equal to $\bar{I}$ while its angle variable is "taken" from the perturbed ray. In the unperturbed waveguide ($V = 0$), Eqs. (4.137) and (4.138) translate to Eqs. (4.128) and (4.130), respectively.

4.2.3 *Statistics of ray travel times*

Thus, we have approximate analytical relation (4.137) connecting variations (caused by internal waves) of geometrical characteristics of the ray path I, θ, p, and z to variations of the ray travel time. Combining this result and stochastic ray theory derived in Sec. 4.1 yields a statistical description of ray travel times [Virovlyansky (2001, 2005a)]. This issue is considered in the present subsection.

4.2.3.1 *Variations of ray travel times caused by internal waves*

Compare a perturbed and unperturbed ray escaping the same point source at the same launch angle. Initial action and angle variables of these rays are denoted by I_0 and θ_0, respectively. In this case Eq. (4.137) can be rewritten in the form

$$\Delta t = \tau_G + \tau_N + \tau_I + \tau_V, \tag{4.139}$$

where

$$\tau_G = \left[G(z, I_0) - G(\bar{z}, I_0) \right] / c_0, \tag{4.140}$$

z and $\bar{z}$ are coordinates of the perturbed and unperturbed rays, respectively, at the observation range;

$$\tau_N = 2\pi \Delta N I_0 / c_0, \tag{4.141}$$

ΔN is the difference between the numbers of minima (numbers of cycles) of the perturbed and unperturbed paths;

$$\tau_I = \frac{\omega_I'(I_0)}{2c_0} \int_0^r \left[I(r') - I_0\right]^2 dr', \tag{4.142}$$

and

$$\tau_V = -\frac{1}{c_0} \int_0^r V(I_0, \theta(r'), r')dr. \tag{4.143}$$

Functions $I(r)$ and $\theta(r)$ in Eqs. (4.142) and (4.143) present the range-dependencies of action and angle variables, respectively, along the perturbed ray path.

At comparatively short ranges $O(100$ km$)$, both rays follow practically the same path. Then $\Delta N = 0$, $z \simeq \bar{z}$, $I(r') \simeq I_0$ and terms τ_G, τ_N and τ_I on the right of Eq. (4.139) are negligible. Equation (4.139) reduces to the well known relation [Flatté *et al.* (1979)]

$$\Delta t \simeq \tau_V \simeq -\frac{1}{c_0^2} \int_0^r \delta c(r, z_0(r))dr, \tag{4.144}$$

where the integration goes along the unperturbed path $z_0(r)$. In this case the travel time variations are caused by sound speed fluctuations crossed by the unperturbed ray. A detailed analysis of statistical characteristics of τ_V have been done in Ref. [Flatté *et al.* (1979)]. The rms variations of τ_V grows with range as $r^{1/2}$.

At ranges $O(1000$ km$)$, perturbed and unperturbed paths with equal starting parameters are no longer close and usually have different numbers of cycles ($\Delta N \neq 0$) [Virovlyansky (2001); Beron-Vera *et al.* (2003); Virovlyansky (2003, 2005a)]. For steep rays with large starting actions I_0 the main contribution to Δt at long ranges comes from the term τ_N [Virovlyansky (2001, 2003, 2005a)]. It follows from Eqs. (1.43) and (1.46) that $2\pi I_0 > G(I_0, z)$ and, hence, $|\tau_G| < |\tau_N/\Delta N|$. Therefore, the term τ_N may be used as a rough estimates of Δt for steep rays. For flat rays (small I_0), both τ_N and τ_G become negligible as compared to τ_I. Since the latter typically exceeds τ_V, the relation

$$\Delta t = \tau_N + \tau_I \tag{4.145}$$

gives a rough estimate of Δt at ranges $O(1000 \text{ km})$ for rays with arbitrary launch angles.

To apply the Wiener process approximation for description of Δt we approximate ΔN by $y/(2\pi)$. This yields

$$\tau_N = y(r)I_0/c_0. \tag{4.146}$$

According to Eqs. (4.33) and (4.142), we have

$$\tau_I = \frac{\omega_I'(I_0)}{2c_0} \int_0^r x^2(r')dr'. \tag{4.147}$$

Equations (4.145)–(4.147) allow one to model the distribution of Δt using the procedure described at the end of Sec. 4.1.5. Then the dependence of Δt on the launch angle enters through the initial action I_0.

The difference in travel times of perturbed and unperturbed rays versus launch angle is shown in the upper panel of Fig. 4.15. In this figure we consider rays with positive launch angles. Results for negative launch angles look similar (not shown).

Making use of Eqs. (4.52) and (4.54), we have evaluated 50000 realizations of $x(r)$ and $y(r)$ for 3000 km propagation. Each pair $(x(r), y(r))$ models a chaotic ray with starting action I_0. A set of ξ_n's for each ray has been generated independently. Then, by formulas (4.146) and (4.147), we have computed τ_N and τ_I for all rays. The lower panel in Fig. 4.15 presents the sum $\tau_N + \tau_I$ as a function of launch angle. It is seen that our simple model based on Eqs. (4.52)–(4.54) and (4.145)–(4.147) gives a reasonable prediction for the spread of Δt. Note that at the 3000 km range, we have $\tau_V = O(5 \text{ ms})$ [Flatté *et al.* (1979); Virovlyansky (2001)], and it is negligible compared to the spread of Δt seen in Fig. 4.15.

An analytical estimate for the rms value of Δt can be easily derived for steep rays. In this case $\Delta t \simeq \tau_N$ and $\langle \Delta t \rangle = 0$. For steep rays we can ignore the reflecting boundary at $x = -I_0$ and exploit Eq. (4.49). Combining Eqs. (4.49) and (4.146) yields a standard deviation of Δt

$$\sigma_{\Delta t} = |\omega_I'(I_0)| \frac{r^{3/2}}{c_0} \left(\frac{B}{3}\right)^{1/2}. \tag{4.148}$$

Two thick solid lines in the upper panel of Fig. 4.15 depict $\pm\sigma_{\Delta t}$ versus launch angle. Equation (4.148) estimates the spread of travel times for a bundle of rays escaping a point source at (steep) launch angles from a narrow angular interval centered at angle χ_0 corresponding to the starting

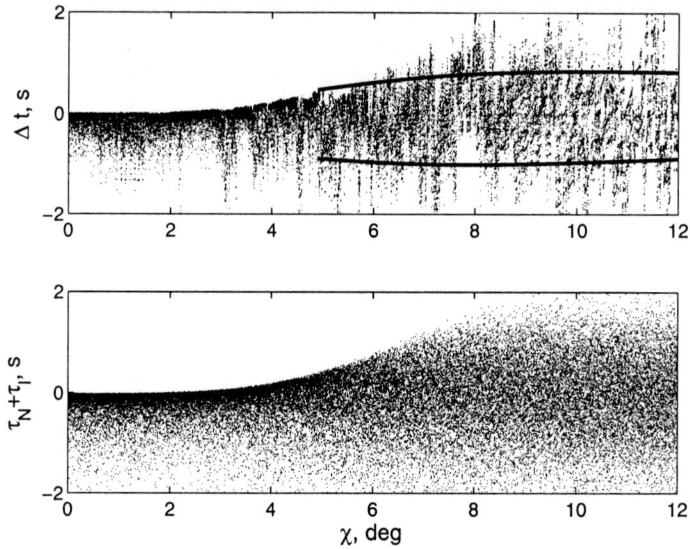

Fig. 4.15 Difference between travel times of perturbed and unperturbed rays with equal launch angles χ. Upper panel: result of direct numerical ray tracing; solid lines show deviations $\pm\sigma_{\Delta t}$ predicted by Eq. (4.148). Lower panel: simulation of sum $\tau_N + \tau_I$ in the Wiener process approximation on the basis of Eqs. (4.52), (4.54), (4.146), and (4.147) [Virovlyansky *et al.* (2007)].

action I_0. According to Eq. (4.148) the rms spread of travel times grows with range like $r^{3/2}$. The same range dependence was found by other authors [Beron-Vera and Brown (2004)].

4.2.3.2 *Spread and bias of fuzzy segments*

Numerical ray tracing demonstrates that a weak perturbation causes not only the spread of a timefront segment but also its bias. This effect is clearly seen in Fig. 4.16 that presents a magnified view of perturbed and unperturbed segments from Fig. 4.12 associated with the identifier +140. In Fig. 4.12 the perturbed segment is marked by thick points (upper panel) and the unperturbed one is marked by a thick solid line (lower panel).

Introduce a quantity τ_i (see Fig. 4.16) that represents a time delay between the arrival of i-th ray contributing to the given fuzzy segment and the unperturbed segment corresponding to the same identifier. In fact, τ_i is a difference between travel times of perturbed and unperturbed rays escaping a point source and arriving at the same observation point ($z = \bar{z}$)

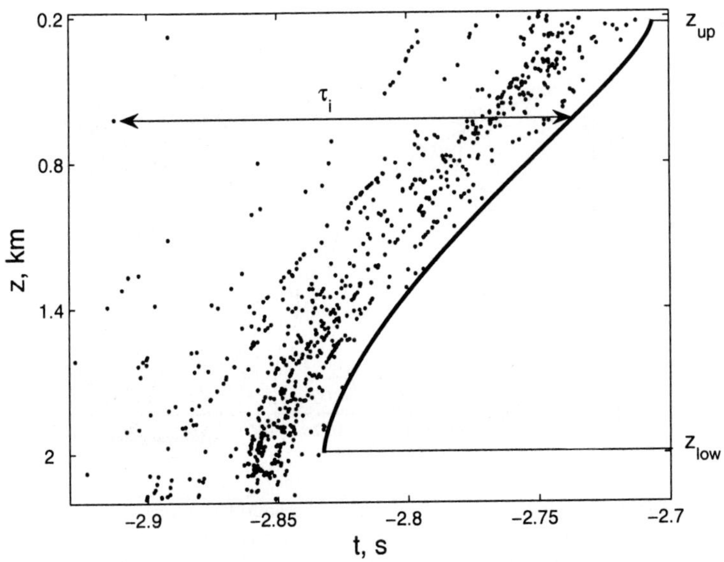

Fig. 4.16 Arrivals of rays with the identifier +140 at the range 3000 km are shown in the time-depth plane. Points and solid lines depict arrivals with and without internal waves present, respectively.

with equal identifiers. Let us apply formulas (4.139)–(4.143) to estimate τ_i. Setting $\tau_G = \tau_N = 0$ and, neglecting τ_V, we get [Virovlyansky (2001, 2003, 2005a)]

$$\tau_i = \frac{\omega_I'(\bar{I}_M)}{2c_0} \int_0^r \left[I(r') - \bar{I}_M\right]^2 dr', \tag{4.149}$$

where $I(r)$ is an action of the i-th perturbed ray as a function of range, and $\bar{I}_M$ is the mean action of rays forming the unperturbed segment. According to Eq. (4.149) the sign of τ_i is determined by the sign of $\omega_I'(\bar{I}_M)$. For typical deep water waveguides the cycle length grows with launch angle $|\chi|$ and therefore $\omega_I'(I)$ is negative. Then $\tau_i < 0$ and the perturbed segment is biased toward early times [Virovlyansky (2001, 2003, 2005a)]. Let us emphasize that this statement is valid only at long enough ranges where $|\tau_I|$ for most rays exceeds $|\tau_V|$. At short ranges, where τ_V dominates, there exists a different mechanism of the bias. It is discussed in Ref. [Codona *et al.* (1985)].

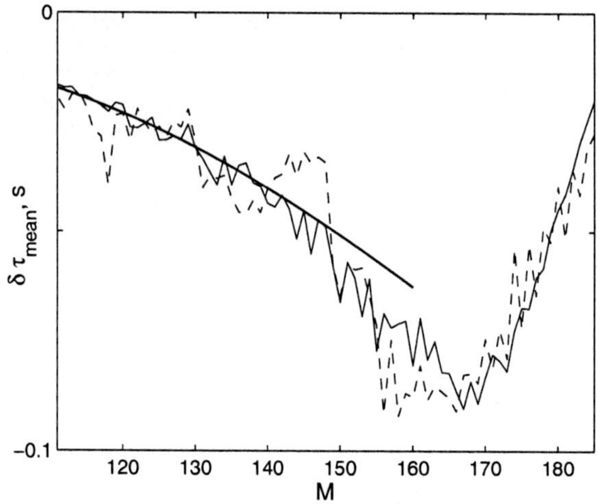

Fig. 4.17 Mean bias of the fuzzy segment at the sound channel axis as a function of number of ray turning points, M. The plot is constructed for rays starting upward. Thin solid line: direct ray tracing. Thick solid line: analytical estimate (4.155). Dashed line: simulation in the Wiener process approximation based on Eqs. (4.52), (4.54), and (4.151) [Virovlyansky *et al.* (2007)].

Since all (or almost all) τ_i have the same sign, both spread and bias of a fuzzy segment are characterized by the quantity

$$\delta\tau_{\mathrm{mean}} = N^{-1}\sum_{i=1}^{N}\tau_i. \tag{4.150}$$

Summation is taken over rays belonging to the selected fuzzy segment and arriving at depths in an interval $z_{\mathrm{low}} < z < z_{\mathrm{up}}$, where z_{low} and z_{up} are depths of the lower and upper ends of the unperturbed segment, respectively. A thin solid line in Fig. 4.17 shows $\delta\tau_{\mathrm{mean}}$ for segments with identifiers $+M$ for $M = 110, 111, \ldots, 185$ obtained by direct ray tracing. Values of $\delta\tau_{\mathrm{mean}}$ for segments with identifiers $+M$ and $-M$ are close for most M (not shown).

In order to estimate τ_i in the Wiener process approximation, we present the action of the i-th ray as $I(r) = I_{0,i} + x(r)$, where $I_{0,i}$ is a starting action. From Eq. (4.48) we get

$$\tau_i = \frac{\omega_I'(\bar{I}_M)}{2c_0}\left[\left(I_{0,i} - \bar{I}_M\right)^2 r + \left(I_{0,i} - \bar{I}_M\right)\int_0^r x(r')dr' + \int_0^r x^2(r')dr'\right]. \tag{4.151}$$

To avoid confusion, note that in Sec. 4.2.3.1 we compared travel times of perturbed and unperturbed rays with equal starting parameters that, generally, arrive at quite different depth points. This explains the difference between the right-hand sides of Eqs. (4.147) and (4.151), obtained on the basis of the same formula (4.142).

The fact that the rays under consideration have the same identifier can be (approximately) accounted for by introducing a constraint [Virovlyansky (2001, 2005a)]

$$\left|\theta_i - \bar{\theta}_M\right| < \pi/2, \tag{4.152}$$

where θ_i and $\bar{\theta}_M$ are angle variables of the i-th perturbed ray and the unperturbed ray with starting action $\bar{I}_M$, respectively. Making use of Eq. (4.46), we rewrite Eq. (4.152) as

$$\left|\omega(I_{0,i})r + y(r) - \omega(\bar{I}_M)r\right| < \pi/2. \tag{4.153}$$

Here we have neglected the difference in initial values of angle variables of perturbed and unperturbed rays. This additional approximation, that slightly relaxes constraint (4.152), is justified by the fact that the values of $|y(r)|$ at 3000 km significantly exceed π for most rays. Employing a one more approximation $\omega(I_{0,i})r - \omega(\bar{I}_M) = \omega_I'(\bar{I}_M)(I_{0,i} - \bar{I}_M)$, we rewrite Eq. (4.153) in the form [Virovlyansky (2001, 2005a)]

$$\left| I_{0,i} - \bar{I}_M + \int_0^r x(r')dr' \right| < \frac{\pi}{2\omega_I'(\bar{I}_M)}. \tag{4.154}$$

Using the numerical procedure described at the end of Sec. 4.1.5, we have computed realizations of $x(r)$ for 50000 fan rays. Selecting rays satisfying the constraint (4.154) and evaluating τ_i by formula (4.151), we have obtained desired estimates of $\delta\tau_{\text{mean}}$ shown in Fig. 4.17 by a dashed line.

Statistical characteristics of τ_i for steep rays, defined by Eq. (4.151) in the presence of the constraint (4.154), can be evaluated analytically. In Refs. [Virovlyansky (2001, 2005a, 2006)] it was shown that

$$\delta\tau_{\text{mean}} = \frac{\omega_I'(\bar{I}_M)}{12c_0}Br^2. \tag{4.155}$$

In Fig. 4.17 this dependence for segments with $M < 160$ is displayed by a thick solid line. The segments with larger M are formed by flat rays. Figure 4.17 demonstrates that predictions (numerical and analytical) obtained in the Wiener process approximation agree fairly well with the results of direct ray tracing.

4.2.3.3 *Pulse signal at the observation point*

Travel times of eigenrays arriving at a point receiver set at a depth z_r are determined by intersections of the timefront and horizontal line $z = z_r$. As in the case of a periodic perturbation, each fuzzy segment gives a compact cluster of travel times representing arrivals of eigenrays with an identifier associated with the segment.

The effect of the internal-wave-induced perturbation on eigenrays may be interpreted in the following way. In the presence of perturbation, the unperturbed eigenray splits into a bundle of new eigenrays with the same identifier as the original one. Travel times of the new eigenarys form a cluster. Correspondingly, the ray pulse coming to the observation point through an unperturbed eigenray turns into what we will call the fuzzy ray pulse representing a superposition of pulses coming through the eigenrays belonging to the bundle. The widening of this pulse may be characterized by the quantity $\delta\tau_{\mathrm{mean}}$. However, the spread of the ray travel times may be considered only as an upper estimate of the pulse widening. The point is that the connection between the ray travel times and ray amplitudes remains, practically, unknown. Rays forming the cluster, generally, have quite different amplitudes and, correspondingly, give quite different contributions to a peak of the received signal associated with the cluster. A numerical example considered in Ref. [Beron-Vera *et al.* (2003)] demonstrates that in spite of the large number of eigenrays forming the cluster, a fuzzy ray pulse may be dominated by only a few pulses coming through eigenrays with maximum amplitudes. The spread of their travel times may be significantly less than the total width of the cluster.

We know from Sec. 1.4 that the delays between successive unperturbed ray pulses propagating along the steep rays usually are larger than for the flat ones. We also know that the steep rays are less chaotic and form fuzzy segments with smaller width than the flat ones. Therefore arrivals of fuzzy ray pulses coming through the steep rays are much better resolved than later arriving pulses propagating along the flat paths.

4.2.3.4 *Stability of the timefront and the Fermat's principle*

The smallness of $\Delta\tau_I$ that ensures stability of timefront segments can be qualitatively interpreted from the viewpoint of Fermat's (Hamilton's) principle. Indeed, present the eikonal in the form

$$S = S_0 - \int V dr, \tag{4.156}$$

where $S_0 = \int (pdz - H_0 dr)$. As we know, a contribution from the last term in the right hand side of Eq. (4.156) to the travel times of steep rays (which form the stable segments) is insignificant. Neglecting this term, we get

$$\Delta\tau_I = \frac{1}{c_0}[S_0(P) - S_0(U)] \tag{4.157}$$

where $S_0(P)$ and $S_0(U)$ are the values of S_0 evaluated over trajectories of perturbed (P) and unperturbed (U) eigenrays with the same identifier, respectively. According to Fermat's principle, the ray U in the unperturbed waveguide is a stationary path of the functional S_0. The absence of linear in ΔI terms in Eq. (4.137) and (4.142) is a consequence of this principle. Since ΔI is our small parameter, the absence of the first order term in ΔI gives a qualitative explanation to the smallness of $\Delta\tau_I$ and, hence, to the stability of the timefront segments.

A much more detailed understanding of the timefront stability in the presence of the sound speed fluctuations can be obtained for relatively short ranges where the travel time clusters do not yet appear [Godin (2007)].

4.2.3.5 *Travel time as a function of launch angle*

Figure 4.18 shows the ray travel time at 3000 km as a function of the launch angle χ in the unperturbed (smooth curve) and perturbed (points) waveguide. Similar distributions of ray arrivals have been observed in a waveguide with a periodic perturbation considered in Sec. 2.6.2 (cf. Fig. 2.23). As in Fig. 2.23, we see dark horizontal stripes located at times corresponding to the "shelves" on the unperturbed $t(\chi)$ curve (analogue of the $t(p_0)$ curve analyzed in Sec. 2.6.2). The stripes are formed by steep rays with launch angles exceeding 5°. The density of points depicting ray arrivals in the $t - \chi$ plane takes maximum values at the stripes. This effect has a qualitative explanation based on the following two properties of chaotic rays.

(i) Connection between the ray depth z at the given observation range r and the action and angle variables of the ray path at this range, I and θ, is determined by the function $z(I, \theta)$ present in the right-hand side of the second of Eqs. (1.49). The derivative $\partial z(I, \theta)/\partial\theta$ vanishes at (wrapped) θ equal to 0 (or 2π) and π, corresponding to the turning depths of a ray path with action I (see Eq. (1.59)). Since the wrapped angle variable θ at a megameter range r is uniformly distributed on $(0, 2\pi)$, this means that the most part of perturbed rays at long ranges have vertical coordinates z close to their turning depths.

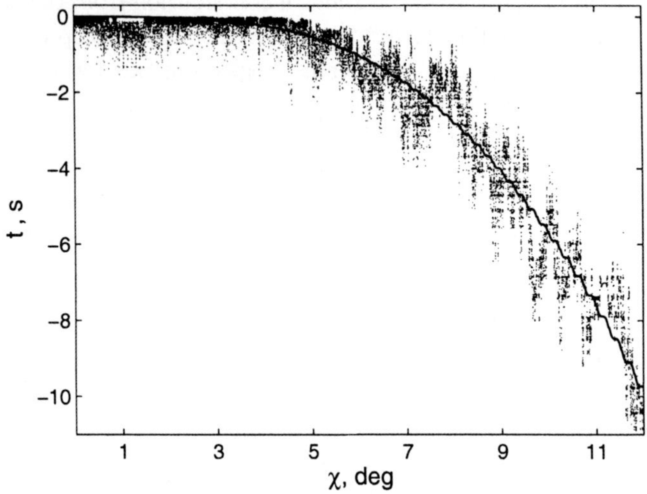

Fig. 4.18 Ray travel time versus launch angle in the unperturbed (solid line) and perturbed (points) waveguides at a range of 3000 km. Rays escape a point source set at the depth $z = 0.7$ km. This plot is constructed using results of the same ray tracings as the timefronts shown in Fig. 4.12.

(ii) Consider a perturbed ray and an unperturbed ray which are steep, connect the same points $(0, z_s)$ and (r, z), and have the same identifier. As we already know, even if $r = O(1000\text{km})$, travel times of these rays are close. Besides, it is shown in Ref. [Virovlyansky (2005a)] that the rms spread of the action along the trajectory of a steep chaotic ray is comparatively small (about $\sqrt{Br/3}$). This means, that not only the mean action variable of the perturbed path, but its final action, as well, should be close to the action of the unperturbed ray. Therefore, if the arrival depth z of the perturbed ray is close to its turning depth, then the same is true for the unperturbed ray, and vise versa.

Thus, at any given range of order 1000 km most rays have travel times close to the travel times of unperturbed rays whose coordinates z at this range are close to their turning depths. Therefore, the density of the travel time distribution has maxima at times corresponding to the "shelves" of the unperturbed $t(\chi)$ curve. This effect manifests itself in the presence of the horizontal dark stripes in Fig. 4.18.

4.3 Modal Structure of the Sound Field in a Waveguide with Random Inhomogeneities

The objective of this section is the derivation of an approximate analytical description of the modal structure in a deep ocean acoustic waveguide. Traditionally, this issue is addressed proceeding from the mode coupling equations [Kohler and Papanicolaou (1977); Dozier and Tappert (1978); Gorodetskaya *et al.* (1999); Voronovich and Ostashev (2006); Morozov and Colosi (2007)]. These equations are complicated and, therefore, usually they should be treated numerically. Here, an alternative approach is discussed. We argue that a simple (albeit rough) analytical description of the modal structure in a fluctuating ocean can be obtained by combining (i) relations (3.118) and (3.123) expressing the mode amplitude through solutions of ray equations and (ii) stochastic ray theory derived in Sec. 4.1.

Under conditions of ray chaos, the number of mode rays contributing to the given mode grows with range exponentially and rapidly becomes very large. Assuming that their phases are independent, we sum up these rays incoherently. The sum is estimated using the stochastic ray theory. As a result an approximate analytical expression for a coarse-grained distribution of acoustic energy between normal modes is obtained.

When considering a transient wave field, our attention is focused on the *mode pulses*, that is sound pulses carried by individual modes. It is shown that the temporal structure of the mode pulse is determined to a significant extent by the travel times of mode rays. The use of stochastic ray theory makes it possible to estimate the spread and bias of the mode pulse caused by internal waves. It turns out that for high modes the spread and bias grow with range like $r^{3/2}$ and r^2, respectively. First these range dependencies were found in Ref. [Colosi and Flatté (1996)] on the basis of numerical simulations. Later an analytical estimate for the spread was derived in Refs. [Virovlyansky *et al.* (2006); Virovlyansky (2006); Udovidchenkov and Brown (2008)]. In this section we follow Ref. [Virovlyansky *et al.* (2009)].

4.3.1 *Redistribution of acoustic energy between normal modes under conditions of ray chaos*

First consider a monochromatic wave field. Its modal representation given by Eq. (3.31) can be written as

$$u(r,z,\Omega)=\sum_m a_m(r,\Omega)\varphi_m(z,\Omega), \tag{4.158}$$

where we have emphasized the dependence of the eigenfuntions φ_m and their amplitudes a_m on the (angular) carrier frequency Ω. Here and in what follows, we consider a decomposition of the wave field into eigenfunctions of the unperturbed waveguide. As it has been shown in Sec. 3.2.4.1, the mode amplitude a_m may be presented as a superposition of contributions from the mode rays, that is, rays whose action variables at the observation range r satisfy the quantization rule (3.114). In what follows we will restrict our attention to modes whose turning points lie within the water bulk. Equation (3.114) can be presented in the form

$$I(r, p_0, z_0) = I_{m\Omega}, \tag{4.159}$$

where

$$I_{m\Omega} = \frac{m - 1/2}{\Omega} c_0, \; m = 1, 2, \ldots. \tag{4.160}$$

If the action of the ray path is expressed by the function $I(r, I_0, \theta_0)$, Eq. (4.159) translates to

$$I(r, I_0, \theta_0) = I_{m\Omega}. \tag{4.161}$$

Equations (4.159) and (4.161) in two equivalent forms give a condition that singles out mode rays contributing to the m-th mode at frequency Ω. It should be emphasized, that at different frequencies these equations singles out different rays. In this sense, rays contributing to the mode with the given number m are frequency dependent.

The right hand sides of Eqs. (4.159) and (4.161) formally depend on two starting parameters of the ray path (p_0 and z_0, or I_0 and θ_0). But these parameters are not independent. Their connection is determined by a particular source exciting the wave field. Therefore, for a given observation range r, we get an equation in one unknown parameter. For example, in the case of a point source all rays escape the same depth z_0 and Eq. (4.159) determines the starting momenta of mode rays.

In this subsection, we will consider a single mode source exciting the m_0-th mode of the unperturbed waveguide. So,

$$u(0, z, \Omega) = \varphi_{m_0}(z, \Omega). \tag{4.162}$$

As it was explained in Sec. 3.2.4.1, in this case all the rays have the same initial action $I_0 = I_{m_0\Omega}$. The condition which singles out mode rays follows from Eq. (4.161) as

$$I(r, I_{m_0\Omega}, \theta_0) = I_{m\Omega}. \tag{4.163}$$

It determines the values of θ_0 corresponding to the mode rays. A contribution from an individual mode ray to the amplitude of the m-th mode denote by δa_m. According to Eq. (3.123), one finds

$$|\delta a_m|^2 = \frac{1}{kD(I_{m_0\Omega})\,|p_0\partial I(r,p_0,z_0)/\partial z_0|}, \tag{4.164}$$

where p_0 and z_0 are determined by Eqs. (4.159) and $p_0 = \sqrt{2\left[H_0(I_{m_0\Omega}) - U(0,z_0)\right]}$. Using Eq. (4.28), the above expression can be rewritten in the form

$$|\delta a_m|^2 = \left.\frac{1}{2\pi k_0\,|\partial I(r,I_0,\theta_0)/\partial\theta_0|}\right|_{I_0=I_{m_0,\Omega},\theta_0=\tilde{\theta}_0}, \tag{4.165}$$

where $\tilde{\theta}_0$ is a solution of Eq. (4.163).

Let us consider squared mode amplitudes $|a_m(r,\Omega)|^2$ which we call the mode intensities. Our task will be to derive an analytical expression for a smoothed mode intensity

$$J_m(r,\Omega) = \sum_{m'} |a_{m'}(r,\Omega)|^2\, e^{-\frac{(m-m')^2}{2\Delta_M^2}} \Big/ \sum_{m'} e^{-\frac{(m-m')^2}{2\Delta_M^2}}, \tag{4.166}$$

where Δ_M is a smoothing scale. Under conditions of ray chaos, the number of mode rays contributing to a given mode at a long range becomes very large. Then it is natural to expect that a rough estimate of $J_m(r,\Omega)$ can be obtained by incoherent summation of these rays. Numerical results presented in Refs. [Virovlyansky and Zaslavsky (2000); Virovlyansky (2000)] support this expectation. Analytical evaluation of an incoherent sum of mode rays expressing $J_m(r,\Omega)$ can be performed on the basis of the approach from Sec. 5.1 that allows one to replace the summation of rays by the statistical averaging.

For a single mode source determined by Eq. (4.162), an incoherent sum of mode rays representing the mode intensity in Eq. (4.165) can be presented in the form

$$|a_m(r,\Omega)|^2 = \frac{1}{2\pi k_0}\sum_j \frac{1}{|\partial I(r,I_{m_0\Omega},\theta_0)/\partial\theta_0|_{\theta_0=\theta_{0,j}}}$$
$$= \frac{1}{2\pi k_0}\int_0^{2\pi} d\theta_0\;\delta\left(I_{m\Omega} - I\left(r,I_{m_0\Omega},\theta_0\right)\right), \tag{4.167}$$

where the index j in the middle expression numbers the contributing mode rays. The integral in the last expression has the form of Eq. (4.55) with $F = \text{const}\ \delta(I_{m\Omega} - I)$ and the integration over θ_0 instead of p_0. It can be treated in the same manner as the integral on the right of Eq. (4.55). Dividing the interval of integration into small subintervals, we note that the integral over each subinterval once again has the form similar to that of Eqs. (4.1), (4.6), and (4.7). For a subinterval $(\theta_0, \theta_0 + \delta\theta)$ we have

$$\frac{1}{\delta\theta} \int_{\theta_0}^{\theta_0+\delta\theta} d\theta_0\ \delta\left(I_{m\Omega} - I\left(r, I_{m_0\Omega}, \theta_0\right)\right) = P_{I|I_0}(I_{m\Omega}, r|I_{m_0\Omega}). \tag{4.168}$$

Since the right hand side does not depend on θ_0 , we get

$$|a_m(r,\Omega)|^2 = \frac{1}{k_0} P_{I|I_0}(r, I_{m\Omega}|I_{m_0\Omega}). \tag{4.169}$$

According to Eq. (3.34), the summation over m in Eq. (4.166) may be approximately replaced by an integration over the action variable. Assuming $m \gg \Delta_M$, we can formally extend the integration over m' to the infinite limits. Then

$$\sum_{m'} e^{-\frac{(m-m')^2}{2\Delta_M^2}} = \int_{-\infty}^{\infty} dm' e^{-\frac{(m-m')^2}{2\Delta_M^2}} = k_0 \int_{-\infty}^{\infty} dI'\ e^{-\frac{k^2(I_{m\Omega}-I')^2}{2\Delta_M^2}} = \sqrt{2\pi}\Delta_M. \tag{4.170}$$

Substituting this in Eq. (4.166) and replacing the summation over m by integration over I, as in Eq. (4.170), we find

$$J_m(r,\Omega) = \frac{1}{\sqrt{2\pi}\Delta_M} \int_{-\infty}^{\infty} dI'\ P_{I|I_0}(r, I'|I_{m_0\Omega}) \exp\left[-\frac{k^2\left(I_{m\Omega} - I'\right)^2}{2\Delta_M^2}\right]. \tag{4.171}$$

Substituting Eq. (4.43) yields

$$J_m(r,\Omega) = \frac{1}{\sqrt{2\pi\left(\Delta_M^2 + k^2 Br\right)}} \exp\left[-\frac{\left(m - m_0\right)^2}{2\left(\Delta_M^2 + k^2 Br\right)}\right]. \tag{4.172}$$

This result is valid only for high modes. At a carrier frequency of 75 Hz the condition (4.45) in our environmental model is met only for mode rays corresponding to $m > 7$. When treating low modes one should take Eq. (4.171) with a finite lower limit of integration and use Eq. (4.38) which gives an expression for $P_{I|I_0}(r, I|I_0)$ valid for both steep and flat rays. Note that according to Eq. (4.172) at very long ranges the number of effectively excited modes grows like $r^{1/2}$.

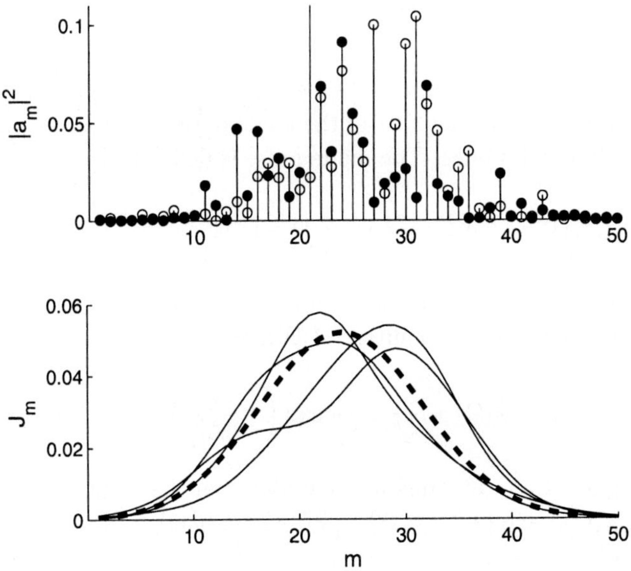

Fig. 4.19 Mode intensities at range 3000 km for a single mode source exciting the 24-th mode ($u(0, z) = \varphi_{24}(z)$) at a carrier frequency of 75 Hz. *Upper panel.* Open and solid circles show mode intensities for two realizations of perturbation (1.17). *Lower panel.* Smoothed mode intensities computed for four realizations of the perturbation (thin solid lines) and prediction of Eq. (4.172) (heavy dashed line). The smoothing scale is $\Delta_M = 4$ [Virovlyansky *et al.* (2009)].

To check the validity of Eq. (4.172) we have solved (numerically) parabolic equation (1.17) at a carrier frequency of 75 Hz with an initial condition $u(0, z) = \varphi_{24}(z)$. Mode intensities at range $r = 3000$ km are shown in the upper panel of Fig. 4.19 for two realizations of random perturbation δc. The lower panel presents results for the smoothed mode intensities evaluated with a smoothing scale $\Delta_M = 4$. Thin solid lines graph J_m for four realizations of perturbation (values of $|a_m|^2$ for two of them are presented in the upper panel). The heavy dashed line shows the prediction of Eq. (4.172). It is seen that the smoothed mode intensity weakly depends on a particular realization of perturbation, and our analytical estimate is in a reasonable agreement with simulations.

In Ref. [Virovlyansky (2006)] it is shown that the wave field excited by a point source can be treated in a similar way.

4.3.2 *Transient wave field*

Turn our attention to studying the modal structure of a narrow-band pulse signal radiated by a point source set at a depth $z_{\rm s}$. Our task is to derive a ray-based description of what we call the mode pulses, that is, sound pulses carried by individual modes. The mode pulse is defined as

$$\psi_m(t,r,\Omega) = \int d\Omega' a_m(r,\Omega')w(\Omega-\Omega')e^{i\Omega'(r/c_0-t)}, \tag{4.173}$$

where $a_m(r,\Omega')$ are mode amplitudes of the wave field satisfying parabolic equation (1.17) with initial condition (1.20), and

$$w(\Omega) = \frac{1}{\sqrt{2\pi}\Delta_\Omega}\exp\left(-\frac{\Omega^2}{2\Delta_\Omega^2}\right) \tag{4.174}$$

is a weighting function. This definition implies that the point source emits a sound pulse $g(t) = \exp\left[-i\Omega t - \Delta_\Omega^2 t^2/2\right]$, whose bandwidth and length are estimated to be $\delta\Omega = \sqrt{2\pi}\Delta_\Omega$ and $\tau_{\rm s} = 2\pi/\delta\Omega = \sqrt{2\pi}/\Delta_\Omega$, respectively. It is assumed that the radiated wave field is registered on a long vertical array and at each frequency is decomposed into a sum of normal modes [Wage *et al.* (2003)]. Then the mode pulses are synthesized in accord with Eq. (4.173). Argument Ω of the function ψ_m indicates the central frequency of an emitted signal. Each mode pulse is a wave packet spreading with range.

4.3.2.1 *Arrival time of an instantaneous frequency*

As in the case of a monochromatic source, we will proceed from the ray-based representation of the mode amplitude described in Sec. 3.2.4. Take a mode ray contributing to the m-th mode at a central frequency of an emitted pulse Ω. Its starting momentum – denote it by p_0^* – satisfies the condition (4.159) $I(r,p_0^*,z_{\rm s}) = I_{m\Omega}$. It is natural to assume that there exists a bundle of rays with starting momenta from a small interval centered at p_0^* which are mode rays for the m-th mode at frequencies close to Ω: for a ray escaping the source with momentum $p_0^*+\delta p$ there exists a frequency Ω' such that $I(r,p_0^*+\delta p,z_{\rm s}) = I_{m\Omega'}$. The contribution from this bundle to the mode pulse – denote it by $\delta\psi_m(r,t)$ – is given by insertion of Eq. (3.118) into Eq. (4.173). This yields

$$\delta\psi_m(r,t) = \int d\Omega' w\left(\Omega-\Omega'\right)A\exp\left\{i\Phi - i\Omega' t + i\tilde{\beta}\right\}, \tag{4.175}$$

where

$$A = \frac{1}{\sqrt{2\pi \left|\partial I / \partial p_0\right|}}, \tag{4.176}$$

$$\Phi = k\left(S + \sigma g(I_{m,\Omega'}, Z_{m\Omega'})\right), \tag{4.177}$$

$Z_{m\Omega'}$ is the arrival depth of the mode ray, and $\tilde{\beta}$ is a frequency independent quantity. The derivative $\partial I/\partial p_0$ and eikonal S should be evaluated at the end point of the mode ray, and, therefore, they are functions of m and Ω' (see remark after Eq. (4.161)). At long ranges, the phase Φ rapidly varies with Ω' and the spreading of the wave packet $\delta\psi_m(r,t)$ can be investigated using the stationary phase technique. The stationary point is determined by the condition

$$t = \partial\Phi/\partial\Omega'. \tag{4.178}$$

Frequency Ω' satisfying this equation is interpreted as an instantaneous frequency of the wave packet at time t. For a given Ω', Eq. (4.178) predicts the arrival time of a constituent of the mode pulse with this instantaneous frequency.

Since $\tilde{\beta}$ does not depend on frequency, we find from Eqs. (4.177) and (3.34)

$$\frac{\partial\Phi}{\partial\Omega'} = \frac{1}{c_0}\left[r + S\left(r, Z_{m\Omega'}\right)\right] + t_1 + t_2, \tag{4.179}$$

where

$$t_1 = \frac{1}{c_0}\left[\frac{\partial S(r,z)}{\partial z} + \sigma\frac{\partial g(z,I)}{\partial z}\right]_{z=Z_{m\Omega'},I=I_{m\Omega'}} \frac{\partial Z_{m\Omega'}}{\partial\Omega'}, \tag{4.180}$$

and

$$t_2 = \frac{\sigma}{c_0}\left[g(Z_{m\Omega'}, I_{m\Omega'}) - I_{m\Omega'}\left.\frac{\partial g(Z_{m\Omega'}, I)}{\partial I}\right|_{I=I_{m\Omega'}}\right]. \tag{4.181}$$

It follows from Eqs. (3.111) and (3.112), that the term t_1 vanishes because the quantity in square brackets in Eq. (4.180) is identically zero at each frequency Ω'. According to Eqs. (1.43), (1.46), and (3.34), $g(z,I)$ is a monotonic function of I and $g(Z_{m\Omega'}, I_{m\Omega'})$ is always less than $mT'c_0/2$, where $T' = 2\pi/\Omega'$. Therefore, the magnitude of t_2 is, typically, a few time less than $mT'/2$. We will see that this is significantly less than the mode pulse length. Therefore, the arrival time of a fragment of the wave packet $\delta\psi_m(r,t)$ with an instantaneous frequency Ω' can be estimated as

$$t(r,\Omega',m) = \frac{1}{c_0}\left[r + S\left(r, Z_{m\Omega'}\right)\right]. \tag{4.182}$$

We will call this quantity the arrival time of an instantaneous frequency Ω'. It is the travel time of a sound pulse through the mode ray corresponding to given r, m, and Ω'.

Two comments concerning this result are noteworthy. First, a geometrical ray may be the mode ray for more than one mode. This occurs if its action I at the observation range satisfies the condition $I = I_{m_1\Omega_1} = I_{m_2\Omega_2}$, where $m_1 \neq m_2$ and both Ω_1 and Ω_2 belong to the frequency band of a radiated signal. Second, generally, at the observation range, there are several rays with equal travel times and different actions. It means that, typically, the instantaneous frequency of the mode pulse cannot be defined: at any moment t, the pulse is a superposition of signals with different instantaneous frequencies.

4.3.2.2 *Mode pulses in a range-independent waveguide*

In the unperturbed (range-independent) waveguide the action variable remains constant along the ray path, so the function $I(r,p_0,z_s)$ does not depend on r. Since I is a monotonically increasing function of $|p_0|$, the condition (4.159) with $z_0 = z_s$ at any range is met for $p_0 = \pm\sqrt{2[H_0(I_{m\Omega'}) - U(z_s)]}$. This means that for any mode m at each frequency Ω' there are two mode rays. Their travel times are close and can be estimated using the approximate formula (1.64). Since we compare ray travel times at the same observation distance, an argument r of the function $\bar{t}(r,I)$, for short, will be omitted. Thus, the arrival time of an instantaneous frequency Ω' may be estimated as

$$\bar{t}(I_{m\Omega'}) = \left[1 + \omega(I_{m\Omega'})I_{m\Omega'} - H(I_{m\Omega'})\right] r/c_0. \tag{4.183}$$

This expression immediately gives the known WKB formula for the group slowness $W_{m\Omega'} = \bar{t}(I_{m\Omega'})/r$ [Munk and Wunsch (1983); Brown *et al.* (1996)].

The quantity $\bar{t}\,(I_{m\Omega})$, with Ω being a central frequency of the emitted signal, gives an estimate for the arrival time of the m-th mode pulse. At short ranges, the duration of this pulse, $\Delta t_{m,\Omega}$, is close to that of the initially emitted signal, τ_s. The quantity $\Delta t_{m,\Omega}$ grows with range because fragments of the mode pulse with different instantaneous frequencies propagate with different velocities. An increase of the mode pulse length due to this mechanism is

$$\tau_{m,\Omega} = \left| \frac{\partial \bar{t}(I_{m\Omega'})}{\partial \Omega'} \right|_{\Omega'=\Omega} \delta\Omega = I_{m\Omega} \left| \gamma(I_{m\Omega}) \right| r \frac{\delta\Omega}{\Omega}, \tag{4.184}$$

where

$$\gamma(I) \equiv \frac{1}{r} \frac{\bar{t}(I)}{dI} = \frac{I}{c_0} \frac{d\omega(I)}{dI}. \tag{4.185}$$

There is a critical distance

$$r_{m,\Omega} = \frac{\tau_s^2 \Omega}{2\pi I_{m\Omega} \left| \gamma(I_{m\Omega}) \right|}, \tag{4.186}$$

where $\tau_{m,\Omega} = \tau_s$. At ranges $r < r_{m,\Omega}$ or $r > r_{m,\Omega}$, the duration of the mode pulse $\Delta t_{m,\Omega}$ can be roughly estimated as τ_s or $\tau_{m,\Omega}$, respectively.

To illustrate the application of the above formulas let us derive a condition of nonoverlapping of mode pulses in the unperturbed waveguide. It follows from Eqs. (4.183) and (3.34) that the difference between arrival times of two consecutive mode pulses at a given observation range is

$$\bar{t}\left(I_{m+1,\Omega}\right) - \bar{t}\left(I_{m\Omega}\right) = \frac{\partial \bar{t}\left(I_{m\Omega}\right)}{\partial m} = \frac{1}{m} \gamma(I_{m\Omega}) I_{m\Omega} r. \tag{4.187}$$

Here and in all the subsequent expressions derived using Eq. (3.34), for short, we replace $m - 1/2$ by m.

At range $r > r_{m,\Omega}$, the m-th mode pulse does not overlap with neighboring pulses if

$$\frac{1}{\tau_{m,\Omega}} \left| \frac{\partial \bar{t}\left(I_{m\Omega}\right)}{\partial m} \right| = \frac{\Omega}{m\delta\Omega} > 1. \tag{4.188}$$

This condition can be rewritten in the form

$$\tau_s > mT, \tag{4.189}$$

where $T = 2\pi/\Omega$ is a period of the central frequency. This criterion was derived in Refs. [Virovlyanskii (1985a)] and [Brown *et al.* (1996)]. It is remarkable that Eq. (4.189) does not depend on the parameters of a waveguide. However, a cautionary remark should be made. The necessary condition of nonoverlapping (4.189) becomes sufficient only at distances exceeding some critical range which does depend on the waveguide and grows with increasing τ_s.

It is interesting and instructive to compare Eq. (4.189) with the condition of nonoverlapping of ray pulses. Take two such pulses arriving at the observation point through paths with actions close to some given value I_0. Then according to Eq. (1.110), a necessary condition of nonoverlapping of the pulses is $\tau_s < 2\pi I_0/c_0$. If $I_0 \simeq I_{m\Omega}$, i.e., the ray pulses have travel times close to the travel time of the m-th mode pulse, this condition translates to

$$\tau_s < mT. \tag{4.190}$$

The conditions of nonoverlapping of ray and mode pulses, Eqs. (4.189) and (4.190), are mutually exclusive because the mode pulse nonoverlapping with its neighbors is formed by a superposition of overlapping ray pulses and vice versa [Munk and Wunsch (1983)].

In conclusion of this subsection note a simple rule of thumb that follows from Eq. (4.189). If an emitted signal includes n periods of a carrier frequency (with a smooth envelope function), then no more than first n mode pulses may be nonoverlapping at any distance in any waveguide.

4.3.2.3 *Mode pulses in the presence of internal waves*

In the presence of internal waves, the above results obtained for the unperturbed waveguide remain valid only at short enough ranges. At long ranges where the chaos is well developed, the mode pulse is formed by many mode rays. We will sum up their contributions incoherently. Our task will be to derive an approximate analytical description of a smoothed mode pulse defined as

$$Y_m(r,\Omega,t) = \frac{1}{\sqrt{2\pi}\Delta_t} \int dt' \, |\psi_m(r,\Omega,t')|^2 \exp\left[-\frac{(t-t')^2}{2\Delta_t^2}\right], \tag{4.191}$$

where Δ_t is a smoothing scale. Substituting Eqs. (4.173) and (4.174) into (4.191), we get

$$Y_m(r,\Omega,t) = \frac{1}{2\pi\Delta_\Omega^2} \int d\Omega_1 d\Omega_2 \; a_m(r,\Omega_1+\Omega_2/2) a_m^*(r,\Omega_1-\Omega_2/2)$$
$$\times \exp\left[-i\Omega_r t - \frac{(\Omega-\Omega_1)^2}{\Delta_\Omega^2} - \frac{1}{2}\left(\frac{1}{2\Delta_\Omega^2} + \Delta_t^2\right)\Omega_2^2\right]. \tag{4.192}$$

Once again, present the mode amplitude $a_m(r,\Omega)$ as a sum of contributions from mode rays expressed by Eq. (3.118). An interval of integration over

Ω_2 is of order $\max(\Delta_\Omega, \Delta_t^{-1})$. Assuming that this interval is sufficiently small, we will use an approximation

$$a_m(r, \Omega_1 + \Omega_2/2) a_m^*(r, \Omega_1 - \Omega_2/2)$$
$$= \frac{1}{2\pi |\partial I(r, p_0, z_s)/\partial p_0|_{p_0 = p_{0\Omega_1}}} e^{i\Omega_2 \partial\Phi(\Omega_1)/\partial\Omega_1}, \qquad (4.193)$$

where the symbol $p_{0\Omega_1}$ denotes the starting momentum of a mode ray contributing to the m-th mode at the frequency Ω_1. The action variable of this ray at the range of observation, according to Eq. (3.34) (see the note after Eq. (4.187)), is

$$I = mc_0/\Omega_1. \qquad (4.194)$$

Equation (4.193) is our main approximation. It implies that (i) the mode pulse is a superposition of (practically independent) pulses associated with bundles of mode rays and (ii) contributions from these pulses are summed up incoherently.

Substituting Eq. (4.193) into (4.192) and integrating over Ω_2 yields

$$Y_m(r, \Omega, t) = \frac{1}{(2\pi)^{3/2} \Delta_\Omega \sqrt{1/2 + \Delta_\Omega^2 \Delta_t^2}} \int \frac{d\Omega_1}{|\partial I/\partial p_0|_{p_0 = p_{0\Omega_1}}}$$
$$\times \exp\left[-\frac{[t - t(r, \Omega_1, m)]^2}{2\Delta_t^2 + \Delta_\Omega^{-2}} - \frac{(\Omega - \Omega_1)^2}{\Delta_\Omega^2}\right]. \qquad (4.195)$$

It follows from Eq. (4.194) that $d\Omega_1 = -mc_0 I^{-2} dI$. Using this relation, we can change the variable of integration in Eq. (4.195) from Ω_1 to p_0. Then

$$Y_m(r, \Omega, t) = \frac{c_0 m}{(2\pi)^{3/2} \Delta_\Omega \sqrt{1/2 + \Delta_\Omega^2 \Delta_t^2}} \int \frac{dp_0}{I^2(r, p_0, z_s)}$$
$$\times \exp\left[-\frac{[t - t(r, p_0, z_s)]^2}{2\Delta_t^2 + \Delta_\Omega^{-2}} - \frac{[\Omega - mc_0/I(r, p_0, z_s)]^2}{\Delta_\Omega^2}\right]. \qquad (4.196)$$

Equation (4.196) presents the smoothed mode pulse as an integral over all paths escaping the point source. This integral has the form of Eq. (4.55) and, therefore, it can be approximately evaluated by means of Eq. (4.58). In order to estimate $t(r, p_0, z_s)$ we will exploit Eq. (4.139) and retain in the right hand side of this equation only the dominant term τ_N. Making use of

Eq. (4.146) yields

$$t(r, p_0, z_s) = \bar{t}(I_0) + y(r)I_0/c_0, \tag{4.197}$$

where I_0 is the starting action of ray escaping the point source with initial momentum p_0. Then Eq. (4.196) may be rewritten in the form

$$Y_m(r,\Omega,t) = \frac{c_0 m}{(2\pi)^{3/2}\Delta_\Omega\sqrt{1/2+\Delta_\Omega^2\Delta_t^2}} \int \frac{dp_0\,dx\,dy}{(I_0+x)^2} P_{x,y|I_0}\left(x, \frac{y}{\omega_I'(I_0)}, r|I_0\right)$$
$$\times \exp\left[-\frac{\left[t-\bar{t}(I_0)-yI_0/c_0\right]^2}{2\Delta_t^2+\Delta_\Omega^{-2}} - \frac{\left(\Omega-\frac{mc_0}{I_0+x}\right)^2}{\Delta_\Omega^2}\right]. \tag{4.198}$$

To simplify this expression, let us consider the function

$$\Psi(I) = \exp\left[-\frac{(\Omega - mc_0/I)^2}{\Delta_\Omega^2}\right] \tag{4.199}$$

and expand its argument in powers about the point $I = I_{m\Omega} = mc_0/\Omega$. This yields

$$\Psi(I) = \exp\left\{-\frac{m^2c_0^2}{I_{m\Omega}^4\Delta_\Omega^2}\Delta I^2 + \frac{2m^2c_0^2}{I_{m\Omega}^5\Delta_\Omega^2}\Delta I^3 + O(\Delta I^4)\right\}, \tag{4.200}$$

where $\Delta I = I - I_{m\Omega}$. If only the first term in the argument of the exponential is retained, then $\Psi(I)$ is of order unity for $|\Delta I| \simeq \Delta I_{\rm cr}$, where

$$\Delta I_{\rm cr} = \frac{I_{m\Omega}^2\Delta_\Omega}{mc_0} = \frac{mc_0\Delta_\Omega}{\Omega^2}. \tag{4.201}$$

Substituting $\Delta I_{\rm cr}$ for ΔI into the second term in the exponential, we conclude that for our narrowband pulses satisfying the condition

$$\Delta_\Omega \ll \Omega, \tag{4.202}$$

this term is small compared to unity, and, therefore, it can be neglected. The same is true for all the terms of higher order. Therefore, when treating narrowband signals, we can retain only the first term in the argument of the exponential in Eq. (4.200).

A further simplification of $\Psi(I)$ is possible if

$$\frac{I_{m\Omega}^2\Delta_\Omega}{mc_0} = \frac{mc_0\Delta_\Omega}{\Omega^2} \ll \sqrt{Br}. \tag{4.203}$$

Then an effective interval of integration over x in Eq. (4.198) is small compared with the corresponding scale of the function $P_{x,y|I_0}(x, y, r|I_0)$, and $\Psi(I)$ can be approximated by the delta-function

$$\Psi(I) = \frac{\sqrt{\pi} I_{m\Omega} \Delta_\Omega}{\Omega} \delta(I - I_{m\Omega}). \tag{4.204}$$

In this approximation Eq. (4.198) reduces to

$$Y_m(r, \Omega, t) = \frac{1}{2\pi\sqrt{1 + 2\Delta_\Omega^2 \Delta_t^2}} \int dp_0 dy P_{x,y|I_0}\left(I_{m\Omega} - I_0, \frac{y}{\omega_I'(I_0)}, r|I_0 \right)$$
$$\times \exp\left[-\frac{\left[t - \bar{t}(I_0) - yI_0/c_0\right]^2}{2\Delta_t^2 + \Delta_\Omega^{-2}} \right]. \tag{4.205}$$

Substituting Eq. (4.51) and evaluating a Gaussian integral over η yields

$$Y_m(r, \Omega, t) = \frac{1}{(2\pi)^{3/2} \Delta_\Omega \sqrt{Br}} \int \frac{dp_0}{\sqrt{2\Delta_t^2 + \Delta_\Omega^{-2} + Br^3\gamma^2(I_0)/6}}$$
$$\times \exp\left\{ -\frac{\left[t - \bar{t}(I_0) - \gamma(I_0) r \left(I_{m\Omega} - I_0\right)/2\right]^2}{2\Delta_t^2 + \Delta_\Omega^{-2} + \gamma^2(I_0) Br^3/6} - \frac{\left(I_{m\Omega} - I_0\right)^2}{2Br} \right\}. \tag{4.206}$$

Since Eq. (4.206) is derived using Eq. (4.51) at frequencies of about 75 Hz, it is valid (like Eq. (4.172)) only for $m > 7$.

Thus, we have two approximate formulas for the smoothed mode pulse given by Eqs. (4.196) and (4.206). Equation (4.196) expresses $Y_m(r, \Omega, t)$ through solutions of the ray equations (ray paths) and it depends on a particular realization of perturbation. In contrast, Eq. (4.206) is an analytical estimate independent of a particular realization of δc. In Fig. 4.20 predictions of Eqs. (4.196) and (4.206) are compared to results of simulations performed by solving (numerically) the parabolic equation (1.17). A point source has been set at the sound channel axis $z = z_a$. The simulations have a center frequency of 75 Hz, and a smoothing scale $\Delta_t = 0.1$ s. The envelope of an emitted signal is determined by Eq. (4.174) with $\Delta_\Omega/(2\pi) = 2$ Hz. The effective bandwidth of the emitted pulse is about 5 Hz. Smoothed mode pulses for $m = 11$, 21, 31, and 41 at the observation range 3000 km are shown by thin solid lines. In order to apply Eq. (4.196) we have traced numerically 50000 rays leaving the point source with initial momenta p_0 uniformly filling an interval corresponding to the launch angles $\pm 14°$. Replacing the integral over p_0 in Eq. (4.196) with a sum over

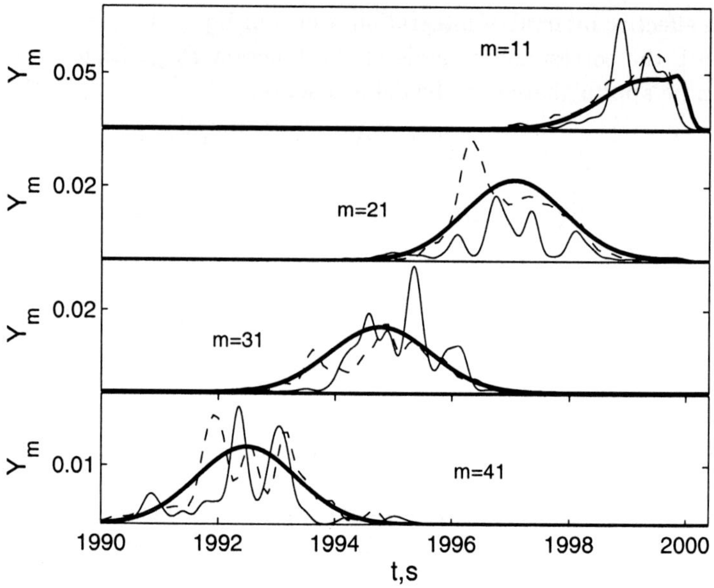

Fig. 4.20 The smoothed mode pulses at 3000 km for $m = 11$, 21, 31, and 41. Parabolic equation based simulations (thin solid lines) are compared to predictions of Eqs. (4.196) (dashed lines) and (4.206) (heavy solid lines) [Virovlyansky *et al.* (2009)].

the computed ray paths yields pulses shown in Fig. 4.20 by dashed lines. Predictions of Eq. (4.206) in Fig. 4.20 are presented by thick solid lines. It is seen that both our approximate formulas are too rough and cannot give a detailed description of the smoothed mode pulse. Nevertheless, they are able to predict the spread of the pulse and its bias caused by the random perturbation. In the following subsection we will consider this issue.

4.3.2.4 *Bias and spread of a mode pulse*

Introduce the symbol $\langle \ldots \rangle_{m\Omega}$ which denotes the averaging over the smoothed mode pulse:

$$\langle g(t) \rangle_{m\Omega} \equiv \int dt\, g(t) Y_m(r, \Omega, t) \Big/ \int dt\, Y_m(r, \Omega, t). \tag{4.207}$$

Our task is to estimate the bias of the mean travel time of the m-th mode pulse relative $\bar{t}\,(I_{m\Omega})$, arrival time of this pulse in the unperturbed waveguide

$$\delta t_{m\Omega} \equiv \langle t - \bar{t}\,(I_{m\Omega}) \rangle_{m\Omega} = \langle t \rangle_{m\Omega} - \bar{t}\,(I_{m\Omega})\,, \tag{4.208}$$

and the spread $\tau_{m\Omega}$ defined as

$$\tau_{m\Omega}^2 \equiv \left\langle \left[t - \bar{t}\,(I_{m\Omega})\right]^2 \right\rangle_{m\Omega} = \left\langle t^2 \right\rangle_{m\Omega} - \bar{t}^2\,(I_{m\Omega})\,. \tag{4.209}$$

The quantities $\langle t \rangle_{m\Omega}$ and $\left\langle t^2 \right\rangle_{m\Omega}$ can be evaluated in two ways: using Eq. (4.196) or (4.206).

Ray tracing. We find from Eq. (4.196)

$$\langle t \rangle_{m\Omega} = \int \frac{dp_0}{I^2(r, p_0, z_s)} t(r, p_0, z_s) \exp\left[-\frac{\left[\Omega - mc_0/I(r, p_0, z_s)\right]^2}{\Delta_\Omega^2}\right] \Bigg/ \Xi_{m,\Omega}, \tag{4.210}$$

and

$$\left\langle t^2 \right\rangle_{m\Omega} = \int \frac{dp_0}{I^2(r, p_0, z_s)} \left[t^2(r, p_0, z_s) + \Delta_t^2 + \Delta_\Omega^{-2}/2\right]$$
$$\times \exp\left[-\frac{\left(\Omega - mc_0/I((r, p_0, z_s))\right)^2}{\Delta_\Omega^2}\right] \Bigg/ \Xi_m, \tag{4.211}$$

where

$$\Xi_{m,\Omega} = \int \frac{dp_0}{I^2(r, p_0, z_s)} \exp\left[-\frac{\left[\Omega - mc_0/I^2(r, p_0, z_s)\right]^2}{\Delta_\Omega^2}\right]. \tag{4.212}$$

Equations (4.210)–(4.212) allow one to calculate the bias and spread of the mode pulse on the basis of numerical ray tracing.

Analytical estimates. They follows from (4.206) as

$$\langle t \rangle_{m\Omega} = \int dp_0\ \left[\bar{t}(I_0) + \gamma(I_0)\,(I_m - I_0)\,r/2\right]$$
$$\times \exp\left[-\frac{\left(I_m - I_0\right)^2}{2Br}\right] \Bigg/ K_m, \tag{4.213}$$

and

$$\left\langle t^2 \right\rangle_{m\Omega} = \int dp_0 \left\{\left[\bar{t}(I_0) + \gamma(I_0)(I_m - I_0)r/2\right]^2 + \Delta_t^2 + \Delta_\Omega^{-2}/2 + \gamma^2(I_0)Br^3/12\right\}$$
$$\times \exp\left[-\frac{\left(I_m - I_0\right)^2}{2Br}\right] \Bigg/ K_m, \tag{4.214}$$

where

$$K_m = \int dp_0 \exp\left[-\frac{\left(I_m - I_0\right)^2}{2Br}\right]. \tag{4.215}$$

Simplified analytical estimates. In order to simplify Eqs. (4.213) and (4.214), change first the integration variable from p_0 to I_0. Then dp_0 is replaced by $q(I_0)dI_0$, where

$$q(I_0) = \frac{dp_0}{dI_0}. \tag{4.216}$$

Due to the exponential present in Eqs. (4.213)–(4.215), the main contributions to these integrals come from I_0 located within a small interval of order $\sqrt{Br}$ centered at I_m. Equations (4.213)–(4.215) can be simplified if variations of $q(I_0)$ and $\gamma(I_0)$ at this interval are negligible, that is, if

$$\left|\sqrt{Br}q'(I_0)/q(I_0)\right| \ll 1, \left|\sqrt{Br}\gamma'(I_0)/\gamma(I_0)\right| \ll 1. \tag{4.217}$$

Then, we get a simple expression for K_m:

$$K_m = \int dI_0 \; q(I_0) \exp\left[-\frac{(I_{m\Omega} - I_0)^2}{2Br}\right] = \sqrt{2\pi Br} q\left(I_{m\Omega}\right). \tag{4.218}$$

It follows from Eqs. (4.208) and (4.213):

$$\delta t_{m\Omega} = \int dI_0 \; q(I_0) \left[\bar{t}(I_0) + \gamma(I_0)\left(I_m - I_0\right) r/2 - \bar{t}(I_{m\Omega})\right]$$

$$\times \exp\left[-\frac{(I_{m\Omega} - I_0)^2}{2Br}\right] \Big/ K_{m\Omega}. \tag{4.219}$$

Using the condition (4.217), we may expand the preexponential term in the above integral in a power series about $I_{m\Omega}$. Using the expansions

$$\bar{t}(I_0) + \frac{r}{2}\gamma(I_0)\left(I_m - I_0\right) = \bar{t}(I_{m\Omega}) + \frac{r}{2}\gamma(I_{m\Omega})\Delta I + O\left(\Delta I^3\right), \tag{4.220}$$

and

$$q(I_0) = q(I_{m\Omega}) + q'(I_{m\Omega})\Delta I + O\left(\Delta I^2\right), \tag{4.221}$$

where $\Delta I = I_0 - I_{m\Omega}$, and neglecting terms $O(\Delta I^3)$, we get

$$\delta t_{m\Omega} = \frac{q'(I_m)\gamma(I_m)B}{2q(I_m)} r^2. \tag{4.222}$$

Now turn our attention to the pulse spread. Substitute Eq. (4.220) into Eq. (4.214), change the integration variable to I_0, and approximately replace $q(I_0)$ and $\gamma(I_0)$ by $q(I_{m\Omega})$ and $\gamma(I_{m\Omega})$, respectively. This yields

$$\langle t^2 \rangle_{m\Omega} = \bar{t}^2(I_{m\Omega}) + \Delta_t^2 + \Delta_\Omega^{-2}/2 + \frac{\gamma^2(I_{m\Omega})Br^3}{3}. \tag{4.223}$$

Correspondingly,

$$\tau_{m\Omega}^2 = \Delta_t^2 + \Delta_\Omega^{-2}/2 + \frac{\gamma^2(I_{m\Omega})Br^3}{3}. \tag{4.224}$$

Due to restrictions imposed by the condition (4.217), the estimates (4.222) and (4.224) in our environmental model at the frequency 75 Hz can be used only for modes with $m > 14$. Note also that at the range 3000 km, the last term in the right hand side of Eq. (4.224) strongly dominates. Therefore, this equation can be further simplified to

$$\tau_{m\Omega} = \frac{|\gamma(I_{m\Omega})|\, B^{1/2} r^{3/2}}{\sqrt{3}}. \tag{4.225}$$

In a slightly different way, this result was derived in Refs. [Virovlyansky *et al.* (2006); Virovlyansky (2006)]. An alternative derivation was given in Ref. [Udovidchenkov and Brown (2008)].

Numerical example. To check the applicability of the above results we have performed the parabolic equation simulation of 100 sound pulses with equal envelope functions $w(\Omega)$ but different central frequencies uniformly filling an interval of 66 to 84 Hz. The bandwidth of each pulse is determined by the same constant $\Delta_\Omega/(2\pi) = 2$ Hz as in Fig. 4.20. This simulation has been performed for four realizations of the perturbation δc. Then we have evaluated the biases, $\delta t_{m\Omega}$, and spreads, $\tau_{m\Omega}$, at the range 3000 km for $m = 1, \ldots, 50$, and for 100 values of Ω corresponding to the above central frequencies. Values of $\delta t_{m,\Omega}$ and $\tau_{m,\Omega}$ averaged over the 100 frequencies we will call the frequency averaged bias and spread and denote by δt_m and τ_m, respectively.

Estimates of $\delta t_{m\Omega}$ and $\tau_{m\Omega}$ for the same m and Ω have been obtained on the basis of ray tracing. In each of the four realizations of our random waveguide, 50000 rays with the same starting parameters as indicated in Sec. (4.3.2.3) have been traced numerically. Then estimates of $\delta t_{m\Omega}$, $\tau_{m\Omega}$, δt_m, and τ_m have been calculated using Eqs. (4.208), (4.210), and (4.212).

Thin solid lines in Fig. 4.21(a) and (b) show the frequency averaged bias δt_m obtained using the parabolic equation simulation and numerical ray tracing, respectively, for four realizations of δc. A heavy solid line, the same in the both plots, depicts an analytical estimates of $\delta t_{m\Omega}$ at the

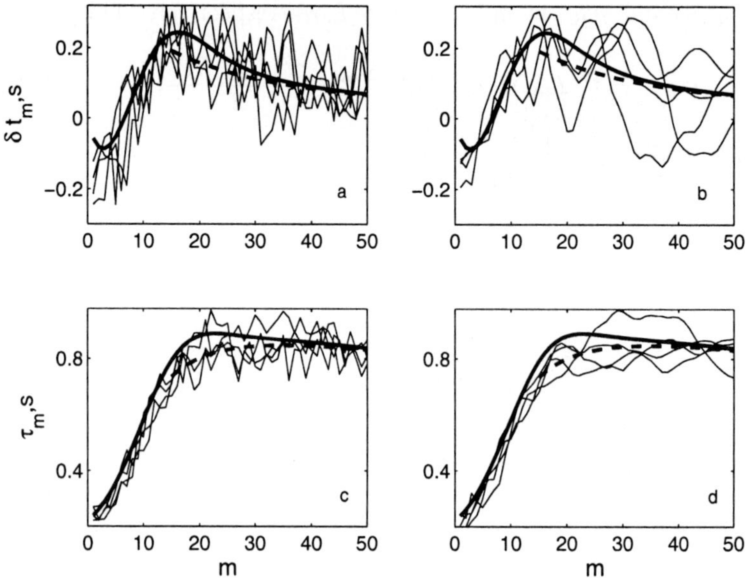

Fig. 4.21 Biases (upper row of plots) and spreads (lower row) of mode pulses at 3000 km. Heavy solid lines present analytical predictions for the bias (a and b) and spread (c and d) at 75 Hz. Heavy dashed line show simplified analytical estimates for the same quantities. Thin solid lines graph frequency averaged biases, δt_m, and spreads, τ_m, obtained by parabolic equation simulation (a and c) and predicted on the basis of a numerical ray tracing (b and d) for four realizations of the perturbation [Virovlyansky *et al.* (2009)].

frequency of 75 Hz predicted by Eqs. (4.208), (4.213), and (4.215). The simplified analytical estimate (4.225) at 75 Hz is shown by a heavy dashed line. Figures 4.21(c) and (d) present similar results for the spread. A heavy solid line is an analytical estimate of $\tau_{m\Omega}$ at 75 Hz given by Eqs. (4.209), (4.213)–(4.215). A heavy dashed line is a simplified analytical estimate given by Eq. (4.225).

In Fig. 4.22 we see the same dependencies as in Fig. 4.21 but the thin solid lines present not frequency averaged results, but results obtained for a single frequency of 75 Hz.

The agreement between simulations and theory based estimates is seen to be good for the spread. It is somewhat surprising that in spite of the restriction indicated after Eq. (4.206) the analytical estimate based on Eqs. (4.208), (4.213), and (4.215) are valid not only for the high modes. The predictions of bias are less accurate. They can be considered only as order-of-magnitude estimates.

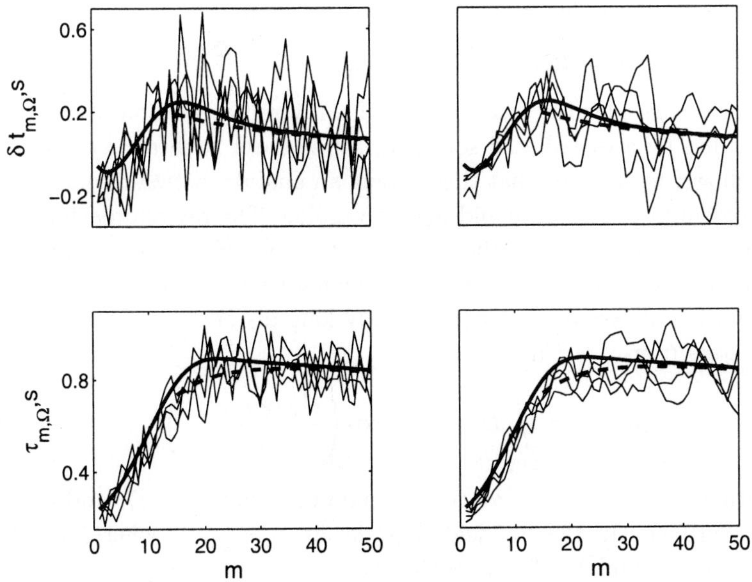

Fig. 4.22 The same as in Fig. 4.21 but thin solid lines in all plots present not frequency averaged biases, δt_m, and spreads, τ_m, but values of $\delta t_{m\Omega}$ and $\tau_{m\Omega}$ computed for $\Omega/(2\pi) = 75$ Hz [Virovlyansky *et al.* (2009)].

4.4 Wave Beam in an Ocean Acoustic Waveguide

So far, we used the ray-based approach for an analysis of the wave field excited by a point source or a single-mode source. A similar description of the wave field excited by a distributed source of a finite size requires special consideration. This issue was examined in Ref. [Virovlyansky *et al.* (2008b)]. Once the ray-based description is introduced, the wave field can be analyzed using the stochastic ray theory in the same way as in the case of a point source or a single-mode source.

In this section, we present the method derived in Ref. [Virovlyansky *et al.* (2008b)] and use it for describing a narrow wave beam radiated by a vertical array. We discuss the method of forming a beam which, in the absence of a random perturbation, diverges slower than other beams with the same initial width. A simple analytical estimate for the widening of the modal spectrum of the beam due to scattering at internal-wave-induced fluctuations of the sound speed is obtained. It generalizes a similar estimate for a single-mode source.

4.4.1 *Ray-based description of a wave beam*

4.4.1.1 *Directivity pattern in a free space*

First, let us consider the introduction of a ray representation of the wave field excited by a distributed source in a free space with $c(r,z) = c_0$. In what follows, we will see that there exists an analogy between solutions of this problem in a free space and in a waveguide. The ray approach can be used only if the wave field in the vicinity of the observation point represents a quasi-plane wave or the superposition of a few such waves. In a free space, this situation always occurs at sufficiently long distances from the source. This follows from the relation

$$u(r,z) = \int dz' u(0,z') \exp\left(\frac{ik_0 (z-z')^2}{2r}\right) \tag{4.226}$$

connecting the wave fields at the observation point, $u(r,z)$, and at the antenna $u(0,z)$. In the far field, that is, at $r \gg l_a^2/\lambda$, where l_a is the aperture length and $\lambda = 2\pi/k_0$ is the wavelength, we have

$$u(r,z) = A \exp(ik_0 S), \tag{4.227}$$

where

$$S = \frac{z^2}{2r}, \tag{4.228}$$

$$A = \int dz' u(0,z') \exp(ik_0 z' \tan\chi), \tag{4.229}$$

$\chi = \arctan(z/r)$ is an angle between the horizontal plane and the straight line connecting the antenna center and the observation point. According to Eq. (4.227), the wave field at the observation point can be interpreted as a contribution from an eigenray with eikonal S and amplitude A. The arrival angle of this ray, $\chi' = \arctan(\partial S/\partial z)$, turns out to be equal to χ. The dependence of A on χ determines the radiation pattern of the source. It is important to emphasize that, however complicated and intricate is the starting field at $r = 0$, at long ranges it always smoothes out, and in the vicinity of an observation point looks as a quasi-plane wave. Therefore, at sufficiently long ranges, the ray representation can be introduced for any source.

In some cases, the ray representation can be introduced at short ranges, as well [Kravtsov and Orlov (1990)]. The simplest example is a linear

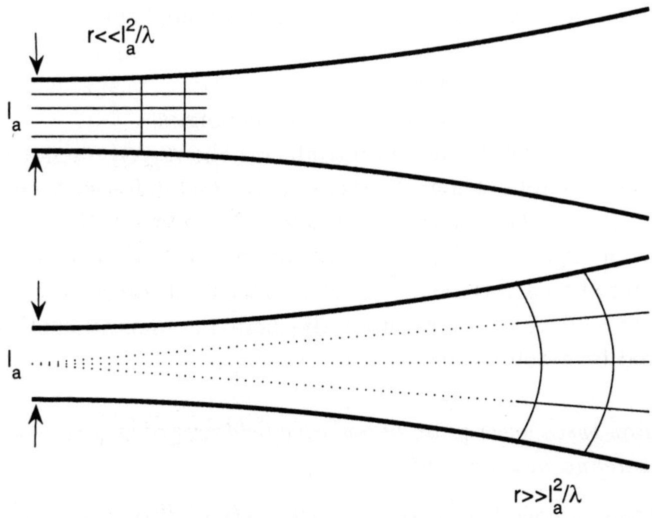

Fig. 4.23 Wave beam generated by a linear antenna in a free space. Top: the ray system in the near-field zone. The constant–phase lines and the ray paths represent segments of the straight lines parallel and perpendicular to the antenna, respectively. Bottom: the ray system in the far-field zone. The constant–phase lines are segments of the parabolas. The continuations of rays toward small distances (dashed lines) intersect at the center of the array.

antenna with the aperture $l_a \gg \lambda$. It radiates a wave beam shown in Fig. 4.23. At ranges $r < l_a^2/\lambda$, that is, in the near field, a beam width is close to l_a, and constant phase lines (far from the beam boundaries) are approximately parallel to the antenna, as is shown in the upper panel of Fig. 4.23. Correspondingly, the rays represent horizontal straight lines. Take the quantity

$$r_* = \frac{l_a^2}{\lambda} \tag{4.230}$$

as a rough estimate of the near field zone length. In the far field zone, that is, at

$$r \gg r_*, \tag{4.231}$$

the wave field is described by another congruence of rays described above and shown in the lower panel of Fig. 4.23. Extensions of these rays into the near field zone depicted by dashed lines cross at the center of the antenna.

In a waveguide, Eq. (4.226) can be used only at sufficiently small ranges, $r < r_{\text{cr}}$, where r_{cr} is an estimate of the maximum range at which the wave field still does not "feel" the dependence of the sound speed on the spatial coordinates. If $r_* < r_{cr}$, then the distributed source can approximately be replaced by a point source with the directivity pattern $A(\chi)$ which assigns different weights to rays with different launch angles.

It is clear that this method of introducing the ray representation has a restricted field of application. A more rigorous and universal approach was derived in Ref. [Virovlyansky *et al.* (2008b)] by applying the Poisson sum formula to the modal representation of the wave field excited by a vertical antenna placed in a waveguide. This approach generalizes the procedure employed in Sec. 3.2.1.5 to transform the mode representation of the sound field excited by a point source into the ray representation.

4.4.1.2 *Ray-based description of the wave field excited by a vertical antenna in a waveguide*

Take a range-independent waveguide with $\delta c(r, z) = 0$. Consider a source representing a vertical thread located along the line $r = 0$. The wave field is governed by parabolic equation (1.17) with $V = 0$ and an initial condition

$$u(0, z) = u_0(z). \tag{4.232}$$

According to Eqs. (3.31) and (3.42), the modal decomposition of the excited wave field is given by the relation

$$u(r, z) = \sum_m b_m \varphi_m(z) e^{-ik_0 E_m r}, \tag{4.233}$$

where

$$b_m = \int dz\; u_0(z)\varphi_m(z). \tag{4.234}$$

Let us assume that the mode amplitude b_m can be presented in the form

$$b_m = \mathcal{A}(I_m) e^{ik_0\alpha(I_m)}. \tag{4.235}$$

Substituting $H_0(I_m)$ for E_m (see Eq. (3.35)), we get

$$u(r, z) = \sum_m \mathcal{A}(I_m)\varphi_m(z) e^{ik_0[\alpha(I_m) - H_0(I_m)r]}. \tag{4.236}$$

Transform this equation using the Poisson summation formula in the same manner as it was done in Sec. 3.2.1.5 for the wave field of a point source. Then

$$u(r, z) = \sum_N \int dm\; \mathcal{A}(I_m)\varphi_m(z) e^{ik_0[\alpha(I_m) - H_0(I_m)r] + 2\pi i N m}. \tag{4.237}$$

Let us split $\varphi_m(z)$ into the sum of two terms in accord with Eq. (3.37). Making use of the quantization rule (3.34), we will change the variable of integration in Eq. (4.237) by replacing m with $k_0 I + 1/2$ (for short, we denote I_m by I). Then the right-hand side of Eq. (4.237) will be presented as a sum of integrals

$$u(r,z) = k_0 \int dI\ \mathcal{A}(I)\varphi_m^\sigma(z) e^{ik_0[\alpha(I) - H_0(I)r] + 2\pi i N m} = \int dI\ \tilde{C} e^{ik_0\Phi}, \tag{4.238}$$

where σ denotes $+$ or $-$,

$$\tilde{C} = k_0 \mathcal{A}(I) D^{-1/2}(I) \left[2\left(H_0(I) - U(z)\right)\right]^{-1/4}, \tag{4.239}$$

$$\Phi = [\alpha(I) + 2\pi N I + \sigma g(I,z) - H_0(I) r] - \pi(N - \sigma/4). \tag{4.240}$$

It follows from Eqs. (1.46), (1.53), and (1.56) that

$$g'(I,z) = \omega(I) L(I,z). \tag{4.241}$$

Here and in what follows, the prime denotes the derivative (ordinary and partial) with respect to action I. For evaluating integral (4.238), let us exploit the stationary phase technique (see Appendix D). This can be done only at sufficiently long ranges where Φ is a much more rapidly varying function of argument I than the amplitude $\tilde{C}$. Using Eqs. (1.47), (1.53), and (4.241) yields

$$\Phi' = \omega(I)\left[R(I,z) - r\right], \tag{4.242}$$

where

$$R(I,z) = \frac{\alpha'(I)}{2\pi} D(I) + N D(I) + \sigma L(I,z). \tag{4.243}$$

The stationary phase condition

$$\Phi' = 0 \tag{4.244}$$

can be represented in the form

$$r = R(I,z). \tag{4.245}$$

For fixed r and z, Eq. (4.245) determines the value of I, which can be interpreted as the action of an eigenray arriving at the observation point (r,z). For fixed I, a pair of equations (4.243) and (4.245) determine the ray path. Comparison of Eq. (4.243) with Eq. (1.61) shows that a ray with action variable I at $r = 0$ starts from the depth z_0 satisfying the condition

$$-\sigma_0 L(I, z_0) = \frac{\alpha'(I)}{\omega(I)}, \tag{4.246}$$

where σ_0 is the sign of the starting launch angle. Taking into account Eq. (1.57) allows one to rewrite the above relation as

$$\alpha'(I) = 2\pi - \theta_0, \tag{4.247}$$

where θ_0 is the initial angle variable of the ray path.

Denote by I_s the action variable satisfying condition (4.244) with fixed r and z. Develop the phase Φ as a series in I (other arguments of Φ are fixed) about the stationary point $I = I_s$:

$$\Phi(I) = \Phi(I_s) + \frac{1}{2}\Phi''(I_s)\,(I - I_s)^2 + \frac{1}{6}\Phi'''(I_s)\,(I - I_s)^3 + \dots. \tag{4.248}$$

The stationary phase technique consists in retaining in this expansion only the first two terms and replacing the amplitude factor $\tilde{C}$ in the rightmost integral in Eq. (4.238) by its value at $I = I_s$. Then our task reduces to the evaluation of a Gaussian integral. Applicability conditions of this approximation will be discussed below.

The second derivative of Φ with respect to I at the stationary point is

$$\Phi''(I_s) = \omega(I)R'(I_s, z). \tag{4.249}$$

Using this expression and evaluating the Gaussian integral yield

$$\begin{aligned} u(r,z) &= \left.\sqrt{\frac{2\pi}{k_0\,|\Phi'|}}\,\tilde{C}e^{ik_0\Phi + i\gamma\pi/4}\right|_{I=I_s} \\ &= \left.\mathcal{A}(I)\sqrt{\frac{k_0}{|R'(I,z)|\tan\chi}}\,e^{ik_0\Phi + i\gamma\pi/4}\right|_{I=I_s}, \end{aligned} \tag{4.250}$$

where γ =sgn(R'), $\tan\chi = \left[2\left(H_0(I) - U(z)\right)\right]^{1/2}$, χ is the grazing angle of a ray path with action I at depth z. This expression presents a contribution from one eigenray arriving at the observation point. The number of contributing eigenrays is determined by the number of stationary phase points, that is, by the number of actions I satisfying condition (4.244) with given r and z. The total wave field at point (r, z) is given by the summation of contributions from all the eigenrays. Equation (4.250) fails at caustics where $R' = 0$.

The above formula was derived for the range-independent waveguide. Nevertheless, it can be used for introducing the ray representation of a wave field excited by a distributed source in the presence of a weak range-dependent perturbation. Let us assume that at short ranges, where the influence of the perturbation is still negligible, Eq. (4.250) is already valid. This allows one to find an explicit expression for the factor C in general formula (1.39). Consider the starting action I_0 as a parameter labeling the ray paths. Combining Eqs. (4.250) and (1.39), it can be shown that in the presence of a perturbation, a contribution from an eigenray is given by the relation

$$u(r, z) = \mathcal{A}(I_0)\sqrt{\frac{k_0}{|\partial z/\partial I_0|}}\, e^{ik_0 S - i\mu\pi/2}, \tag{4.251}$$

where $\partial z/\partial I_0$ is a derivative of the ray depth at the observation point with respect to the starting action, S is the ray eikonal, and μ is the Maslov index. This equation generalizes Eq. (4.250).

The above result can be used in the case where the mode amplitude is expressed as the sum of a few terms of the same form as that in Eq. (4.235). Then each term gives rise to a congruence of rays whose contributions are described by Eq. (4.251). The point source falls into this category. Indeed, according to Eq. (3.44), we have $b_m = \varphi_m(z_s) = \varphi_m^+(z_s) + \varphi_m^-(z_s)$. The terms $\varphi_m^\pm(z_s)$ have the form of Eq. (4.235).

4.4.1.3 *Applicability conditions of the stationary phase technique*

The applicability conditions of our ray-based approach are determined by the applicability conditions of the stationary phase technique used for the evaluation of the integral (4.238). Let us proceed from the fact that the main contribution to integral (4.238) comes from the actions I satisfying the inequality

$$\frac{1}{2}\left|\Phi''(I_s)\right|^2 (I - I_s)^2 \lesssim \pi. \tag{4.252}$$

The effective width of this interval can be estimated as

$$\Delta I_{\text{eff}} = \sqrt{2\pi/\left|\Phi''(I_s)\right|}. \tag{4.253}$$

Within this interval, the first neglected term in the expansion (4.248) should be small compared to π. This means that

$$\frac{1}{6}\left|\Phi'''(I_\mathrm{s})\right| \Delta I_\mathrm{eff}^3 \ll \pi. \tag{4.254}$$

Besides, the effective width ΔI_eff should be small on the scale of the width of the amplitude factor $\mathcal{A}(I)$, that is,

$$\Delta I_\mathrm{eff} \ll \Delta_I, \tag{4.255}$$

where Δ_I is a characteristic scale of $\mathcal{A}(I)$.

At ranges greater than a few cycle lengths, an approximate formula

$$R' \approx \frac{D'}{D} r = -\frac{\omega_I'}{\omega} r \tag{4.256}$$

can be applied far from caustics. Making use of this approximation, we find from Eq. (4.242) the order-of-magnitude estimates

$$|\Phi''| \approx \frac{2\pi r}{D \delta I}, \; |\Phi'''| \approx \frac{2\pi r}{D \delta I^2}, \tag{4.257}$$

and

$$\Delta I_\mathrm{eff} \approx \sqrt{\frac{D^2}{k_0 r \left|D'\right|}} \approx \delta I^{1/2} k_0^{-1/2} D^{1/2} r^{-1/2}, \tag{4.258}$$

where

$$\delta I = D / \left|D'\right| \tag{4.259}$$

is a characteristic scale of the function $D(I)$. Now condition (4.254) can be rewritten in the form

$$r \gg \frac{D}{10 k_0 \delta I}. \tag{4.260}$$

Plots presented in Fig. 4.3 allow one to estimate the quantity δI for our environmental model. It varies from 0.2 km for flat rays to 0.4 km for steep ones. At frequencies $O(100\text{ Hz})$, the condition (4.260) is met at practically any range. Then the applicability of the stationary phase technique is determined by the condition (4.255) which can be rewritten as

$$r \gg \frac{D \delta I}{k_0 \Delta_I^2}. \tag{4.261}$$

Equations (4.260) and (4.261) (simplified versions of Eqs. (4.254) and (4.255)) show that the ray-based description in a waveguide, as in a free space, becomes valid only at sufficiently long ranges. Let us demonstrate

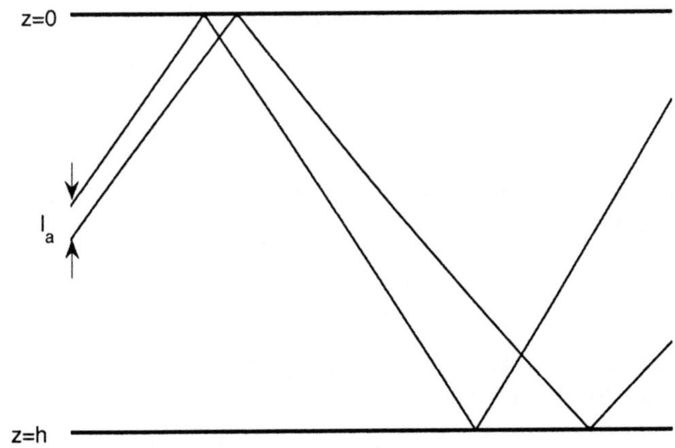

Fig. 4.24 Sketch of a wave beam in an ideal waveguide with perfectly reflecting boundaries.

that the estimates given by Eqs. (4.260) and (4.261) agree with the well-known condition (4.231) derived for a free space. Take a wave beam propagating in an ideal waveguide, i.e., in a range-independent, isovelocity water layer of a width h with a perfectly reflecting hard bottom and a free surface. Assume that the beam is emitted by a linear antenna of length $l_a \ll h$ (see Fig. 4.24). The wave field in this waveguide is easily described using the well-known image method [Jensen *et al.* (1994)]. It superimposes the free-field solution with the fields produced by the image sources. As a result, the wave field is expressed through the solution of the wave equation in a free space. Therefore, the applicability conditions of the ray-based approach in this waveguide should be the same as in a free space. Let us check that in this example, Eqs. (4.254) and (4.255) indeed give the same result as Eq. (4.231).

Making use of Eqs. (1.43), (1.53), and (1.54), we find that in the ideal waveguide

$$H_0(I) = \frac{\pi^2 I^2}{2h^2}, \ \omega(I) = \frac{\pi^2 I}{h^2}, \ D(I) = \frac{2h^2}{\pi I}. \tag{4.262}$$

In this case $U = 0$ and the momentum of a ray path and its action variable are connected by $p^2/2 = H_0(I)$. It means that $p = \pi I/h$. Quantization rule (3.34) yields

$$E_m = \frac{1}{2}\frac{\pi^2}{k_0^2 h^2}\left(m - \frac{1}{2}\right)^2, \ m = 1, 2, \ldots. \tag{4.263}$$

The m-th eigenfunction in this waveguide has the form [Ahluwalia and Keller (1977)]

$$\varphi_m(z) = \sqrt{\frac{2}{h}} \sin\left(\frac{\pi\,(m-1/2)\,z}{h}\right). \tag{4.264}$$

Assume that the center of a vertical antenna is set at a depth z_c and an initial wave field at $r=0$ is

$$u(0,z) = \begin{cases} e^{i\kappa z}, & \text{if } z_c - l_a/2 < z < z_c + l/2 \\ 0, & \text{otherwise} \end{cases}, \tag{4.265}$$

where $\kappa = k_0 \tan\chi_0$. The emitted beam propagates along a reference ray escaping point $(0, z_c)$ at a launch angle χ_0. The mode amplitudes are

$$b_m = \int\limits_0^h dz\; u(0,z)\varphi_m(z) = \sqrt{\frac{2}{h}} e^{i(\kappa-\kappa_m)z_c} \frac{\sin\left[(\kappa-\kappa_m)\,l_c/2\right]}{\kappa-\kappa_m}, \tag{4.266}$$

where $\kappa_m = \pi\,(m-1/2)\,/h$. This expression has the form of Eq. (4.235) with

$$\mathcal{A}(I) = \frac{\sin\left(k_0 l_c\,(p_0 - \pi I/h)\,/2\right)}{l_c\,(p_0 - \pi I/h)}. \tag{4.267}$$

Estimating the width of the main peak of function $\frac{\sin x}{x}$ as π, we get the characteristic scale of the function $\mathcal{A}(I)$

$$\Delta_I = \frac{2h}{k_0 l_c}. \tag{4.268}$$

In the ideal waveguide Eq. (4.256) translates to

$$R' = -\frac{r}{I}. \tag{4.269}$$

Substituting this in Eq. (4.249) yields

$$\Phi'' = -\frac{k_0\pi^2 r}{h^2}. \tag{4.270}$$

This expression does not depend on I, which means that in our approximation Φ''' and all higher derivatives with respect to I vanishes. Therefore, the condition (4.254) is met at all ranges and the applicability of the ray-based description is determined solely by the condition (4.255). According to Eq. (4.253), one finds

$$\Delta I_{\text{eff}} = \sqrt{\frac{2h^2}{\pi k_0 r}}. \tag{4.271}$$

Substituting Eqs. (4.268) and (4.271) into Eq. (4.255), we find that the ray-based approach in an ideal waveguide is valid at ranges r satisfying the condition

$$r \gg \frac{l_a^2}{\lambda}, \tag{4.272}$$

which, consistent with our expectation, coincides with Eq. (4.231).

4.4.1.4 *A weakly divergent beam*

Consider a narrow acoustic beam propagating in a range-independent waveguide along a reference ray starting from the sound channel axis z_a. Action variable of the ray and starting value of its angle variable denote $\bar{I}$ and $\bar{\theta}_0$, respectively. Let the starting wave field $u_0(z)$ be determined by Eq. (4.235) with the function $\mathcal{A}(I)$ concentrated within a small interval of actions centered at $\bar{I}$. Assume that the width of this interval Δ_I, determined as

$$\Delta_I^2 = \int dI(I-\bar{I})^2\mathcal{A}^2(I) \Big/ \int dI\, \mathcal{A}^2(I), \tag{4.273}$$

satisfies the condition

$$\Delta_I \ll \bar{I} \tag{4.274}$$

and

$$\alpha(I) = (2\pi - \bar{\theta}_0)(I - \bar{I}). \tag{4.275}$$

This choice of function $\alpha(I)$ is suggested by Eq. (4.247). We also assume that the initial beam width Δ_z, determined as

$$\Delta_z^2 = \int dz(z-z_a)^2 \left|u_0(z)\right|^2 \Big/ \int dz\, |u_0(z)|^2, \tag{4.276}$$

is small compared with an interval between turning depths of the reference ray.

The number of normal modes forming the beam, Δ_m, can be estimated as

$$\Delta_m^2 = \int dm(m-\bar{m})^2\mathcal{A}^2(I_m) \Big/ \int dm(m-\bar{m})^2\mathcal{A}^2(I_m), \tag{4.277}$$

where $\bar{m} \simeq k_0\bar{I}$ is the number of central modes. If $\mathcal{A}(I)$ is a smooth function of its argument, it follows from the quantization rule (3.34) that at high frequencies we have $\Delta_m \simeq k_0 \Delta_I$. In what follows, we will assume that

$$\bar{m} \gg \Delta_m \gg 1, \tag{4.278}$$

which is consistent with the condition (4.274).

In this subsection, our task will be to find the amplitude function $\mathcal{A}(I)$ corresponding to a beam which diverges with range slower than other beams with the same initial width. First, note that at ranges where the ray-based description is already valid, the spatial spread of the beam can be estimated using Eqs. (4.243) and (4.245). The spread in range is given by

$$\Delta_r = |R'|_{I=\bar{I}}\, \Delta_I. \tag{4.279}$$

At the sound channel axis z_a (far from caustics), the approximate formula (4.256) is valid and the above relation reduces to

$$\Delta_r = \frac{\left|\omega_I'(\bar{I})\right|}{\omega(\bar{I})} r \Delta_I. \tag{4.280}$$

The spreading will be especially slow if $\omega_I'(\bar{I})$ vanishes. In this case, rays forming the beam have close cycle lengths. Even if the wave field is excited by a point source, weakly divergent beams may appear in the vicinities of rays with the actions $\bar{I}$ satisfying the condition $\omega_I'(I) = 0$. This phenomenon was observed in numerical simulations and field experiments [Petukhov (1994); Goncharov and Kurtepov (1994); Brekhovskikh *et al.* (1995)]. In order to estimate the beam spread in range in the case where $\omega_I'(\bar{I}) = 0$, Eq. (4.280) should be modified by including the term $O(\Delta_I^2)$. This yields

$$\Delta_r = \frac{\left|\omega''(\bar{I})\right|}{2\omega(\bar{I})} r \Delta_I^2. \tag{4.281}$$

In the far field, where the ray representation is valid, the beam width at the sound channel axis, i.e., far from the ray turning points, can be estimated as $\Delta_z + \Delta_r / \tan\chi_a$, where χ_a is the grazing angle of the reference ray at $z = z_a$. According to estimates (4.280) and (4.281), the dependence of Δ_r on the function $\mathcal{A}(I)$ comes only through the parameter Δ_I. Therefore, a beam with minimal divergence is defined by the function $\mathcal{A}(I)$ corresponding to a minimum value of Δ_z for a given Δ_I. In order to find this $\mathcal{A}(I)$ we will consider the starting wave field $u_0(z)$ as a superposition of the Brillouin waves.

Take a beam with a positive launch angle which represents a superposition of functions $\varphi_m^+(z)$. We assume that the initial wave field $u_0(z)$ is negligible everywhere except a small area centered at the sound channel axis z_a. The size of this area is much less than the interval between the turning points of modes forming the beam. Within this area, the Brillouin waves can be approximately represented as

$$\varphi_m^+(z) = Q_m(z_a)\exp\left[ik_0 g(I_m, z_a) + p_m(z_a)(z - z_a)\right]. \tag{4.282}$$

It follows from Eqs. (1.47), (3.36), and (3.41) that

$$g'(I_{\bar{m}}, z_a) = \bar{\theta}_0, \quad \left.\frac{\partial p_m(z_a)}{\partial m}\right|_{m=\bar{m}} = \frac{\gamma}{k_0}, \; \gamma = \frac{2\pi}{D(I_{\bar{m}})p_{\bar{m}}(z_a)}. \tag{4.283}$$

We find from Eq. (4.282) the following expression for normal modes with numbers m close to $\bar{m}$ (see Eq. (4.278)):

$$\varphi_m^+(z) \simeq \varphi_{\bar{m}}^+(z_a)\exp\left[i\bar{\theta}_0(m - \bar{m}) + i\gamma(m - \bar{m})(z - z_a)\right]. \tag{4.284}$$

This yields

$$u_0(z) = \varphi_{\bar{m}}^+(z_a)\sum_m \mathcal{A}(I_m)\exp\left[i\gamma(m - \bar{m})(z - z_a)\right]. \tag{4.285}$$

Assuming that $\mathcal{A}(I_m)$ is a smooth function of m and taking into account Eq. (4.278), we approximately replace the summation in Eq. (4.285) by integration over an infinite interval. Note also, that according to the quantization rule (3.34), I_m is a linear function of m. Therefore, it turns out that the functions $u_0(z)$ and $\mathcal{A}(I)$ are connected by a Fourier transform.

Thus, in order to find a beam with a minimum divergence we should minimize the width of $u_0(z)$, keeping constant the width of its Fourier transform. The solution of this problem is well known. The width of any function and that of its Fourier transform satisfy the uncertainty relation [Landau and Lifshitz (1977)]. In our case, the latter takes the form

$$\Delta_I \Delta_z \geq \frac{1}{2k_0\gamma}. \tag{4.286}$$

For a fixed Δ_I, an initial beam width Δ_z takes its minimum value $(2k\gamma\Delta_I)^{-1}$ for

$$\mathcal{A}(I) = \exp\left[-\frac{(I - \bar{I})^2}{4\Delta_I^2}\right]. \tag{4.287}$$

If we consider wave beams with a given width Δ_z, then the optimal starting field is

$$u_0(z) = 2\Delta_m \pi^{1/2}\varphi_{\bar{m}}^+(z_a)\exp\left[-\frac{(z - z_a)^2}{4\Delta_z^2} + ik_0 p_m(z_a)(z - z_a)\right]. \tag{4.288}$$

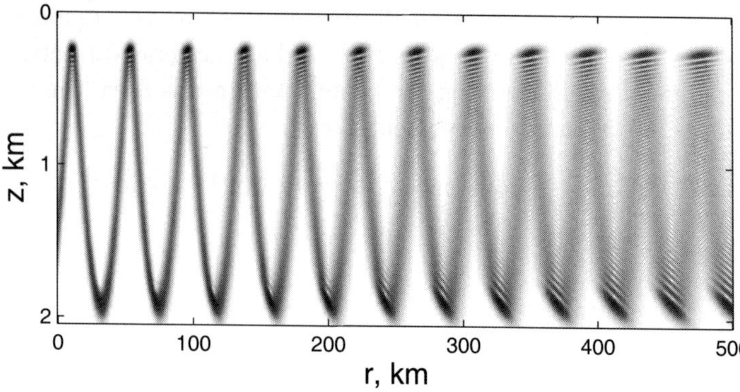

Fig. 4.25 Weakly divergent beam in the unperturbed waveguide. Carrier frequency is 250 Hz.

In quantum mechanics a similar expression describes the wave function of a state with the minimum uncertainty which is called the coherent state [Landau and Lifshitz (1977)]. A more general method of constructing a beam with minimum divergence is discussed in Ref. [Virovlyansky and Nefedova (2009)].

The above result means that the wave beam formed by normal modes with the amplitudes

$$b_m = \exp\left[-\frac{(m-\bar{m})^2}{4\Delta_m^2} - i\bar{\theta}_0 (m-\bar{m})\right] \tag{4.289}$$

spreads slower than other beams with the same initial width $\Delta_z = (2\gamma\Delta_m)^{-1}$. Figure 4.25 shows a wave beam, computed at a carrier frequency of 250 Hz. Its initial wave field is defined by Eq. (4.289) with $\bar{\theta}_0 = \pi/2$, $\bar{m} = 60$ and $\Delta_m = 2\sqrt{2}$. A grazing angle of this beam at the sound channel axis $z_a = 0.7$ km is about 8°. As we know, the applicability condition of the ray-based description is determined by Eq. (4.261). In this example the right hand side of Eq. (4.261) is about 200 km. Figure 4.26 presents magnified views of two parts of the wave beam in Fig. 4.25. Five ray paths with equal starting angle variables $\theta_0 = \bar{\theta}_0$ and actions uniformly distributed within an interval $\bar{I} - 1.5\Delta_I < I < \bar{I} + 1.5\Delta_I$, where $\bar{I} = \bar{m}/k_0$ and $\Delta_I = \Delta_I/k_0$, are superimposed on the beam. Consistent with our expectation, in the far field (right panel) the trajectories fill the beam. It is clearly seen that the wave field increases at caustics located in the vicinities of the beam turning depths. In the near field (left panel), the ray-based

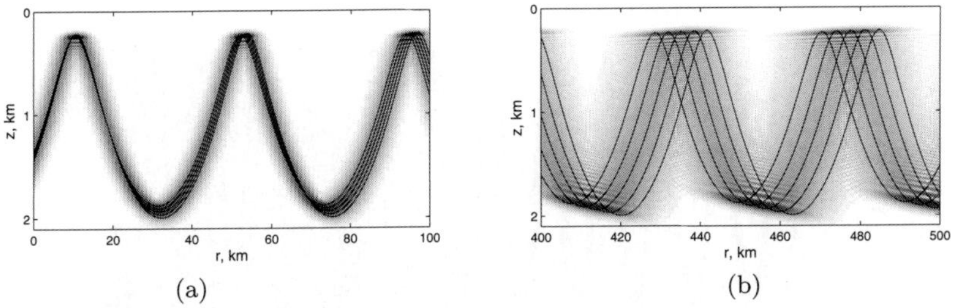

Fig. 4.26 Magnified views of two parts of the wave beam shown in Fig. 4.25. Trajectories of five ray paths with equal starting angle variables are superimposed. At long ranges, they uniformly fill the beam. At short ranges, they are concentrated in the center of the beam. (a) First 100 km. (b) Last 100 km.

description introduced above does not work and the spread of the ray paths is significantly less than the beam width.

In Ref. [Virovlyansky *et al.* (2008b)] it is shown that an approximate analytical description of the wave field determined by Eq. (4.289) can be constructed not only in the far field by in the near field, as well.

Alternative approaches for constructing wave beams with a weak divergence are discussed in Refs. [Morozov and Colosi (2005)] and [Petukhov and Khil'ko (2008)].

4.4.2 *Wave beam and its modal structure in the presence of random internal waves*

In this subsection we examine the beam propagation in the presence of the sound speed fluctuations δc. Possible manifestations of the chaotic ray behavior in the beam structure were initially considered by Tappert [Tappert (2003)]. He hypothesized that the width of a narrow beam propagating through a statistically homogeneous single-scale random medium would increase exponentially (explosively) with range at a rate predicted by the Lyapunov exponent. However, Tappert did not perform numerical simulations to confirm this prediction. Such simulations were done in Ref. [Morozov and Colosi (2005)], where it was demonstrated that acoustical beams in an ocean waveguide with realistic sound-speed perturbations from internal waves do not expand exponentially (or explosively), but they expand diffusively. Nevertheless, the numerical results exhibit a finite-frequency manifestation of ray chaos. It was shown that the wave field has a positive Kolmogorov-Sinai entropy (see 2.1.1) which means that it evolves chaotically. We think that the explosive behavior of a wave beam may be

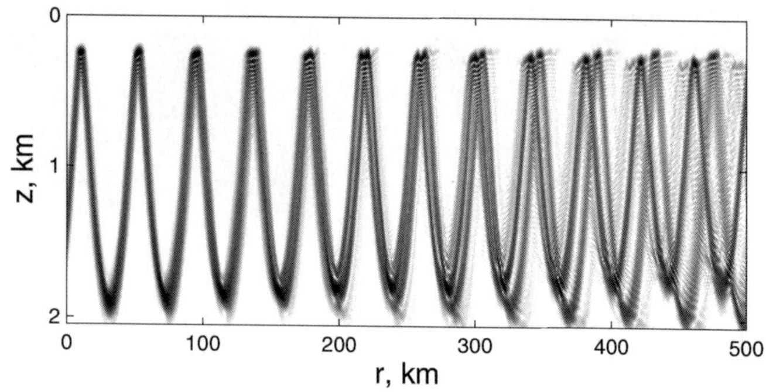

Fig. 4.27 Wave beam with the same starting field as in Fig. 4.25, but in the presence of random internal waves.

observed only at very high carrier frequencies, but not at frequencies of order 100 Hz used in experiments on the long range sound propagation.

Figure 4.27 presents a beam with the same starting field as in Fig. 4.25, but propagating in the presence of random internal waves. Initial parts of the beams in the unperturbed and perturbed waveguides are practically the same. The distinction appears at ranges where unperturbed and perturbed paths with the same starting parameters deviate significantly from each other. At short ranges, the action variables of the perturbed and unperturbed paths remain close and the deviation is determined mainly by the difference between their angle variables. The absolute value of this difference can be roughly estimated as the rms spread of angle variables σ_θ given by Eq. (4.49). Assuming that the influence of the random perturbation becomes significant at a range where $\sigma_\theta = \pi/6$, we estimate this range from the relation

$$r = \frac{\sigma_\theta^{2/3}}{|\omega_I'(I_0)|^{2/3}} \left(\frac{3}{B}\right)^{1/3}, \tag{4.290}$$

that follows from Eq. (4.49). For a starting action I_0 corresponding to our beam, this yields $r \simeq 300$ km which is consistent with what we see in Fig. 4.27.

At megameter ranges, the beam diverges completely in the sense that the wave field fills the whole interval between the turning depths of the reference ray. Due to scattering at the sound speed fluctuations, the modal spectrum of the wave field is widening with range. Once the ray-based description is introduced, the widening can be evaluated analytically in the same manner as it was done in Sec. 4.3.1 for a single mode source. In the case of a single mode source, all the rays have the same initial value of the action variable and the mode intensity obtained by the incoherent summation of mode rays is given by Eq. (4.167). In the case of a weakly divergent beam determined by Eq. (4.289), all the rays have the same starting angle variable θ_0, and the representation of the modal intensity as an incoherent sum of mode rays (analogue of formula (4.167)) is

$$|a_m|^2 = \sum_j \mathcal{A}^2(I_{0,j}) \left| \frac{\partial I}{\partial I_0} \right|^{-1}_{I_0 = I_{0,j}} = \int dI_0 \; \mathcal{A}^2(I_0) \delta\left(I_m - I(r, I_0)\right), \quad (4.291)$$

where the subscript j in the middle expression numbers the mode rays. An analogue of Eq. (4.168) has the form

$$\frac{1}{\Delta I} \int\limits_{I_0 - \Delta I/2}^{I_0 + \Delta I/2} dI_0 \; \delta\left(I_m - I(r, I_0)\right) = P_{I|I_0}(I_m|I_0), \quad (4.292)$$

where ΔI is a small subinterval. Substituting this in Eq. (4.291), we get an analogue of Eq. (4.169)

$$|a_m|^2 = \int dI_0 \; \mathcal{A}^2(I_0) P_{I|I_0}(I_m|I_0). \quad (4.293)$$

Once again let us use the definition of the smoothed mode intensity (4.166) from Sec. 4.3

$$J_m(r, \Omega) = \sum_{m'} |a_{m'}(r, \Omega)|^2 \, e^{-\frac{(m-m')^2}{2\Delta_M^2}} \Bigg/ \sum_{m'} e^{-\frac{(m-m')^2}{2\Delta_M^2}} \quad (4.294)$$

and approximately replace the summation over the mode number m by integration over I. It is assumed that

$$\bar{m} \gg \Delta_M \gg 1. \quad (4.295)$$

Substituting Eq. (4.293) into Eq. (4.294) and taking into account Eq. (4.170) yield

$$J_m = \frac{k_0}{\sqrt{2\pi}\Delta_M} \int dI dI_0 \; \mathcal{A}^2(I_0) \exp\left[-\frac{k_0^2 (I_m - I)^2}{2\Delta_M^2} \right] P_{I|I_0}(I|I_0). \quad (4.296)$$

Let us assume that $\bar{m}$ is so large that the beam is formed by predominantly steep rays satisfying the condition (4.45). The statistical characteristics of

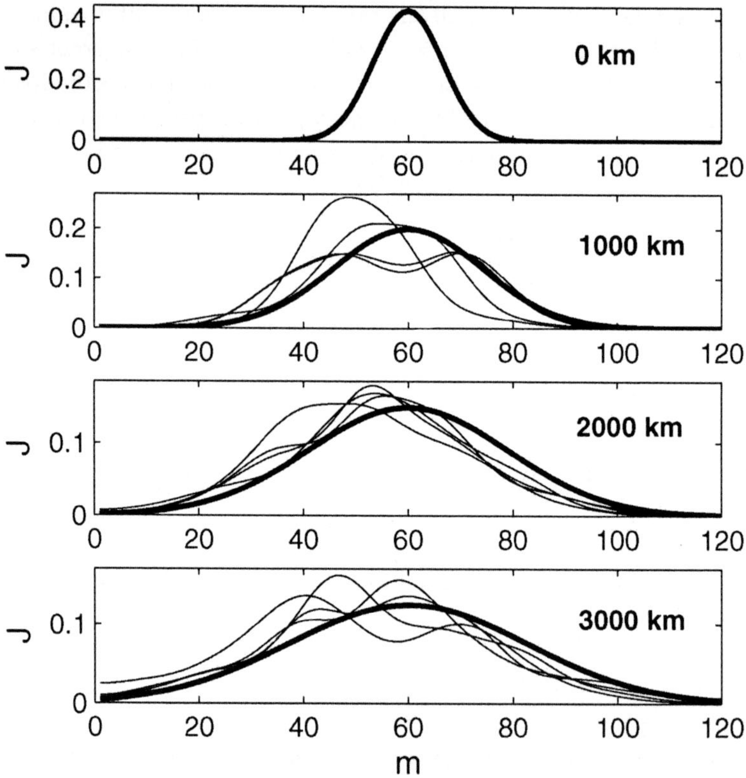

Fig. 4.28 Smoothed mode intensity as a function of mode number m at ranges 0, 1000, 2000, and 3000 km. Thick solid lines: predictions of Eq. (4.297). Thin solid lines: parabolic equation simulations for four realizations of the perturbation δc. The smoothing scale is $\Delta_M = 7$.

these rays are described by the PDF (4.43). Substituting Eqs. (4.287) and (4.43) into Eq. (4.296) and evaluating Gaussian integrals, we obtain

$$J_m = \frac{\Delta_m}{\sqrt{2\left(Brk_0^2 + \Delta_m^2 + \Delta_M^2\right)}} \exp\left[-\frac{(\bar{m} - m)^2}{2\left(Brk^2 + \Delta_m^2 + \Delta_M^2\right)}\right]. \quad (4.297)$$

This expression can be considered as a generalization of Eq. (4.172). Figure 4.28 compares the predictions of Eq. (4.297) and the results obtained by direct (numerical) solution of the parabolic equation (1.17) for four realizations of the internal-wave-induced perturbation δc. It is seen that

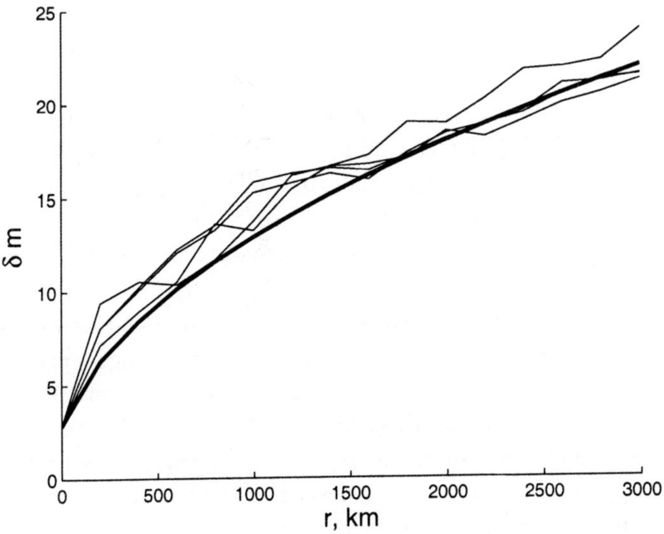

Fig. 4.29 Number of effectively excited modes δm as a function of range r. Thick solid line: prediction of Eq. (4.299). Thin solid lines: results of parabolic-equation simulations for four realizations of perturbation δc.

Eq. (4.297) gives a correct order-of-magnitude estimate for the widening of the modal spectrum. However, the results obtained for different realizations differ significantly even for a large smoothing scale $\Delta_M = 7$. The number δm of effectively excited modes defined by the relation

$$\delta m^2 = \sum_m (m - \bar{m})^2 |a_m|^2 \Big/ \sum_m |a_m|^2, \tag{4.298}$$

can be predicted better. Substituting Eq. (4.293) into this relation and using the same expressions for $\mathcal{A}(I)$ and $P_{I|I_0}(I|I_0)$ as in the derivation of Eq. (4.297), we find

$$\delta m = \sqrt{k_0^2 B r + \Delta_m^2}. \tag{4.299}$$

At long ranges, δm grows as $r^{1/2}$. In Ref. [Morozov and Colosi (2007)] this range dependence was established on the basis of numerical simulations. Figure 4.29 shows that the value of δm depends only weakly on a particular realization of δc, and the prediction of Eq. (4.299) is in reasonable agreement with the results of numerical simulations.

4.5 Arrival Times of Sound Pulses in the Presence of Internal Waves and a Mesoscale Inhomogeneity

As was shown in Sec. 4.2, internal waves weakly affect the early part of the arrival pattern which is formed by fuzzy ray pulses coming through steep paths. The case is similar if the perturbation is a superposition of random internal waves and a mesoscale inhomogeneity. In this case, the early part of the arrival pattern also remains approximately the same both with and without internal waves.

Below, we will consider a mesoscale inhomogeneity induced by a synoptic (mesoscale) eddy. Such eddies, frequently called "rings", are generated by splitting off intensive currents such as the Gulf Stream [Brekhovskikh and Lysanov (1991)]. Alone, the eddy does not cause chaotic ray motion. Its presence leads to changes in the ray travel times which can be described using a simple perturbation theory described in Chapter 1 (Sec. 1.6). Addition of random internal waves gives rise to ray chaos and causes spread and bias of the ray pulses. However, the pulses coming through steep rays are only weakly affected by internal waves. The delays of their arrival times caused by the eddy can be predicted using the simplest environmental model that does not account for internal waves. This fact is very important from the viewpoint of solving inverse problems of the ocean acoustic tomography and thermometry [Munk and Wunsch (1979); Spiesberger and Metzger (1991); The ATOC Consortium (1998)]. It allows one to apply analytical relations presented in Sec. 1.6 that relate the travel time delays to the environmental parameters to be reconstructed.

Unfortunately, fuzzy ray pulses coming through flat ray paths overlap. Therefore, researchers in acoustic tomography traditionally use the information in the early part of the arrival pattern where arrivals of individual steep rays can be resolved and identified.

It might be expected that the use of a vertical receiving array with a narrow directivity pattern allows one to increase the number of resolvable ray pulses. However, at frequencies of about 100 Hz used in the experiments on long range sound propagation, the width of the directivity pattern usually exceeds the difference between the arrival angles of the neighboring eigenrays, especially the flat ones. Therefore, the beamforming is of little help in resolving signals coming through rays with close travel times and close arrival angles.

In this section we discuss an alternative method of the space-time processing that may allow one to measure delays of signals arriving not only via steep paths, but via relatively flat ones, as well [Virovlyansky *et al.* (2008a)]. Our approach is based on the fact that the distribution of ray arrivals in the time-depth plane has characteristic features remaining stable and predictable even under conditions of ray chaos.

4.5.1 *Variation of the timefront in the presence of synoptic eddy*

Let us generalize the environmental model described in Sec. 4.1.2 by introducing an additional constituent of the sound speed field caused by a synoptic eddy. The total sound speed field will be presented in the form

$$c(r,z) = \bar{c}(z) + \delta c_m(r,z) + \delta c_i(r,z), \tag{4.300}$$

where $\delta c_m(r,z)$ and $\delta c_i(r,z)$ are the perturbations induced by the eddy and internal waves, respectively. The background sound speed profile $\bar{c}(z)$ and perturbation $\delta c_i(r,z)$ (in preceding sections it was denoted by δc) were determined in Sec. 4.1.2. The new constituent δc_m will be modeled as

$$\delta c_m(r,z) = c_1 \exp\left(-\frac{(r-r_1)^2}{\Delta r^2} - \frac{(z-z_1)^2}{\Delta z(r)^2}\right), \tag{4.301}$$

where

$$\Delta z(r) = \Delta z_c - \Delta z_v \exp\left(-\frac{(r-r_v)^2}{\Delta r_v}\right), \tag{4.302}$$

$c_1 = -0.005$ km/s, $r_1 = 500$ km, $z_1 = 1$ km, $\Delta r = 80$ km, $\Delta z_c = 0.5$ km, $\Delta z_v = 0.25$ km, $r_v = 520$ km, $\Delta r_v = 20$ km. The isolines of the sound speed field $\bar{c}(z) + \delta c_m(r,z)$ are shown in Fig. 4.30.

In what follows we will consider the wave field excited by a point source set at depth $z = 0.8$ km. The source radiates a pulse signal with a spectrum (see Eq. (1.15))

$$s(\Omega) = \frac{T}{\sqrt{2\pi}} \exp\left(-\frac{T^2\left(\Omega - \bar{\Omega}\right)^2}{2}\right). \tag{4.303}$$

In the numerical simulations presented below, the center frequency of the pulse and constant T defining its initial duration were set to be $\bar{\Omega}/2\pi = 75$ Hz and 0.017 s, respectively.

Numerical ray tracing demonstrates that the distribution of ray arrivals in the time-depth plane in the presence of the mesoscale inhomogeneity δc_m remains qualitatively the same. This is seen in Fig. 4.31 presenting

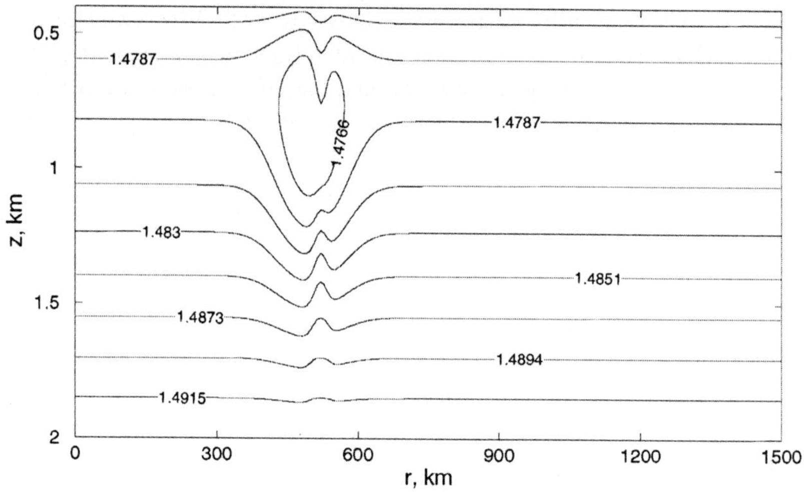

Fig. 4.30 Sound speed field modeling a synoptic eddy in a deep sea. Contour labels are of sound speed in km/s.

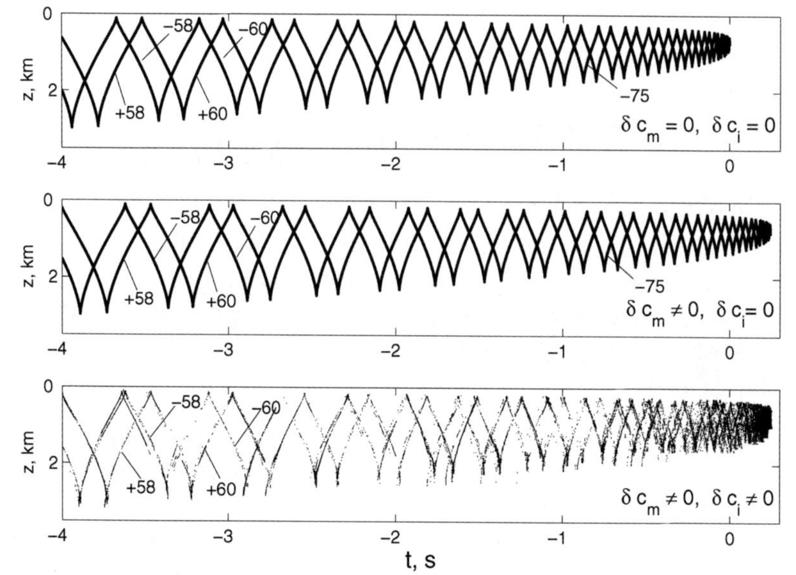

Fig. 4.31 Distributions of ray arrival times in the time-depth plane (timefronts) at a distance of 1500 km. An unperturbed channel (top), a channel with a synoptic eddy (middle), and a channel with a synoptic eddy and internal waves (bottom) [Virovlyansky *et al.* (2008a)].

timefronts at the range $r = 1500$ km. In these and subsequent plots, the time is reckoned from $r/c_0 = 1014.7$ s, which is the arrival time of an axial ray. Starting momenta p_0 of the fan rays traced numerically fill up uniformly an interval corresponding to the launch angles $\pm 12.5°$. The upper panel plots a timefront in an unperturbed waveguide ($\delta c_m = 0$, $\delta c_i = 0$). The middle and lower panels show timefronts in the presence of the synoptic eddy alone ($\delta c_m \neq 0$, $\delta c_i = 0$) and in the presence of both the eddy and internal waves ($\delta c_m \neq 0$, $\delta c_i \neq 0$), respectively. It is seen that the presence of the eddy does not destroy the timefront segments, but only shifts them. Different segments are shifted by different amounts. The mesoscale inhomogeneity under consideration causes a decrease in the sound speed (an eddy with a cold core) which leads to an increase in ray travel times. Therefore, in the presence of perturbation, the segments are shifted to the right relative the unperturbed segments with the same identifiers. Comparison of the middle and lower panels in Fig. 4.31 shows that in the presence of internal waves the segments (i) become fuzzy and (ii) they shift toward the early times. The same effects was observed in Sec. 4.2 in a waveguide without the mesoscale perturbation.

Figure 4.32 shows the intensity distribution of the sound field in the time–depth plane. It is calculated by numerically solving parabolic

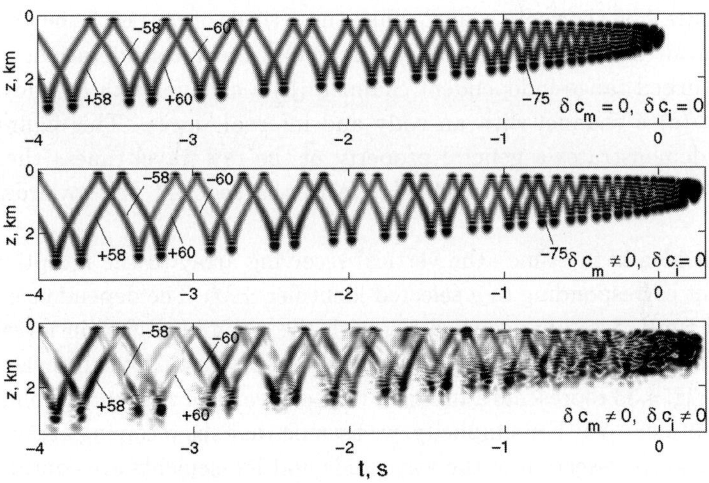

Fig. 4.32 Intensity distribution of the sound field in the time-depth plane at a distance of 1500 km. An unperturbed channel (top), a channel with a synoptic eddy (middle), and a channel with a synoptic eddy and internal waves (bottom) [Virovlyansky *et al.* (2008a)].

equation (1.17) at discrete frequencies with the pulse signal synthesized according to Eq. (1.22). The upper, middle, and lower panels present the data computed for the same models of the medium as the data presented in the corresponding panels of Fig. 4.31. In the absence of internal waves, the evaluation of the timefront allows one to predict the distribution of sound energy in the time–depth plane, including the shifts of segments in this distribution under the effect of a mesoscale inhomogeneity. In the presence of internal waves (the lower panel of Fig. 4.32), only early segments corresponding to steep rays are well resolved.

4.5.2 *The use of a vertical receiving array for measuring travel time delays*

Comparison of the upper and middle panels in Fig. 4.31 demonstrates that in the presence of an eddy the segments shift in time almost without changing their shapes. The same can be seen in Fig. 4.32. The results of numerical simulation suggest that this property of the timefronts may take place for different mesoscale inhomogeneities [Virovlyansky *et al.* (2007)]. This means that it may be not necessary to monitor the arrivals of single rays. It may be sufficient to measure only the time variations of the whole segments.

Figure 4.33 shows (in a single plot) the arrivals of rays with the identifier -75 in all the three models of the medium under consideration: (a) an unperturbed range-independent channel, (b) a channel with an eddy, and, finally, (c) a channel with an eddy and internal waves. This figure once again demonstrates a general property of the ray travel times: the fuzzy segment (c) is slightly biased relative to the segment (b) towards early times.

Our idea is to "tune" the vertical receiving array to the reception of a segment corresponding to a selected identifier $\pm M$. The dependence of the arrival time of an unperturbed ray with the given identifier on its vertical coordinate z at the observation range r denote $\tau(z)$. Note that the curve (a) in Fig. 4.33 represents a function that is inverse of $\tau(z)$ corresponding to the identifier -75. For simplicity, we assume that the receiving array covers the whole cross-section of the waveguide and its elements are continuously distributed over the aperture. Let us introduce a delay $\tau(z)$ for a signal received by an element located at depth z and then sum up the signal

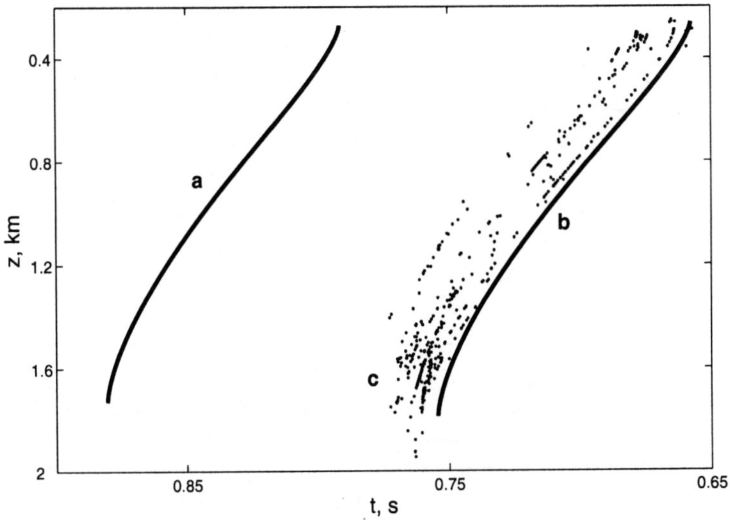

Fig. 4.33 Arrival times of rays with the identifier −75 at a distance of 1500 km in the time-depth plane: (a) an unperturbed channel, (b) in the presence of a synoptic eddy, and (c) in the presence of both a synoptic eddy and internal waves [Virovlyansky *et al.* (2008a)].

intensities from all the elements. This procedure yields an output signal

$$g(t) = \int_{z_{\min}}^{z_{\max}} dz \; |\tilde{v}(r, z, t - \tau(z))|^2 , \tag{4.304}$$

where $z_{\min}$ and $z_{\max}$ are the depths of the upper and lower points of the unperturbed segment associated with the selected identifier.

Figure 4.34 shows three signals $g(t)$ at a range of 1500 km computed by applying the above procedure to the wave field obtained by solving numerically the parabolic equation. (The plots presented in Fig. 4.32 were constructed on the basis of the same numerical simulations.) The receiving array was tuned to the segment with the identifier −75. The upper panel of Fig. 4.34 demonstrates the signal $g(t)$ in an unperturbed channel with $\delta c_m = 0$ and $\delta c_i = 0$. The purpose of introducing the delays $\tau(z)$ is to equalize the arrival times of the rays with the identifier −75 at all array elements. The contributions from pulses arriving via these rays form a peak of the signal $g(t)$ in the upper panel centered at $t = 0$ and indicated by an arrow. Apart from this peak (let us call it the main peak), $g(t)$ has maxima corresponding to other (neighboring) segments of the timefront

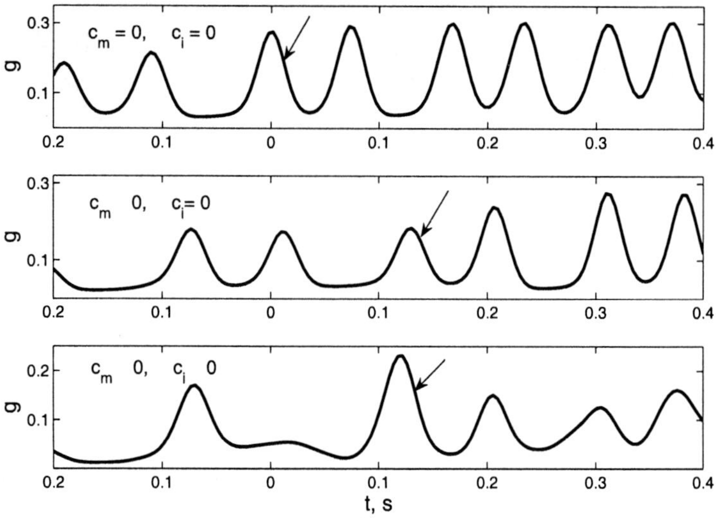

Fig. 4.34 Output signal $g(t)$ at a distance of 1500 km obtained by tuning to the segment with the identifier -75 in an unperturbed channel (top), in the presence of a synoptic eddy (middle), and in the presence of both synoptic eddy and internal waves (bottom). The arrows indicates the main peaks [Virovlyansky *et al.* (2008a)].

which correspond to similar functions $\tau(z)$. The position of each maximum on the t axis shows the time delay between the corresponding segment of the timefront and the selected segment with the identifier -75.

It was noted above that in the presence of a synoptic eddy, the segment shifts almost as a whole. Therefore, the introduction of the same delays $\tau(z)$, as in the unperturbed channel, equalizes the arrival times of the rays with the selected identifier in the perturbed waveguide as well. However, now their contributions produce a peak of $g(t)$ (we continue to call it the main peak) at the time equal to the delay of the selected segment caused by the eddy. This peak is clearly seen in the middle plot.

The lower plot displays the signal $g(t)$ in the presence of both eddy and internal waves. The position of its main peak shows the mean bias of the fuzzy segment. Let us emphasize that in this example the antenna is tuned to a segment formed by relatively flat rays with grazing angles at the channel axis of about 7°.

Thus, the main peak corresponding to the identifier -75 is clearly seen in the presence of internal waves. However, this is not always so. The upper panel in Fig. 4.35 demonstrates the output signal $g(t)$ obtained in

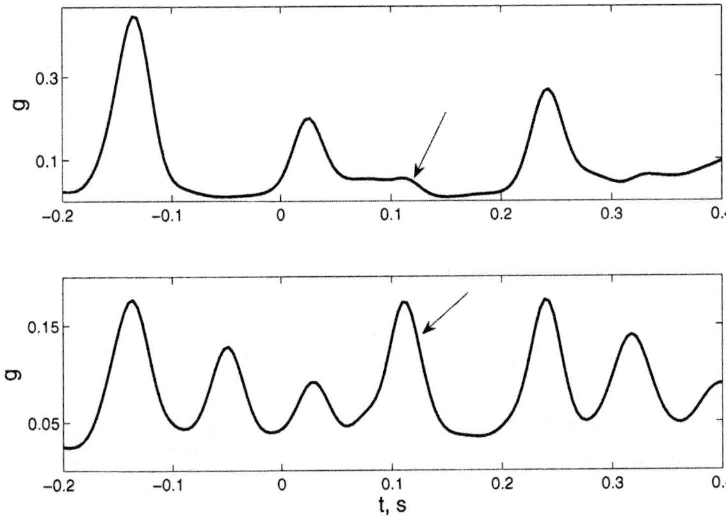

Fig. 4.35 Output signal $g(t)$ at 1500 km obtained in the case of array tuning to the segment with the identifier +73 (top) and the result of averaging over six signals $g(t)$ obtained in the case of array tuning to the segments with identifiers ±72, ±73, and ±74 (bottom) [Virovlyansky *et al.* (2008a)].

the case where the array is tuned to a segment with the identifier +73. The main peak is poorly seen here. To resolve it, we use the fact that the neighboring segments shift by close time intervals. Accordingly, the positions of the main peaks corresponding to them are also close. Using this fact, we averaged six signals $g(t)$ obtained by tuning the array to segments with identifiers ±72, ±73, and ±74. In the averaged signal shown in the lower plot, the main peak is clearly seen.

The stars in Fig. 4.36 show the estimates for the shifts of segments with identifiers $-M$, where $M = 53, \ldots, 79$, which were obtained using the discussed procedure. We took the signal obtained by averaging over six $g(t)$ calculated for $\pm(M-1)$, $\pm M$, and $\pm(M+1)$ as the output signal $g(t)$ for each M. The shifts of the timefront segments in the absence of internal waves are shown in Fig. 4.36 by a solid line. It connects the points indicating the segment shifts at the channel axis ($z = z_{\mathrm{a}}$) at $\delta c_m \neq 0$ and $\delta c_i = 0$. Again, as in Fig. 4.33, we see that the presence of internal waves leads to small shifts of segments towards early times.

Thus, we have demonstrated that the use of a vertical receiving array can significantly extend the number of resolved arrivals in comparison with

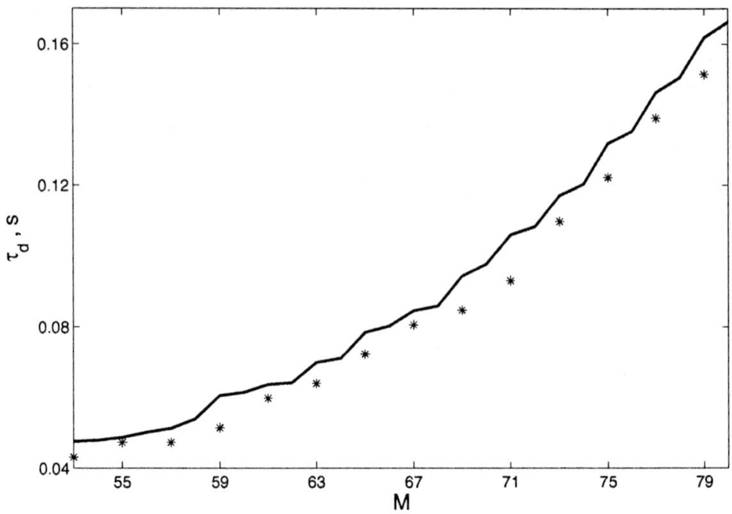

Fig. 4.36 Shifts of timefront segments τ_d under the effect of a synoptic eddy in the presence (stars) and absence (solid line) of internal waves [Virovlyansky *et al.* (2008a)].

the case where signals are recorded on a point receiver. In the lower panel in Fig. 4.32, we see that at $t < -1.5$ s the fuzzy ray pulses do not overlap and they can be resolved in the output signal of a point receiver. These pulses come via relatively steep rays with grazing angles at the sound channel axis greater than 9° and with $M < 67$. The use of the proposed procedure allows one to measure the travel time delays not only for these rays, but also for fairly flat rays with grazing angles at the channel axis of up to 6°, with M reaching a value of 80, and with arrival times of up to -0.45 s. Our approach is based on the fact that in the presence of a mesoscale inhomogeneity a segment shifts almost as a whole. This statement is based on the results of numerical simulations, and its applicability range needs additional research.

In Ref. [Virovlyansky *et al.* (2007)], it was demonstrated that the delays of timefront segments carry information on the temperature field perturbation averaged over distance. Therefore, it is possible to reconstruct climatic variations of the mean oceanic temperature on the basis of measured delays. We considered the simplest realization of the proposed approach to avoid overloading the presentation with unimportant details. The processing algorithm can be improved by taking into account the fact that the rays arriving at the observation point with the same identifiers have grazing angles that differ only slightly [Virovlyansky (2005a)].

Before summing according to Eq. (4.304), it seems to be expedient to apply the corresponding spatial filtering in order to suppress the signals arriving at angles that widely differ from the arrival angles of the rays forming the segment to which the array is tuned.

The above results of numerical simulation once again confirm that the internal waves cause not only the diffusion of the timefront segments, but also their additional small delays. This fact should be taken into account in solving inverse problems.

4.6 Summary

In this chapter we considered the ray and wave chaos in a realistic model of the USC in a deep ocean with the sound speed fluctuations induced by random internal waves.

Chaotic ray motion in a single realization of the random medium can be described using a statistical approach introduced in Sec. 4.1. In the scope of this approach, any statistical characteristic of a ray path starting from a given point (p_0, z_0) of the phase space is determined by averaging over a bundle of rays starting from a small area centered at this point. The bundle plays the role of a statistical ensemble. Statistical description is greatly simplified due to the fact that the angle variable θ rapidly randomizes and already at comparatively short ranges (hundreds of kilometers) the wrapped θ is more or less uniformly distributed on $(0, 2\pi)$. Diffusion of the action variable I is described by the Fokker-Planck equation.

It turned out that in our environmental model the diffusivity weakly depends on the ray launch angle and, therefore, can be approximated by a constant. Then, the Hamilton equations are approximated by Langevin equations (4.34) and (4.47) whose solution are readily expressed through a random Wiener process representing the simplest model of diffusion. In the Wiener process approximation, the PDFs of ray parameters are obtained analytically or using a simple numerical procedure described at the end of Sec. 4.1.5.

At megameter ranges, where the ray chaos is well developed, the wave field at any observation point is formed by contributions from a large number of chaotic rays. Incoherent summation of their intensities gives a coarse-grained distribution of the wave field intensity. The sum over chaotic rays can be expressed through integration over their starting parameters. In our approach, this integration is interpreted as a statistical averaging. This yields an analytical estimate for the smoothed intensity of the sound field.

Section 4.1.8 represents a purely deterministic approach to ray motion in a randomly-inhomogeneous waveguide. Specific Poincaré map described in this section allows one to find domains of finite-range stability in the phase space. Presence of such domains infers intermittent ray behavior, that is, irregular pieces of a ray trajectory alternate with regular ones.

In Sec. 4.2 we studied the properties of ray travel times. It was demonstrated that, as in a waveguide with a periodic perturbation, rays with equal identifiers have close travel times. This property manifests itself in the effect of clustering. A weak perturbation causes an eigenray to split into a cluster of eigenrays whose travel-time spread is small and whose centroid has a travel time that is close to that of the eigenray in the background environment. All the eigenrays belonging to the cluster have the same identifier as the original unperturbed one. Their travel times are determined by intersections of the timefront segment with the line $z = z_r$, where z_r is the depth of an observation point. Due to the effect of clustering, the perturbation, even at megameter ranges, does not destroy the timefront completely. It causes only diffusion of the timefront segments and their bias toward early times. Segments, formed by the early arriving steep rays, remain almost unaffected. Quantitative explanation of the clustering effect is based on Eq. (4.137) for the difference in travel times of perturbed and unperturbed rays. This formula, combined with the stochastic ray theory derived in Sec. 4.1, yields analytical estimates for the spread and bias of the timefront segment due to the sound speed fluctuations. Our results also explain why segments formed by steep rays are significantly more stable compared to those formed by flat rays. It was shown that the stability of ray travel times is associated with the Fermat's principle.

Manifestations of chaotic ray dynamics in the modal structure of the wave field is studied using the ray-based formalism for description of mode amplitudes derived in Sec. 3.2.4. In the scope of this approach the amplitude of the m-th mode is formed by contributions from rays – we call them the mode rays – whose action variables at the observation range up to a (frequency dependent) multiplicative constant are approximately equal to m. Under conditions of ray chaos, the number of mode rays contributing to the given mode exponentially grows with range. At long ranges, the coarse-grained distribution of mode intensities can be evaluated by incoherent ray summation. Analytical estimate for the coarse-grained distribution of acoustic energy between normal modes is obtained by combining the ray-based formalism and the stochastic ray theory based on the Wiener process approximation.

The ray-based formalism derived for the monocromatic wave field can be applied for studying the transient wave field, as well. In Sec. 4.3 our attention was focused on the mode pulses, that is, sound pulses carried by individual modes. It turned out that the arrival time of a fragment of the m-th mode pulse with the instantaneous frequency Ω is approximately equal to the travel time of a mode ray contributing to the m-th mode at this frequency. This result is a key point for the analysis of the transient wave field. It allows one to apply the stochastic ray theory for the description of mode pulses. This approach yields estimates for the spread and bias of the mode pulse caused by random internal waves.

In Sec. 4.4, it was shown how the ray-based description can be introduced for a wave field excited by a vertical antenna. In the preceding sections we dealt only with a point source and a single-mode source. First, the ray approach was used to examine the spatial spread of a narrow acoustic beam propagating in a range-independent waveguide. We found an initial distribution of the wave field at the antenna corresponding to a beam which diverges slower than other beams with the same initial width. Then, using the stochastic ray theory we estimated the range interval at which the influence of the internal-wave-induced perturbation on the beam is still negligible. Finally, we evaluated the widening of the modal spectrum of the wave beam due to scattering at the sound speed fluctuations.

In Sec. 4.5 we considered the propagation of sound pulses through a synoptic eddy representing an example of the mesoscale inhomogeneity. It was assumed that the random internal waves are also present and give rise to ray chaos. In this environmental model, only pulses propagating via steep rays can arrive at the observation point without overlapping and be resolved. In the method of ocean acoustic tomography, their delays are used as input parameters in reconstructing a perturbation of the temperature field caused by the eddy. We showed that the use of a vertical array allows one to increase the number of resolved pulses. Procedure of the space-time processing discussed in Sec. 4.5 makes it possible to resolve pulses coming through not only steep rays but through relatively flat ones, as well. The procedure is based on the fact that the timefront segments remains relatively stable even at megameter ranges where the ray chaos is well developed.

Most results discussed in this chapter were derived using the ray-based description of the sound field combined with the stochastic ray theory. A major limitation of our approach is an insufficient understanding of its applicability conditions. This is a common problem for practically all ray-based approaches. But in our case an additional difficulty arises due to the conjecture that the averaging over ray starting parameters may be replaced by the statistical averaging. Obviously, this may be done only when evaluating smoothed characteristics of the wave field. However, the question of selecting proper smoothing scales remains open and requires a further investigation. The smoothing scales used in our calculations were selected empirically.

Chapter 5

Glossary of Some Concepts and Notations in Hamiltonian Chaos Theory

A brief and simple description of basic concepts, methods, and approaches in Hamiltonian chaos theory in the alphabetic order is given with a number of illustrations. Rigorous results and their proofs can be found in a number of textbooks and monographs on dynamical systems theory and Hamiltonian chaos [Arnold *et al.* (1988); Lichtenberg and Liberman (1983); Zaslavsky (2005)].

Bernoulli Shift

Measuring or computing a trajectory, we in fact obtain sequences of numbers. Discrete maps, producing such sequences, can be simple generators of dynamical chaos. We demonstrate that with a one-dimensional map known as *a Bernoulli shift*:

$$x_{n+1} = 2x_n \pmod 1,$$

where (mod 1) means that one omits an integer part of a number x at each iteration step. In spite of the fact that the Bernoulli shift has the analytic solution $x_j = 2^j x_0$, it may generate random sequences. A binary representation of an initial number x_0 is

$$x_0 = \sum_{j=1}^{\infty} a_j \cdot 2^{-j} : (0.a_1 a_2 a_3 \ldots),$$

where a takes the values 0 or 1. For example,

$$x_0 = \frac{1}{2} + \frac{1}{8} + \frac{1}{16} + \frac{1}{64} + \ldots = 0.101101\ldots.$$

In the binary representation, the operator $\hat{f}(x) = 2x \pmod 1$, when acting on x, shifts the point to the right. Taking as a initial condition, for example, the above mentioned number x_0, we get: $x_1 = 0.01101\ldots$, $x_2 = 0.1101\ldots$, $x_3 = 0.101\ldots$. Iterations of the Bernoulli shift are sensitive exponentially to small variations in initial conditions: $\Delta x_n = 2^n \Delta x_0 = e^{n \ln 2} \Delta x_0$, where $\nu = \ln 2$ is a positive *Lyapunov exponent* (see the corresponding article in the Glossary). In order to obtain a number with the accuracy of N digits after n iterations, one needs to know the initial number with the accuracy of $N + n$ digits. The horizon of predictability

$$n_p = \frac{1}{\ln 2} \ln \frac{1}{\Delta x(0)}$$

depends strongly on the Lyapunov exponent and only logarithmically on the uncertainty in initial data $\Delta x(0)$.

With $n > n_p$ only statistical predictions are possible. Sequences of iterations in the Bernoulli shift have the same statistical properties as classical random processes like a coin toss or a roulette. Whereas there is a random mechanism in the classical random processes (for example, a roulette wheel), randomness arises in the Bernoulli shift from a fully deterministic law.

The operator $2\times$ provides stretching of a segment of initial data, and the operator $\pmod 1$ provides its folding and returns iterations of initial points back into the unit interval. It is a universal mechanism generating dynamical chaos not only in discrete maps but in differential equations as well. If the initial point x_0 is a rational number, then the sequence of iterations will repeat itself beginning with, say, a number m. Such a point is a m-periodic point of the map or a m-periodic cycle, the analogue of a periodic trajectory of continuous systems. The Bernoulli shift contains a countable set of periodic points of all the periods. The set of initial conditions (rational numbers), generating periodic sequences, is infinite, countable, and of a zero measure. Random (chaotic) sequences are generated under the Bernoulli map by irrational numbers, the set of which is infinite, uncountable, and of unit measure.

Bifurcation

Bifurcation means a qualitative change in the topology in *the phase space* (see the corresponding article in the Glossary) under varying control parameters of a dynamical system under consideration. The number of *stationary points* (see the corresponding article in the Glossary) and their stability may change when varying the parameters. Those values of the parameters, under which bifurcations occur, are called *critical* or *bifurcation values.*

Cantorus

Some *invariant tori* (see the corresponding article in the Glossary) in typical unperturbed Hamiltonian systems break down under a perturbation. Suppose that an invariant torus with the frequency f breaks down at a critical value of the perturbation frequency ω. If f/w is a rational number, then a chain of resonances or *islands of stability* (see the corresponding article in the Glossary) appears at its place. If f/w is an irrational number, then *a cantorus* appears at the place of corresponding invariant torus. Cantorus is a Cantor-like invariant set [Percival (1980); MacKay *et al.* (1984)] the motion on which is unstable and quasiperiodic (see the article "Fractal" in the Glossary). Cantorus looks like a closed curve with an infinite number of gaps. Its topological measure is equal to zero, therefore, cantori are *fractal* (see the corresponding article in the Glossary). Since the motion on a cantorus is unstable, it has stable and unstable *manifolds* (see the corresponding article in the Glossary). All the points on a cantorus belong to the same quasiperiodic *trajectory* (see the corresponding article in the Glossary) if its initial point belongs to the same cantorus. Cantori are singular objects that do not occupy a volume of a finite measure in *the phase space* (see the corresponding article in the Glossary). However, they form an infinite hierarchy around *islands of stability.* The closer a cantorus is to the island's boundary, the narrower are its gaps. Cantori influence essentially the transport in the phase space because it may take a long time for particles and their trajectories to percolate through cantori gaps. Since the smallest gaps appear in the cantori which are close to the very boundaries of the islands, one observes at those places an increased density of phase points on Poincaré sections. The island's boundaries are called *dynamical traps* (see the corresponding article in the Glossary) and the long stay of trajectories there is called *a stickiness.*

Cantori are not a single reason for stickiness and dynamical traps. Hyperbolic trajectories along with their stable and unstable *manifolds* (see the corresponding article in the Glossary) form such a complicated tangle where particles and their trajectories may be trapped for a long time.

Dynamical Trap

A dynamical trap is a domain in the *phase space* (see the corresponding article in the Glossary) where a particle (and its trajectory) may spend an arbitrary long but finite time [Zaslavsky (2002, 2002a, 2005)] in spite of the fact that the trajectory is chaotic in any relevant sense. Strictly speaking, it is the definition of a quasitrap. Absolute traps, where particles could spend an infinite time, are not possible in Hamiltonian systems which have no attractors. The dynamical traps are caused by *stickiness* of trajectories, mainly, to the boundaries of *islands of stability* (see the corresponding article in the Glossary) where *cantori* (see the corresponding article in the Glossary) are situated. Up to now there is no full classification and description of dynamical traps. Dynamical traps influence significantly transport in Hamiltonian systems specifying its anomalous statistical properties.

Fractal

The name "*fractal*" was coined by [Mandelbrot (1982)] to describe irregular and self-similar structures, i.e., objects small parts of which are similar in a sense to big parts and those in turn are similar to whole object. Such objects in mathematics as *Cantor sets* with zero topological measure, Weierstrass functions, which are everywhere continuous but not differentiable anywhere, Julia sets, etc. [Mandelbrot (1982)] have been known for many years. However, they have been considered as exotic objects not existing in real world. One of the definitions of the fractal is the following: fractal is a set whose Hausdorff-Besikovich measure is larger than its topological measure [Mandelbrot (1982)].

Hamiltonian chaos produces different kinds of fractals. Stochastic layers and hierarchies of *islands of stability* (see the corresponding article in the Glossary) and of *cantori* (see the corresponding article in the Glossary) are fractal. Typical chaotic trajectories are fractal in a sense [Zaslavsky (2005)]. Chaotic scattering, exit-time functions [Ott (1993); Gaspard (1998)], and Poincaré recurrences [Zaslavsky (2005)] are fractal as well. Chaotic invariant sets in Hamiltonian systems are fractal and have a Cantor-like structure (see the article "Horseshoe mapping" in the Glossary).

Fractal sets appear in dynamical systems in a natural way if there exists a mechanism removing phase points out off a given region in the phase space. In dissipative systems it is a dissipation which shrinks in course of time an initial phase volume to a set called "*an attractor*". In chaotic dissipative systems such attractors may be fractal sets (strange attractors). In Hamiltonian systems the phase volume is conserved (Liouville theorem), and fractal sets may appear when a scattering problem is formulated.

In the long-range underwater sound propagation, the scattering problem can be formulated as follows. One computes travel times or travel length for the rays arriving to a given range as functions of their initial grazing angle. It has been shown by [Zaslavsky and Abdullaev (1997)] that a sensitive dependence of ray trajectories in the mixed phase space on initial conditions produces the scattering functions with intermittent smooth and wildly oscillating fragments. Magnification of the intervals of the grazing angle, where the travel times (travel lengths) oscillate in a complicated way, shows a self-similarity that is typical with other scattering problems [Gaspard (1998); Budyansky *et al.* (2004)]. Another manifestation of fractal properties of rays has been found by [Abdullaev and Zaslavsky (1988)] when computing the spatial frequency as a function of the initial grazing angle. The corresponding plot is a kind of a devil's staircase, a multistep function with a self-similarity.

The famous Cantor fractal is formed as follows [Mandelbrot (1982)]. Take the closed interval $[0,1]$ and remove the open middle third interval $(1/2,2/3)$, leaving the two intervals $[0,1/3]$ and $[2/3,1]$. Then remove the middle open thirds of each of these two intervals, leaving four closed intervals of length $1/9$ each, etc. The total length of remaining segments is $\lim_{n\to\infty} 2^n r = \lim_{n\to\infty}(2/3)^n \equiv \lim_{n\to\infty} e^{-n\ln(3/2)}$ (one gets $N = 2^n$ segments with the length $r = (1/3)^n$) each after n iterations). Though the set of remaining segments is infinite, its total length or the Lebesque measure is zero (see the article "Measures in dynamical systems" in the Glossary). Topological measure of the classical Cantor fractal d_t is zero. Other measures have been introduced to characterize such dust-like objects called *Cantor sets*. The Hausdorff-Besikovich measure is a common used one

$$d_{HB} = \lim_{r\to 0} \frac{\ln N}{\ln(1/r)}.$$

In the case of the classical Cantor fractal one gets: $d_{HB} = \ln 2/\ln 3 = 0.63\ldots$, i.e., the fractal measure is not an integer. It is larger then the topological measure of a point ($d_t = 0$), but smaller then the topological measure of an interval ($d_t = 1$). The chaotic invariant set Λ in *the horseshoe mapping* (see the corresponding article in the Glossary) with zero Lebesque measure is an example of a Cantor-like fractal.

Dynamical systems may produce another kinds of fractals with nonzero Lebesque measure. Let us, for example, take the open interval $[0,1]$ and remove the middle third at the first step, then remove $(1/3)^2$ part of the middles in the two remaining segments at the second step, and remove $(1/3)^n$ part of the middles of the remaining segments at the n-th step. The total length of the remaining segments are $d_f = \prod_{n=1}^{\infty}(1-3^{-n}) = 0.56\ldots$, i.e. the topological measure is not an integer but the Hausdorff-Besikovich measure is equal to one. Such fractal sets with positive Lebesque measure are called *fat fractals* [Ott (1993)]. They need one or more scaling exponents to be characterized.

Typical chaotic Hamiltonian systems having the mixed phase space with *islands of stability* (see the corresponding article in the Glossary) and *dynamical traps* (see the corresponding article in the Glossary) produce, as a rule, fat fractals with the probability density functions (for example, distributions of number of rays over exit times) having power-law "tails". Hyperbolic chaotic Hamiltonian systems, that do not have *KAM tori* (see the corresponding article in the Glossary) and *cantori* (see the corresponding article in the Glossary), produce Cantor-like fractals with exponential probability density functions.

Hamiltonian Chaos

Hamiltonian chaos is dynamical chaos in Hamiltonian systems (see the article "Hamiltonian dynamics" in the Glossary). A deterministic dynamical system is called chaotic if it is locally unstable, i.e. it has at least one positive *Lyapunov exponent* (see the corresponding article in the Glossary)) and is globally mixing.

The mixing is defined as follows. Let B is a region with dye in a waterpool A with a circulation. The volume of B at $t=0$ is $V(B_0)$. Let C is another region in A. The amount of dye in C is $V(B_t \cap C)$ at the moment of time t and its concentration in C is $V(B_t \cap C)/V(C)$. The definition of global mixing is: $V(B_t \cap C)/V(C) - V(B_t)/V(A) \to 0$ at $t \to \infty$, i.e., the concentration of dye in any region C in the waterpool A is the same as in the entire waterpool. Recall that in Hamiltonian systems the phase fluid is incompressible, i.e. $V(B_t) = V(B_0)$.

Local instability (see the article "Stable and unstable motion" in the Glossary) produces an exponential sensitivity of trajectories to small variations of initial conditions and/or control parameters. Generally speaking, it is difficult to prove analytically existence of chaos in a system under consideration. Manifestations of dynamical chaos become evident when computing *Poincaré sections* (see the corresponding article in the Glossary), Melnikov integrals (see the article "Melnikov method" in the Glossary), intersections of stable and unstable *manifolds* (see the corresponding article in the Glossary), and maximal *Lyapunov exponents* (see the corresponding article in the Glossary).

Theory of Hamiltonian chaos is presented in a number of monographs and textbooks [Arnold *et al.* (1988); Lichtenberg and Liberman (1983); Zaslavsky (2005)]. *The phase space* in a typical Hamiltonian system is mixed, i.e., the regions with regular motion coexist with chaotic ones. In Fig. 5.1 we show *a Poincaré section* (see the corresponding article in the Glossary) of a Hamiltonian system with the mixed phase space with chains of *islands of stability* (see the corresponding article in the Glossary) separated by stochastic layers (where motion is chaotic) and confined between *invariant KAM tori* (see the corresponding article in the Glossary). KAM tori are stable invariant *manifolds* with boundaries that are impenetrable to particle's transport. There are an infinite number of cantori around the nested islands of stability (not shown in the figure). They are Cantor-like unstable invariant sets with gaps (see the article "Fractal" in the

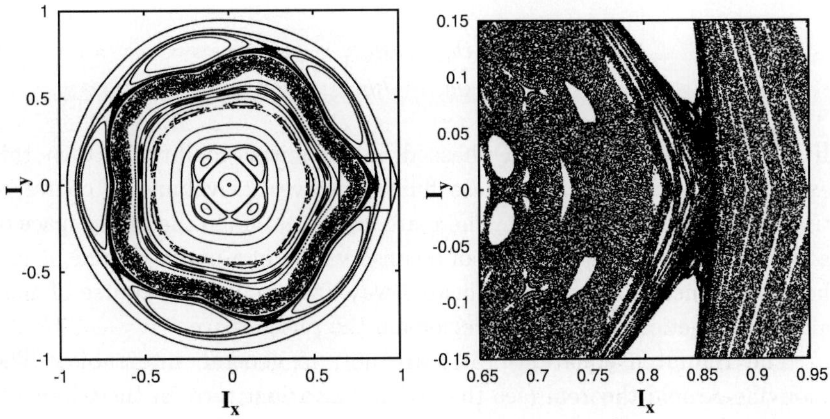

Fig. 5.1 Left panel — Poincaré section of a Hamiltonian system with *islands of stability* (see the corresponding article in the Glossary) (closed solid curves) and a stochastic layer between confining *invariant tori* (see the corresponding article in the Glossary). Right panel — zoom of the small region of the stochastic layer indicated in the left panel.

Glossary) transport throw which is possible but difficult. Hamiltonian chaos is a special type of motion with the properties both of regular motion (due to determinism of equations of motion) and stochastic motion (due to local instability of trajectories).

Hamiltonian Dynamics

Hamiltonian dynamics is a geometry in *phase space* (see the corresponding article in the Glossary) [Arnold (1989)]. Each state of a Hamiltonian system with N degrees of freedom in the phase space is described by N generalized positions $(q_1, \ldots, q_N)$ and momenta $(p_1, \ldots, p_N)$ which are pairwise canonically conjugated variables. The equations of motion of a Hamiltonian system are specified with the help of a Hamiltonian function of the generalized positions and momenta

$$\dot{\mathbf{q}}_i = \frac{\partial H}{\partial \mathbf{p}_i}, \quad \dot{\mathbf{p}}_i = -\frac{\partial H}{\partial \mathbf{q}_i}. \tag{5.1}$$

If the Hamiltonian $H(p, q, t)$ depends on time, then the corresponding system can be studied in the enlarged $(2N + 1)$-dimensional phase space $(q_1, \ldots, q_N; p_1, \ldots, p_N; t)$ where it has $N + 1/2$ degrees of freedom.

The equations (5.1) satisfy to the incompressibility condition

$$\sum_{i=1}^{N} \left(\frac{\partial \dot{q}_i}{\partial q_i} + \frac{\partial \dot{p}_i}{\partial p_i} \right) = 0. \tag{5.2}$$

If one specifies a volume (a phase drop) of initial conditions, then this expression means that the phase fluid conserves its volume in course of time (the Liouville theorem). There are no attractors in the phase space of a Hamiltonian system. In spite of a conservation of a phase volume, it can be transformed in a very complicated way and occupies in course of time all the energetically accessible regions in the phase space.

The Hamilton equations (5.1) are not, in general, integrable. The Liouville-Arnold theorem (see the article "Invariant tori" in the Glossary) states that a Hamiltonian system with N degrees of freedom is *fully integrable* if there exist N linearly independent first integrals of motion C_i in involution, i.e. with zero Poisson brackets $\{C_i, C_j\} \equiv 0, i, j = 1, 2, \ldots, N$. Equations of motion (5.1) of a fully integrable system can be always transformed to the following form:

$$\dot{I}_i = -\frac{\partial H}{\partial \theta_i} = 0, \quad \dot{\theta}_i = \frac{\partial H}{\partial I_i} \equiv \omega_i \quad (I_1, \ldots, I_N), \tag{5.3}$$

where I_i and θ_i are pairwise canonically conjugated variables known as *action* and *angle*, respectively.

Trajectories of a Hamiltonian system with N integrals of motion lie on N dimensional invariant *manifolds* (see the corresponding article in the Glossary) in the $2N$ dimensional phase space. The manifolds have torus topology and are called *invariant tori* (see the corresponding article in the Glossary). Any trajectory, starting on a given torus, stays on it all the time. If a Hamiltonian system is fully integrable, then the representation in terms of I_i and θ_i is global, i.e., the phase space is partitioned to invariant tori, and any trajectory is located on some torus. If a system is not integrable, then some trajectories do not lie on invariant tori. Up to now, there is no complete theory of behavior of nonintegrable Hamiltonian systems. However, there exists very important *Kolmogorov-Arnold-Moser theorem* (see the corresponding article in the Glossary) about the behavior of Hamiltonian systems under weak perturbations.

Heteroclinic (Homoclinic) Structure

In an integrable Hamiltonian system *separatrices* (see the corresponding article in the Glossary) connect either two hyperbolic *stationary points* (see the corresponding article in the Glossary) in such a way that the stable separatrix of one point $W_s^{(0)}(h_1)$ coincides with the unstable separatrix of the other point $W_u^{(0)}(h_2)$ and vice versa (see Fig. 5.2a) or $W_s^{(0)}$ and $W_u^{(0)}$ of the same hyperbolic point coincide (see Fig. 5.2b). In the former case one gets *heteroclinic connection*, whereas in the latter one — *homoclinic connection*.

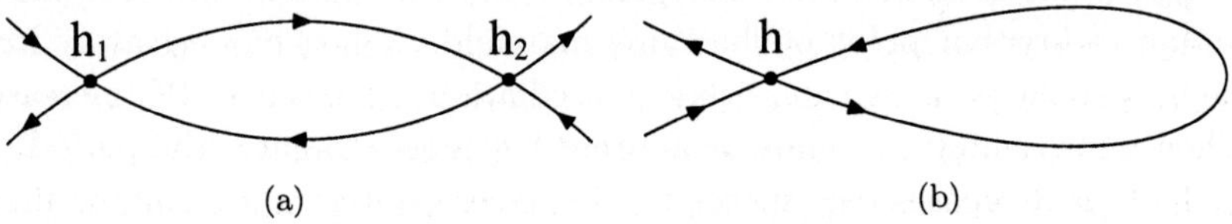

Fig. 5.2 (a) A heteroclinic connection of nonperturbated separatrices of two hyperbolic points h_1 and h_2 and (b) a homoclinic connection with one hyperbolic point h on a phase plane.

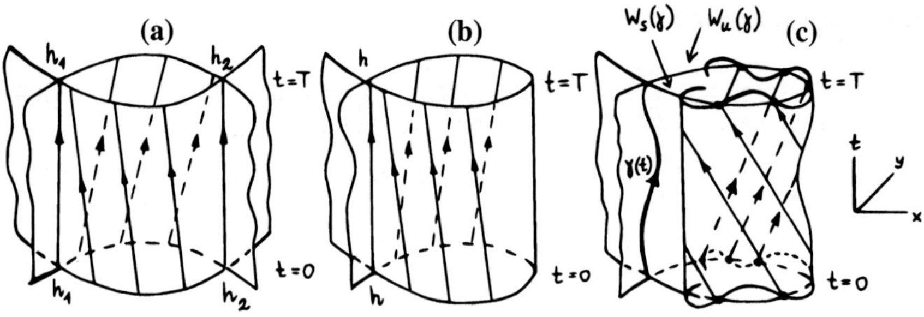

Fig. 5.3 Schematic representation of (a) heteroclinic and (b) homoclinic connections of a nonperturbed one-degree-of-freedom system in the extended phase space (x, y, t). (c) Under a perturbation with the period T, a saddle point h becomes a periodic hyperbolic trajectory $\gamma(t)$ whose stable $W_{\rm s}(\gamma)$ and unstable $W_{\rm u}(\gamma)$ manifolds are surfaces intersecting in the extended phase space. The lines with arrows on those manifolds represent typical trajectories.

In Fig. 5.2 those connections are shown in the phase plane (x, y) of a Hamiltonian system with one degree of freedom, and in Fig. 5.3(a) and (b) they are shown in the enlarged phase space (x, y, t). Under a perturbation with a period T, hyperbolic (saddle) points of the unperturped system become unstable periodic *trajectories* (see the corresponding article in the Glossary) $\gamma(t)$ with stable and unstable separatrix branches which are called stable $W_{\rm s}(\gamma)$ and unstable $W_{\rm u}(\gamma)$ *manifolds* (see the corresponding article in the Glossary) of the corresponding hyperbolic trajectory $\gamma(t)$. In the enlarged phase space the manifolds $W_{\rm s}(\gamma)$ and $W_{\rm u}(\gamma)$ are two-dimensional surfaces which do not coincide but intersect (see Fig. 5.3(c)). The corresponding curves $W_{\rm s}(\gamma)$ and $W_{\rm u}(\gamma)$ intersect each other on *a Poincaré section* (see the corresponding article in the Glossary) in *homoclinic points.* H. Poincaré proved that there are an infinite number of homoclinic points of intersections of stable and unstable manifolds of a hyperbolic trajectory.

Any point of an invariant manifold maps, by definition, on the Poincaré section to another point of the same manifold. When moving away from the hyperbolic point the amplitude of oscillations of the curve $W_{\rm u}$ increases. When approaching the same or another hyperbolic point, "the period" of oscillations decreases (the successive distances between the points of intersections of $W_{\rm s}$ and $W_{\rm u}$ decrease when approaching to h) because of slowing down of the motion nearly h. It results in a complicated *heteroclinic* or *homoclinic structure.*

Horseshoe Mapping

Phase volumes stretch and fold nearby hyperbolic trajectories. The map, known as *a horseshoe map* or Smale's horseshoe [Smale (1967)] is a geometric model of unstable motion. Figure 5.4 demonstrates schematically evolution of the small phase volume 1–2–3–4 chosen on the Poincaré section nearby a saddle point h in such a way that the side 1–2 is a fragment of its unstable manifold $W_{\rm u}$ and the sides 1–3 and 2–4 are fragments of its stable manifold $W_{\rm s}$ (see Fig. 5.4(a)). When iterating points of this volume forward in time, the volume stretches along the unstable manifold (the side 1′–2′ as before is on $W_{\rm u}$), shrinks along the stable manifold (the sides 1′–3′ and 2′–4′ are on $W_{\rm s}$) and the deformed volume looks like a horseshoe (Fig. 5.4(b)). When iterating the phase points backward in time, the volume shrinks along $W_{\rm u}$ and stretches along $W_{\rm s}$ in such a way that the horseshoe is formed in the perpendicular direction (see Fig. 5.4(c) where only the first phase of folding is shown). After a finite number of iterations, the forward and backward images of the phase volume intersect many times.

Following to S. Smale (1967), let us consider mapping of a square with the unit side when the square shrinks in horizontal direction, stretches along the vertical direction and folds (Fig. 5.5). The square is not mapped into itself, and the iterated points, which leave the square under the map, are not considered further. Let the horizontal strip H_0 in the upper panel of Fig. 5.5 contain points with the coordinates $(0 \le x \le 1,\ 0 \le y \le \lambda_h^{-1})$, and the horizontal strip H_1 contains points with $(0 \le x \le 1,\ 1 - \lambda_h^{-1} \le y \le 1)$.

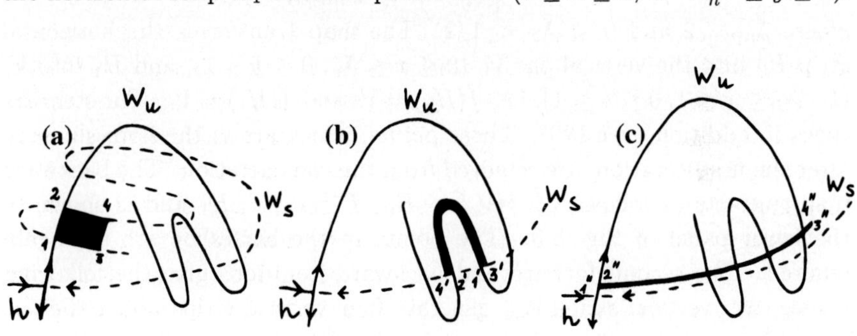

Fig. 5.4 "A horseshoe" appears in *the homoclinic structure* (see the corresponding article in the Glossary) on the *Poincaré section* (see the corresponding article in the Glossary) as a result of (b) forward and (c) backward iterations of the phase volume 1–2–3–4 (a). Segments of the stable $W_{\rm s}$ (dashed line) and unstable $W_{\rm u}$ manifolds of the saddle point h are shown.

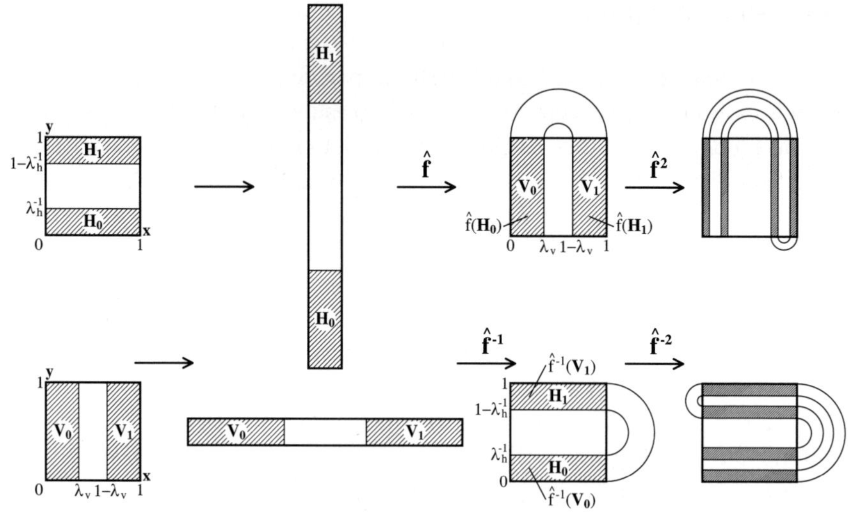

Fig. 5.5 Forward (upper panel) and backward (lower panel) iterations of the horseshoe map.

Then the linear horseshoe mapping is given by transformation

$$\hat{f}(H_0) : \begin{pmatrix} x \\ y \end{pmatrix} \to \begin{pmatrix} \lambda_v & 0 \\ 0 & \lambda_h \end{pmatrix} \begin{pmatrix} x \\ y \end{pmatrix} ;$$
$$\hat{f}(H_1) : \begin{pmatrix} x \\ y \end{pmatrix} \to \begin{pmatrix} -\lambda_v & 0 \\ 0 & -\lambda_h \end{pmatrix} \begin{pmatrix} x \\ y \end{pmatrix} + \begin{pmatrix} 1 \\ \lambda_h \end{pmatrix} , \tag{5.4}$$

where $\lambda_h > 2$ and $0 < \lambda_v < 1/2$. The map transforms the horizontal strip H_0 into the vertical one V_0 ($0 \le x \le \lambda_v$, $0 \le y \le 1$), and H_1 into V_1 ($1 - \lambda_v \le x \le 1, 0 \le y \le 1$), i.e. $\hat{f}(H_0) = V_0$ and $\hat{f}(H_1) = V_1$. The strip H_1 turns in addition over 180°. Those points, which are in the horseshoe arc after the first iteration, are removed from the consideration. The backward mapping acts as follows: $\hat{f}^{-1}(V_0) = H_0$, $\hat{f}^{-1}(V_1) = H_1$ and is shown in the lower panel of Fig. 5.5. The points in the horseshoe arc are again ignored. The second forward and backward iterations give the following result: two vertical strips $V_{0,1}$ generate four vertical strips under the the forward iteration and two horizontal strips $H_{0,1}$ generate four horizontal ones under the the backward iteration. The forward n-th iteration produces 2^n vertical strips with the width λ_v^n each and the backward n-th iteration produces 2^n horizontal strips with the width λ_h^{-n} each. As a result of the forward iterations, at $n \to \infty$ there remain *a Cantor set* (see the article

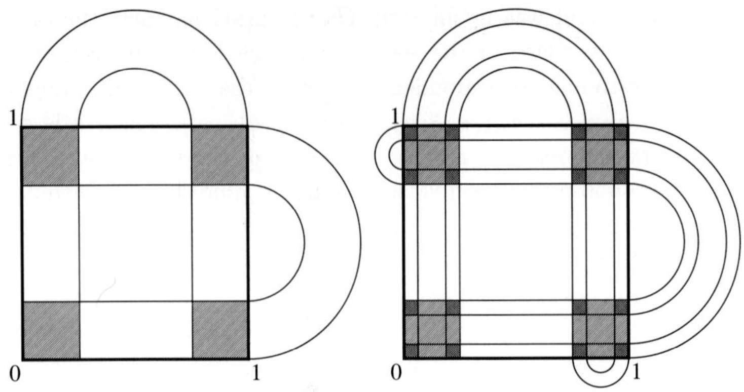

Fig. 5.6 Points in shadowed squares remain in the initial unit square as a result of $n = 1$ (left) and $n = 2$ (right) forward and backward iterations. At $n \to \infty$ the set of such points is an invariant set Λ.

"Fractal" in the Glossary) of vertical strips. In order to know where the points of this set are situated in the initial square, it is necessary to perform a backward iteration that produces at $n \to \infty$ a Cantor set of horizontal strips.

An invariant set Λ of the horseshoe map is formed at the intersection of all the vertical and horizontal strips. The black squares in Fig. 5.6 contain the points remaining in the initial unit square after $n = 2$ forward and backward iterations. After n iterations, one gets 4^n squares with side length going to zero exponentially at $n \to \infty$. The set Λ consists of the points which stay forever in the initial square after forward and backward iterations.

Each point of the set Λ is a hyperbolic one because by definition it is located at the intersection of stretching and shrinking directions. Therefore, orbits in Λ with close initial positions will diverge exponentially under the iterations. It is the main property of dissipative and *Hamiltonian chaos* (see the corresponding article in the Glossary) that can be proved in a simple way [Smale (1967)]. Let us consider the horizontal strips H_0 and H_1 in the upper panel Fig. 5.5. Each point in the set Λ corresponds to a doubly infinite sequence of 0 and 1 constructed as follows: if after n iterations the image of a given point is in the strip H_0, we put 0 at the n-th place of the sequence to the right of its center. If it is in the strip H_1, then put 1 at that place. After each iteration the image is either in H_0 or H_1, otherwise it would be out of the square. The same is valid for backward

iterations: if a point was in the strip H_0 m iterations back, then it should be 0 at the m-th place of the sequence to the left of its center. Thus, there is a one-to-one correspondence between the points in Λ and the sequences $(\ldots, a_{-m}, \ldots, a_{-2}, a_{-2}; a_0, a_1, a_2, \ldots, a_n, \ldots)$, where a is equal to 0 or 1 and the symbol ";" means the center of the sequence. Therefore, the horseshoe mapping is equivalent (homomorphic) to the *Bernoulli shift* (see the corresponding article in the Glossary) for doubly infinite digit sequences for which the *Lyapunov exponent* (see the corresponding article in the Glossary) is positive and equal to $\nu = \ln 2$.

The Smale's result can be formulated as follows: the horseshoe map (5.4) has an invariant set Λ consisting of a countable set of unstable periodic orbits with all the periods, an uncountable set of aperiodic orbits and a dense orbit. *A dense orbit* of an invariant set is a trajectory which comes infinitely closely in course of time to each point of the set. The horseshoe mapping is a paradigm of chaotic dynamics as in dissipative and Hamiltonian systems because stretching and folding of phase volumes nearby hyperbolic trajectories produce the dynamics similar to that arising under iterations of the horseshoe map (see Fig. 5.4).

Invariant Torus

We give a brief description of Hamiltonian systems in the article "Hamiltonian dynamics". The Liuville-Arnold theorem specifies that: 1) all the trajectories of a fully integrable Hamiltonian system with N degrees of freedom and N first integrals in involution C_i lie on N dimensional invariant *manifolds* (see the corresponding article in the Glossary) in the $2N$ dimensional *phase space* (see the corresponding article in the Glossary) which are *invariant tori*; 2) the corresponding *trajectories* (see the corresponding article in the Glossary) are quasiperiodic and specified by N incommensurate frequencies $\omega_i = \omega_i(C_1, \ldots, C_N)$, and 3) satisfy to the equations of motion (5.3). In fully integrable Hamiltonian systems the action I_i is a constant on the corresponding invariant torus, and the angle variable has a simple solution $\theta_i = \omega_i t + \text{const}$ [Arnold *et al.* (1988)].

Invariant torus is *a resonant one* if its eigenfrequences are commensurate, i.e., if $k_1\omega_1 + k_2\omega_2 + \ldots + k_N\omega_N = 0$ for nonzero k_i. If it is not the case, the torus is called *a nonresonant one.* In the former case a trajectory is closed on the torus, and the motion is *(multi)periodic.* In the latter one trajectories are not closed, and the corresponding motion is *quasiperiodic.*

In a nondegenerate fully integrable system, i.e. if

$$\det\left|\frac{\partial\omega_i(\mathbf{I})}{\partial I_j}\right| = \det\left|\frac{\partial^2 H(\mathbf{I})}{\partial I_i \partial I_j}\right| \neq 0, \tag{5.5}$$

each invariant torus has its own frequency. In a nondegenerate system the set of nonresonant tori is more powerful than the set of resonant tori (however, the latter one is dense) because in the set of real numbers the rational ones constitute the subset of zero measure.

Island of Stability

Island of stability is a region on *a Poincaré section* (see the corresponding article in the Glossary) filled with regular trajectories. The islands of stability appear as a result of *nonlinear resonances* (see the corresponding article in the Glossary) between natural frequencies of a nonlinear dynamical system under consideration and perturbation frequencies. Rotational islands correspond to finite regular motion in a bounded region in *the phase space* (see the corresponding article in the Glossary). Ballistic islands correspond to infinite regular motion, i.e. they all filled with ballistic regular trajectories.

Kolmogorov-Arnold-Moser Theorem and KAM Torus

The Kolmogorov-Arnold-Moser theorem (KAM theorem) states that under a sufficiently small conservative Hamiltonian perturbation a majority of *nonresonant invariant tori* does not disappear but they are slightly deformed in such a way that there appear invariant tori (called *KAM tori*) in *the phase space* (see the corresponding article in the Glossary) of a nonintegrable Hamiltonian system filled up everywhere densely with (quasi)periodic *trajectories* (see the corresponding article in the Glossary) [Arnold *et al.* (1988)]. The KAM theorem says nothing about the fate of *resonant tori*. It has been shown in numerous studies that they may break down and produce seeds of *Hamiltonian chaos* (see the corresponding article in the Glossary). Since resonant tori is a set of zero measure (the probability to find such a torus under a random choice of initial conditions is equal to zero), the KAM theorem can be formulated as follows: under a sufficiently small perturbation, almost all invariant tori of the integrable system under consideration are conserved. When proving the theorem, it is stated what is it "sufficiently small" and "almost all" [Arnold *et al.* (1988)].

Lyapunov Exponent

Chaotic motion is characterized by an exponential sensitivity to small variations in initial conditions. It means that initially closed trajectories may diverge exponentially fastly in time (not linearly as in the case with regular trajectories). *A Lyapunov exponent* is a measure of mean velocity of exponential divergence (convergence) of close trajectories.

Equations of motion of an autonomous dynamical system are

$$\dot{x}_i = F_i(x_1, \ldots, x_n), \quad i = 1, \ldots, n. \tag{5.6}$$

Linearizing Eqs. (5.6) nearby a given trajectory $x(t) = (x_1, x_2, \ldots, x_n)$ with the initial condition $X(0)$, we get the equations of motion for small deviations

$$\delta\dot{x}_i = \sum_{j=1}^{n} \delta x_j \left(\frac{\partial F_i}{\partial x_j}\right)_{\mathbf{X}=\mathbf{X}(t)}. \tag{5.7}$$

The norm

$$|\Delta(t)| = \sqrt{\sum_{i=1}^{n} \delta x_i^2(t)} \tag{5.8}$$

is a measure of divergence between the chosen trajectory X and a neighbor trajectory with close initial condition $x(0) + \delta x(0)$. Let us introduce the mean velocity of exponential divergence of trajectories

$$\nu(\mathbf{X}(0)) = \lim_{\substack{t\to\infty \\ \Delta(0)\to 0}} \frac{1}{t} \ln \frac{|\Delta(t)|}{|\Delta(0)|}, \tag{5.9}$$

where $|\Delta(0)| = \sqrt{\sum_{i=1}^{n} \delta x_i^2(0)}$. The limit exists and is finite. The $n \times n$ Jacobian matrix of the linearized system with the elements $(\partial F_i/\partial x_j)_{x=x(t)}$ has a set of n basis vectors. The velocities of divergence (or convergence) in each direction, characterized by its own eigenvector, specify a set of *Lyapunov exponents* which can be ordered by their values: $\nu_1 \geq \nu_2 \geq \cdots \geq \nu_n$. A Hamiltonian system with N degrees of freedom has $2N$ Lyapunov exponents with the following symmetry properties [Lichtenberg and Liberman (1983)]: $\nu_i = -\nu_{2N-i+1}$. In other words, a divergence of trajectories in one direction ($\nu > 0$) is compensated exactly by their convergence in the other direction ($\nu < 0$). It is a consequence of the Liouville theorem on conservation of the phase volumes in Hamiltonian systems. Thus, the sum of all Lyapunov exponents in a Hamiltonian system is zero.

In course of time a small initial phase volume stretches mostly in the direction corresponding to a maximal Lyapunov exponent. Computation with the expression (5.9) gives namely that value which is known as *a maximal Lyapunov exponent.* When computing ν in a chaotic system, the norm $|\Delta(t)|$ increases exponentially, and the distance between close trajectories cannot be treated as small. To bypass this obstacle, the quantity $|\Delta(t)|$ is renormalized to the initial value $|\Delta(0)|$ at each integration step. Generally speaking, values of Lyapunov exponents depend on the choice of a test trajectory $x(t)$. It is not the case in hyperbolic chaotic systems, but the choice of a test trajectory, say, inside *an island of stability* (see the corresponding article in the Glossary) gives obviously $\nu = 0$. The limit (5.9) can be achieved in chaotic systems with the bounded phase space for a reasonable computation time. In open systems $\nu \to 0$ at $t \to \infty$, and chaos in such systems is transient. So-called *finite-time and finite-size Lyapunov exponents* may serve measures of transient chaos.

Manifold

Manifold is a fundamental notion in topology. Its rigorous definition can be found in all textbooks on this subject. It is sufficient here to define a manifold as a smooth subspace in *the phase space* (see the corresponding article in the Glossary). Smoothness means that it is possible to introduce a local coordinate system in a neighbourhood of each point of the manifold. An infinite line and a circle are examples of one-dimensional manifolds, surfaces of a sphere and a torus are two-dimensional manifolfds, the three-dimensional linear space R^3 is an example of a three-dimensional manifold. A segment with its limit points and the surface of a cone are not manifolds because the limit points and the top of the cone do not satisfy to the smoothness criterion.

Nondegenerated hyperbolic *invariant tori* in Hamiltonian systems have *stable* W_s and *unstable* W_u *invariant manifolds* filled up with trajectories asymptotic to quasiperiodic trajectories on an hyperbolic torus at $t \to \infty$ (W_s) and $t \to -\infty$ (W_u). In integrable Hamiltonian systems, the manifolds W_s and W_u coincide, as a rule, pairwisely. In nonintegrable systems they may intersect each other transversally forming a complicated *homoclinic* or *heteroclinic* structures (see the corresponding article in the Glossary).

To give a visual picture of these abstract objects, let us consider stable and unstable manifolds of a periodic saddle trajectory $\gamma(t)$ appearing in

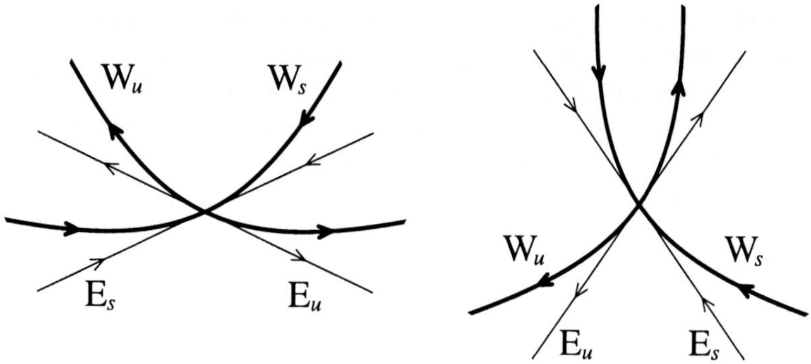

Fig. 5.7 Stable $W_s(\gamma(t))$ and unstable $W_u(\gamma(t))$ manifolds of a saddle trajectory $\gamma(t)$ are shown on the phase plane at t_1 (left) and $t_2 > t_1$ (right) along with the corresponding linear invariant submanifolds $E_s(t)$ and $E_u(t)$.

a plane flow of incompressible fluid under a periodic perturbation from a stationary saddle point of the corresponding integrable system. The manifolds $W_s(\gamma(t))$ and $W_u(\gamma(t))$ are collections of points through which those trajectories of fluid particles go at the moment of time t that are asymptotic to the saddle trajectory $\gamma(t)$ at $t \to \infty$ and $t \to -\infty$, respectively. These manifolds depend on time and the trajectory $\gamma(t)$. In Fig. 5.7 we show schematically a geometry of stable and unstable manifolds nearby a periodic saddle trajectory $\gamma(t)$ at different time moments. Both $W_{s,u}(\gamma(t))$ and the corresponding linear invariant sets $E_{s,u}(t)$, which are specified with the help of a linearization of the velocity field nearby the corresponding saddle point, move in space and time. Under a periodic perturbation, stable and unstable manifolds are periodic functions of time. What happens with W_s and W_u far away from $\gamma(t)$ is discussed in the article "Heteroclinic (homoclinic) structure".

The existence of these manifolds and their structural stability follow from the corresponding theorems which can be found, for example, in the textbook by [Guckenheimer and Holmes (1983)]. Unstable manifolds can be seen with a naked eye in laboratory experiments on chaotic advection in fluids with a dye [Ottino (1989); Koshel and Prants (2008)]. Theoretically, they are curves of an infinite length and complicated form. In real experiments with dyes, they are, of course, diffusive lines owing to molecular diffusion and a technical noise. Since the stable and unstable manifolds are material lines in two-dimensional flows, trajectories of fluid particles cannot cross them, i.e. W_s and W_u are transport barriers. Any material line in a flow is, of course, a transport barrier. Exclusiveness of stable and unstable manifolds is defined by the fact that they divide topologically different regions in flows.

Measure in Dynamical Systems

When studying chaotic dynamical systems, we are forced to leave attempts to describe chaotic dynamics in terms of individual trajectories. In spite of the fact that equations of motion are fully deterministic, their solutions may be so unstable that it is practically impossible to predict the behavior of chaotic trajectories, and one needs to use statistical methods. It is reasonably, for example, to ask how frequently "a typical" trajectory visits a specified domain in the phase space. If a dynamical system is a discrete map, for example, *the Bernoulli shift* (see the corresponding article in the Glossary), the computational algorithm may be the following. Choose in a random way "a typical" initial point, equipartition the unity interval, and compute how many times iterations of the initial point visit a specified segment of the unity interval. Dividing the number obtained to the total number of iterations (it should be large enough), one gets a number meaning the frequency of visiting the segment by "a typical" trajectory of the Bernoulli shift.

The phase space (see the corresponding article in the Glossary) of a continuous dynamical system is partitioned by cells with an equal volume $\Delta\Gamma$. One computes an amount of time "a typical" trajectory spends in the i-th cell $\Delta\Gamma_i$, in other words, one computes the ratio of an amount of time t_i the trajectory spends in the i-th cell to the total amount of time t: $\rho_i(t) = \lim_{t\to\infty}(t_i/t)$. It is *a probability density*, and its integral $\mu(\Delta\Gamma_i) = \int_{\Delta\Gamma} \rho d\Gamma$ is a probability for trajectories to come to a given region in the phase space $\Delta\Gamma_i$. A real nonnegative number μ is *a measure* of that region. A dynamical system can be prescribed a measure in the phase space just like in probability theory one introduces a measure in the space of random events that is equal to a probability of an event in the case of the discrete space and to a probability density in the continuous space.

The notion of a measure can be introduced in another way as well. Let us choose N initial points (particles) and compute their positions at the moment of time t. Suppose we have found $N_i(t)$ particles in the i-th cell. The number $\rho_i'(t) = N_i(t)/N$ corresponds to the density of particles in the i-th cell at time t. Generally speaking, the measures introduced by the two different ways, may be different. They are equal, i.e., an averaging over time is equal to an averaging over the measure (or an ensemble), if the system under consideration is ergodic.

A method of prescribing a real nonnegative number μ to every subset A in the phase space, may be called "a measurement", and the very number $\mu(A)$ is called *a measure of the set* A. It is possible to introduce a large number of different measures. A measure is "good" if almost each initial condition generates a trajectory which "measures" in the limit $t \to \infty$ every subset in the phase space in an equal way. There are, however, initial conditions generating periodic trajectories which have no such a property. If such and other similar initial conditions (of zero measure) are ignored, then a limit value of the amount of time, a trajectory spends in a given region on the phase space, does not depend on the initial conditions. Measures with such a property is called *natural* or physical ones. If the measure of all the phase space is equal to 1, it is called *a probabilistic* one. If it is not changed under time shifts, the corresponding measure is *invariant*. Assuming the existence of a natural probabilistic invariant measure, we can use statistical methods to specify dynamical systems, namely, to introduce and compute different distribution functions, mean values, variations and high moments, correlation function, entropies, etc.

Thus, measure is a map that put a real nonnegative number $\mu(A)$ in a correspondence to a set A. The number $\mu(A)$ is a measure of the set A. The common used measure in the Euclidean space is a Lebesque measure μ_L. The Lebesque measure of a set A is the Lebesque integral $\mu_L(A) = \int_A d\mathbf{x}$. For examples the length of an interval $[a, b]$ is $\int_a^b dx$. Measure of a set is a generalization of the notion of the length of an interval, the square of a plane figure, and the volume of a body to more abstract sets. The invariant measure can be computed, as a rule, for simple discrete mappings only. There is a natural invariant measure in Hamiltonian systems, called a Liouville one, that satisfies to the Liouville equation. The existence of an invariant measure allows to compute *Lyapunov exponents* (see the corresponding article in the Glossary) because time averages along a given direction are the same with all the initial conditions excepting a zero-measure set.

Melnikov's Method

The Melnikov's method provides an analytical study of the motion nearby the unperturped separatrix of a dynamical system close to an integrable

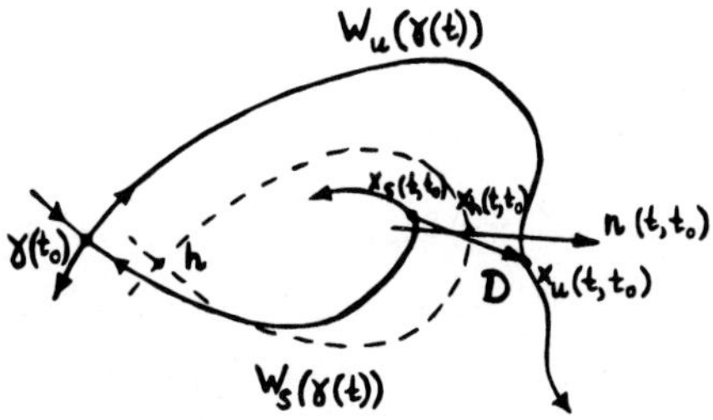

Fig. 5.8 Melnikov's separatrix splitting. The dashed curve is an unperturped separatrix of the hyperbolic point h displaced under a perturbation to the point γ at a time moment t_0.

one. This method allows to prove analytically intersections of stable and unstable *manifolds* (see the corresponding article in the Glossary) and calculate a distance between them (see the article "Heteroclinic (homoclinic) structure" in the Glossary) in the first order of the perturbation amplitude. The idea and the results obtained by [Melnikov (1963)] are based on using global solutions of the corresponding unperturped system when calculating perturbation solutions. We describe a simple variant of the Melnikov's method following to [Lichtenberg and Liberman (1983); Guckenheimer and Holmes (1983)].

Let us consider a two-dimensional dynamical system

$$\dot{\mathbf{x}}_i = \mathbf{f}(\mathbf{x}) + \varepsilon \mathbf{g}(\mathbf{x}), \tag{5.10}$$

where $\mathbf{f} = (f_1(\mathbf{x}), f_2(\mathbf{x}))^T$ is a Hamiltonian vector field, $f_1 = \partial H_0/\partial x_2, f_2 = -\partial H_0/\partial x_1$, $\mathbf{x} = (x_1, x_2)^T$ (T means a vector transpose), and $\mathbf{g} = (g_1(\mathbf{x}), g_2(\mathbf{x}))^T$ is a small periodic perturbation (not necessary to be a Hamiltonian one). The system (5.10) has a separatrix loop at $\varepsilon = 0$ (see Fig. 5.8) which splits under a perturbation. The saddle point of the unperturped system h becomes a saddle periodic trajectory $\gamma(t)$. The aim is to calculate a distance between the stable $W_{\rm s}(\gamma(t))$ and unstable $W_{\rm u}(\gamma(t))$ branches of the unperturped separatrix which not only changes in time but may change the sign in course of time.

In the first order of a small ε, both the manifolds $W_{\rm s}(\gamma(t))$ and $W_{\rm u}(\gamma(t))$ are close to the unperturped separatrix $\mathbf{x}_h(t)$. Moreover, the corresponding trajectories $\mathbf{x}_{\rm s,u}(t, t_0)$ admit equiconvergent estimates in the intervals $[t_0, \infty)$ and $(-\infty, t_0]$ for $W_{\rm s}(\gamma(t))$ and $W_{\rm u}(\gamma(t))$, respectively (see the proof in the book by Guckenheimer and Holms (1983)):

$$\mathbf{x}_{\rm s,u}(t, t_0) = \mathbf{x}_h(t - t_0) + \varepsilon \delta \mathbf{x}_{\rm s,u}(t, t_0) + O(\varepsilon^2), \tag{5.11}$$

where t_0 is an initial time moment which is a parameter specifying a distance passed by a phase point along a trajectory. Substituting (5.11) into (5.10), we get in the first order

$$\delta\dot{\mathbf{x}}_{\mathrm{s,u}} = \hat{F}(\mathbf{x}_h)\delta\mathbf{x}_{\mathrm{s,u}} + \varepsilon\mathbf{g}(\mathbf{x}_h(t-t_0), t), \tag{5.12}$$

where

$$\hat{F}(\mathbf{x}_h) = \begin{pmatrix} \dfrac{\partial f_1}{\partial x_1} & \dfrac{\partial f_1}{\partial x_2} \\ \dfrac{\partial f_2}{\partial x_1} & \dfrac{\partial f_2}{\partial x_2} \end{pmatrix}$$

is a Jacobian matrix for the vector function $\mathbf{f}$ taken at the unperturped separatrix $\mathbf{x}_h(t-t_0)$. One needs to find such solutions of (5.12) for $\mathbf{x}_{\mathrm{s}}$ when $t > t_0$ and for $\mathbf{x}_{\mathrm{u}}$ when $t < t_0$ that satisfy to the condition

$$\lim_{t\to+\infty} \mathbf{x}_{\mathrm{s}} = \lim_{t\to-\infty} \mathbf{x}_{\mathrm{u}} = \mathbf{x}_\gamma,$$

where $\mathbf{x}_\gamma$ is a location of the hyperbolic point of the perturbed system. Those solutions differ by the vector $\mathbf{D}(t, t_0) = \mathbf{x}_{\mathrm{s}}(t_1, t_0) - \mathbf{x}_{\mathrm{u}}(t, t_0) = \delta\mathbf{x}_{\mathrm{s}}(t, t_0) - \delta\mathbf{x}_{\mathrm{u}}(t, t_0)$. *Separatrix splitting* at t_0 at the point $\mathbf{x}_h(0)$ (or a distance between the manifolds $W_{\mathrm{s}}(\gamma(t))$ and $W_{\mathrm{u}}(\gamma(t))$) is specified as a projection of the vector $\mathbf{D}$ on the normal $\mathbf{n}$ to the unperturped separatrix at the point $\mathbf{x}_h(0)$:

$$d(t_0) = \varepsilon\frac{\mathbf{n}(\mathbf{x}_h(0)) \cdot \mathbf{D}(t_0)}{|\mathbf{n}(\mathbf{x}_h(0))|} + O(\varepsilon^2). \tag{5.13}$$

The vector of the normal is specified with the help of (5.10) (at $\varepsilon = 0$) as:

$$\mathbf{n}(\mathbf{x}_h(0)) = \left(-f_2(\mathbf{x}_h(0)), f_1(\mathbf{x}_h(0))\right)^T,$$

where the dot in (5.13) symbolizes an inner vector product. We can rewrite the expression (5.13) introducing the outer vector product $\mathbf{x} \wedge \mathbf{y} \equiv x_1 y_2 - x_2 y_1$ as follows:

$$d(t_0) = \varepsilon\frac{\mathbf{f}(\mathbf{x}_h(0)) \wedge \mathbf{D}(t_0)}{|\mathbf{f}(\mathbf{x}_h(0))|} + O(\varepsilon^2).$$

It is a technical problem to find $d(t_0)$ in an explicit form [Melnikov (1963)] and to obtain the Melnikov's formula

$$d(t_0) = \varepsilon\frac{M(t_0)}{|\mathbf{f}(\mathbf{x}_h(0))|} + O(\varepsilon^2), \tag{5.14}$$

where $M(t_0)$ is *a Melnikov's function* or *a Melnikov's integral* which in a general case of a non-Hamiltonian perturbation $\mathbf{g}$ takes the form

$$M(t_0) = \int_{-\infty}^{\infty} \left(\mathbf{f}(\mathbf{x}_h(t - t_0)) \wedge \mathbf{g}(\mathbf{x}_h(t - t_0), t)\right) dt.$$

If the perturbation is a Hamiltonian one, i.e., if $g_1 = \partial H_1/\partial x_2$, $g_2 = \partial H_1/\partial x_1$, then we get

$$M(t_0) = \int_{-\infty}^{\infty} \{H_0(\mathbf{x}_h(t - t_0)), H_1(\mathbf{x}_h(t - t_0), t)\} dt,$$

where $\{,\}$ is a Poissonian bracket. If the function $M(t_0)$ has simple zeroes, then stable and unstable manifolds of a saddle trajectory of the system under consideration intersect each other transversally.

It should be noted that a separatrix splitting does not imply an intersection of the stable and unstable manifolds. The Melnikov's method not only proves the transversal intersections of the manifolds and existence of chaos but it provides, as well, a measure for the separatrix splitting and, therefore, a measure for the width of a *stochastic layer.*

Nonlinear Resonance

In linear systems the resonance occurs at the perturbation frequency close to a natural frequency of the system under consideration. In a nonlinear system with sufficiently strong nonlinearity, the resonance may occur practically at any frequency of the excitation Ω. Since nonlinear systems possess, in general, infinitely many natural frequencies ω_i, the resonance condition $m\omega_i = n\Omega$ is satisfied with an infinite number of positive integers m and n.

An isolated nonlinear resonance is represented on *the Poincaré section* (see the corresponding article in the Glossary) by nested invariant curves, forming *an island of stability* (see the corresponding article in the Glossary) or a resonant island, with an elliptic point (see the article "Stationary points" in the Glossary) in its center. Elliptic points on the Poincaré section are images of periodic trajectories in *the phase space* (see the corresponding article in the Glossary).

There are nonlinear resonances of different orders. *A primary nonlinear resonance*, $\omega_1 = \Omega$, in a system with one-and-half degrees of freedom

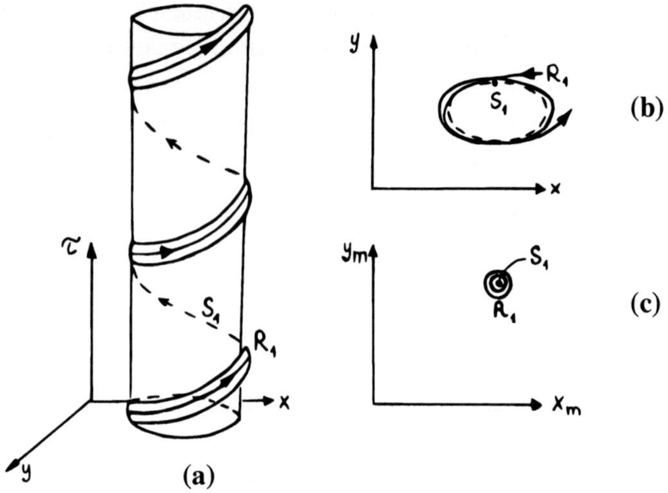

Fig. 5.9 Schematic illustration of a primary nonlinear resonance. (a) In the extended phase space (x, y, τ) a tube, filled by quasiperiodic trajectories R_1 of the primary resonance, winds the surface that contains the periodic trajectory S_1 of that resonance. (b) The periodic trajectory (dashed closed curve S_1) and one of the quasiperiodic trajectories of the primary resonance R_1 (solid open curve) are shown on the phase plane (x, y). (c) Stationary elliptic point S_1 and the invariant resonant curves of the quasiperiodic trajectories R_1 are shown on the Poincaré section (x_m, y_m).

is illustrated in Fig. 5.9 as it looks in the extended phase space (x, y, τ), on the phase plane (x, y), and on the Poincaré section (x_m, y_m). In the extended space (Fig. 5.9(a)), a tube with quasiperiodic trajectories winds around the cylindrical surface that contains the periodic trajectory S_1 of that resonance. The quasiperiodic trajectories lie on the surfaces of the nested cylinders which are densely filled with those trajectories. The periodic S_1 and one of the quasiperiodic trajectories R_1 are shown in Fig. 5.9(b) by the dashed closed and solid open curves, respectively. The periodic trajectory is represented on the Poincaré section (Fig. 5.9(c)) by the point S_1, whereas the family of the quasiperiodic trajectories is mapped onto the corresponding nested resonant invariant curves.

The periodic trajectory of *a secondary nonlinear resonance* S_2 winds the surface of the tube filled with the quasiperiodic trajectories R_1 of the corresponding primary resonance (see Fig. 5.10(a)). The tubes with the quasiperiodic trajectories of the secondary resonance R_2 (not shown in the figure) wind around S_2. This complicated motion is simplified on the Poincaré section (x_m, y_m) in Fig. 5.10(b) demonstrating schematically

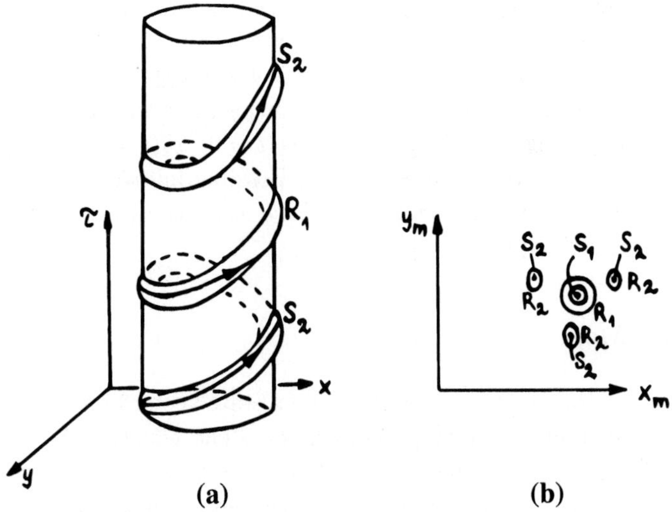

Fig. 5.10 Schematic illustration of a secondary nonlinear resonance. (a) In the extended phase space (x, y, τ) the periodic trajectory of the secondary resonance S_2 winds the surface of a tube filled by quasiperiodic trajectories of the primary resonance R_1. (b) On the Poincaré section (x_m, y_m) an island of a primary resonance with the elliptic point S_1 is surrounded by three islands of a secondary resonance filled by invariant curves of the secondary resonance R_2.

the island of the primary resonance with the elliptic point S_1 surrounded by three islands of the secondary resonance with the elliptic and invariant curves of the corresponding quasiperiodic trajectories R_2. The phase point on the periodic trajectory of the secondary resonance S_2 turns around the elliptic point of the primary resonance S_1 and returns to its initial position for three perturbation periods.

There are infinitely many nonlinear resonances of different orders in typical chaotic Hamiltonian systems. They are represented on Poincaré sections (see Fig. 5.1) by chains of islands of a different size. Islands of primary resonances are surrounded by chains of smaller islands of secondary resonances which, in turn, are surrounded by islands of higher-order resonances of smaller sizes, etc.

Phase Space

The phase space is a n-dimensional abstract space with the coordinates being components $x_i (i = 1, 2, \ldots, n)$ of a state vector of the dynamical system under consideration $\dot{x}_i = F_i(x_1, x_2, \ldots, x_n, t)$. In mechanical systems generalized positions and momenta are coordinates in the phase space. A state of a dynamical system at each time moment is *a phase point* in the phase space. In course of time the phase point moves along a curve which is called *a phase trajectory* beginning at an initial point $[x_1(t=0), x_2(t=0), \ldots, x_n(t=0)]$. A set of the phase trajectories with all possible initial conditions constitutes *a phase portrait. The extended phase space* is a phase space with an additional coordinate (time or phase).

Poincaré Map

Let $\mathbf{X}_n = (x_1(t_n), x_2(t_n), \ldots, x_N(t_n))$ be a state vector of a N-dimensional dynamical system at the time moment t_n. The equation $\mathbf{X}_n = \hat{f}\mathbf{X}_{n-1}$, specifying the states of the system at discrete time moments, is called *a Poincaré map* if the moments (t_n) are chosen accordingly to one of the rules specified below. The Poincaré idea was to fix coordinates of a phase point at specified time moments or when it crosses a given surface in the phase space. The corresponding trajectory is represented by a set of points in the phase space or on one of its projections. If the system is perturbed by an external force with the period T, it is convenient to fix coordinates of phase points at the moments of time multiplied by the period $t_n = nT$ $(n = 1, 2, \ldots)$ and project them to a plane. If the dynamical system under consideration is autonomous, then one chooses a surface in the phase space with the dimension which is smaller by one then the phase space dimension and fixed the moments t_n when the corresponding trajectory intersects transversally the surface in one and the same direction. That surface is called *a Poincaré section surface.* Projecting the intersection points onto a specified plane, one gets *a Poincaré section.* The Poincaré sections are usually computed with a large number of initial conditions and values of the control parameters. When studying specific properties of chaotic motion, for example, *stickiness*, it is sufficient to compute a Poincaré section with a single typical chaotic trajectory for a long time.

Separatrix

In the systems with one degree of freedom *a separatrix* is a special, singular trajectory connecting hyperbolic *stationary points* (see the corresponding article in the Glossary) and separating topologically different regions of motion. The period of the phase point motion along a separatrix is infinite because the velocity at stationary points is zero by definition. One gets *a homoclinic connection* if a separatrix connects the same hyperbolic point (see Fig. 5.3(b)). If a separatrix connects different saddle point, one gets *a heteroclinic connection* (Fig. 5.3(a)). Perturbed separatrices may arise in dynamical systems under a perturbation. They are stable and unstable *manifolds* (see the corresponding article in the Glossary) of the corresponding hyperbolic points (trajectories).

Stable and Unstable Motion

Stability and *instability* are fundamental properties of motion that are manifested not only nearby *stationary points* (see the corresponding article in the Glossary). *A trajectory* (see the corresponding article in the Glossary) $\mathbf{X}(t)$ with the initial condition $\mathbf{X}_0$ is called *stable by Lyapunov* if for any number ε there exists a number $\delta(\varepsilon)$ such that for all $\tilde{\mathbf{X}}(t)$ the inequality $\|\mathbf{X}(t) - \tilde{\mathbf{X}}(t)\| < \varepsilon$ is satisfied for any trajectory $\tilde{\mathbf{X}}(t)$ such that $\|\mathbf{X}_0 - \tilde{\mathbf{X}}_0\| < \delta$. It means that the diameter of a phase drop with the center at $\mathbf{X}_0$ when $t = 0$ does not exceed in course of time a given value ε if it was smaller than δ at $t = 0$ (see Fig. 5.11). Besides the Lyapunov stability specifying a closeness of phase points at the same time moment, one can define *an orbital stability.* Any trajectory which is stable by Lyapunov is orbitally stable but not vice versa. If a trajectory is stable by Lyapunov, then the corresponding phase drop is forever compact in a stream tube. Whereas an initially compact drop in a stream tube with an orbitally stable trajectory, stays forever in that tube but spreads along that trajectory. In other words, two initially close points in the drop may diverge from each other in course of time.

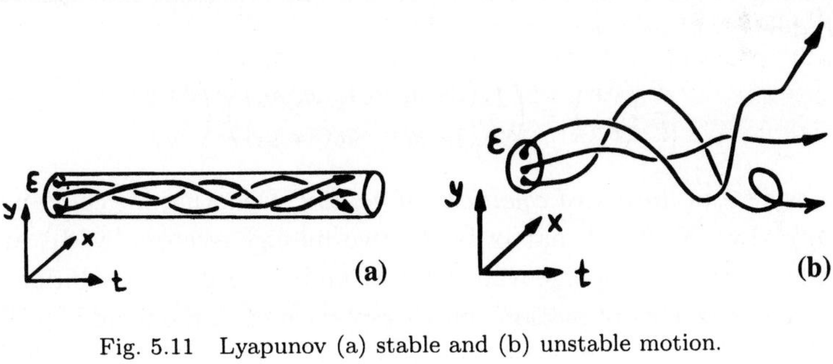

Fig. 5.11 Lyapunov (a) stable and (b) unstable motion.

Stability Points

Analysis of the dynamical system under consideration begins with finding *the phase portrait* (see the article "Phase space" in the Glossary) of an autonomous version of the system, finding its stationary points, and studying motion in a small neighbourhood of each of them. Let us represent the equations of motion in the form of a set of the first-order differential equations

$$\dot{\mathbf{X}} = \mathbf{F}(\mathbf{X}),$$

where $\mathbf{X}$ is a vector with components being state variables of the system. *Stationary points* (the other names: equilibrium, rest, special, singular or fixed points) are specified as $\dot{\mathbf{X}} = 0$ or $\mathbf{F}(\mathbf{X}_\mathrm{s}) = 0$. To study a character of motion nearby a stationary point let us expand the function $\mathbf{F}(\mathbf{X})$ in a Taylor series and analyze the corresponding linearized equations of motion. As an illustrative example, we consider a set of two equations

$$\begin{aligned} \dot{x} &= f(x, y), \\ \dot{y} &= g(x, y), \end{aligned}$$

with the coordinates of their stationary points satisfying to the equations: $f(x_\mathrm{s}, y_\mathrm{s}) = 0$ and $g(x_\mathrm{s}, y_\mathrm{s}) = 0$. Let us introduce small deviations δx and δy nearby one of the points $x = x_\mathrm{s} + \delta x$ and $y = y_\mathrm{s} + \delta y$ and expand f and g in a series in powers of δx and δy:

$$\begin{aligned} \delta\dot{x} &= f_x(x_\mathrm{s}, y_\mathrm{s})\delta x + f_y(x_\mathrm{s}, y_\mathrm{s})\delta y + f_{xy}(x_\mathrm{s}, y_\mathrm{s})\delta x\delta y + \ldots, \\ \delta\dot{y} &= g_x(x_\mathrm{s}, y_\mathrm{s})\delta x + g_y(x_\mathrm{s}, y_\mathrm{s})\delta y + g_{xy}(x_\mathrm{s}, y_\mathrm{s})\delta x\delta y + \ldots. \end{aligned}$$

Neglecting terms above the first order, one can represent the equations for small deviations as a set

$$\frac{d}{dt}\begin{pmatrix} \delta x \\ \delta y \end{pmatrix} = \begin{pmatrix} f_x(x_\mathrm{s}, y_\mathrm{s}) & f_y(x_\mathrm{s}, y_\mathrm{s}) \\ g_x(x_\mathrm{s}, y_\mathrm{s}) & g_y(x_\mathrm{s}, y_\mathrm{s}) \end{pmatrix}\begin{pmatrix} \delta x \\ \delta y \end{pmatrix},$$

which is a set of *linearized equations of motion.* Denoting the column vector $(\delta x, \delta y)^T$ by $\delta\mathbf{X}$, 2×2 matrix by $\hat{F}$, two its eigenvectors by $\mathbf{d}_1$, and $\mathbf{d}_2$, and the corresponding eigenvalues by λ_1 and λ_2, a general solution of the linearized equations of motion can be represented in the form

$$\delta\mathbf{X} = c_1\mathbf{d}_1 e^{\lambda_1 t} + c_2\mathbf{d}_2 e^{\lambda_2 t},$$

where $c_{1,2}$ are integration constants. The eigenvalues $\lambda_{1,2}$ are roots of the equation

$$\det[\hat{F} - \lambda \hat{I}] = 0,$$

where $\hat{I}$ is a unit matrix.

The full classification of stationary points can be found in any textbook on dynamical systems. Here, we reproduce general statements about Hamiltonian systems and area-preserving maps. There are the following three possibilities:

(1) λ_1 and λ_2 is a complex-conjugated pair $\lambda_1 = e^{i\alpha}$ and $\lambda_2 = e^{-i\alpha}$ on a unit circle. Then small deviations δx and δy rotate around the corresponding point and the corresponding phase trajectories are ellipses. Such a stationary point is called *a stable* or *an elliptic* one.

(2) λ_1 and λ_2 are real numbers with the condition $\lambda_2 = \lambda_1^{-1}$. The motion nearby such a point is unstable, and it is called *a hyperbolic* or *a saddle* point.

(3) There is a special case when $\lambda_1 = 1$ and $\lambda_2 = -1$. Such a stationary point is called *parabolic*.

Trajectories

We give below definitions of the types of trajectories of dynamical systems in *a phase space* (see the corresponding article in the Glossary). The phase point, specifying a state of the dynamical system at a given time moment, changes its position in the phase space in course of time. The corresponding curve is called *a phase trajectory*. Because of uniqueness of solutions of differential equations, phase trajectories cannot cross each other. If the motion is periodic, then the corresponding trajectory is called *a periodic* one. Circle is the simplest image of a periodic trajectory. Trajectories with a long period usually have more complicated forms. Multi-frequency motion can be periodic if the frequencies are commensurate, i.e., if there exists a set of nonzero integers (positive or negative) $k_1, k_2, \ldots$ such that $k_1\omega_1 + k_2\omega_2 + \ldots = 0$. If such a set does not exist, the motion is called *quasiperiodic*. In the two-dimensional case, a quasiperiodic trajectory winds the surface of a torus without self-intersections and is not closed. The motions along and across the torus have different frequencies, and their ratio is an irrational number. *Aperiodic* or *chaotic* trajectories do not lie on the surfaces of tori in the phase space.

Periodic and quasiperiodic trajectories can be stable and unstable. The latter ones are called *hyperbolic trajectories.* Chaotic trajectories are, generally speaking, unstable, however, their fragments of an arbitrary but finite length may demonstrate a stability. There are special types of chaotic trajectories. Let $\mathbf{X}(t)$ be an unstable (hyperbolic) trajectory. A trajectory which asymptotically approaches the hyperbolic trajectory at $t \to -\infty$ and $t \to \infty$, is called *homoclinic.* Let $\mathbf{X}(t)$ and $\tilde{\mathbf{X}}(t)$ be two hyperbolic trajectories. A trajectory which asymptotically approaches $\mathbf{X}(t)$ at $t \to -\infty$ and $\tilde{\mathbf{X}}(t)$ at $t \to \infty$ is called *heteroclinic.*

A periodic trajectory in the extended phase space winds a surface of the (deformed) cylinder or torus, and its projection onto a phase plane is a smooth close curve (perhaps, with self-intersections). The phase point along a periodic trajectory returns to its initial position for the time nT, where T is a period of the trajectory and $n = 1, 2, \ldots$. Figure 5.12(a) demonstrates the periodic trajectory on the surface of the straight cylinder. Projection of the quasiperiodic trajectory onto a phase plane is a smooth open curve winding in a comparatively narrow strip (Fig. 5.12(b)). Projection of the aperiodic trajectory onto a phase plane is a smooth open curve wandering in an energetically accessible regions of the plane (Fig. 5.12(c)).

Figure 5.13 illustrates Poincaré sections of the main types of trajectories in the phase space. A trajectory with the period equal to T is represented by a point on the Poincaré section with the same period (Fig. 5.13(b)). A periodic trajectory with another value of the period is represented by a finite set of points on the Poincaré section with the period T (Fig. 5.13(c)). A quasiperiodic trajectory never returns to the starting point. If we wait long enough, the quasiperiodic orbit will be as close as we want to returning at some point in time. Therefore, it covers a continuous invariant curve

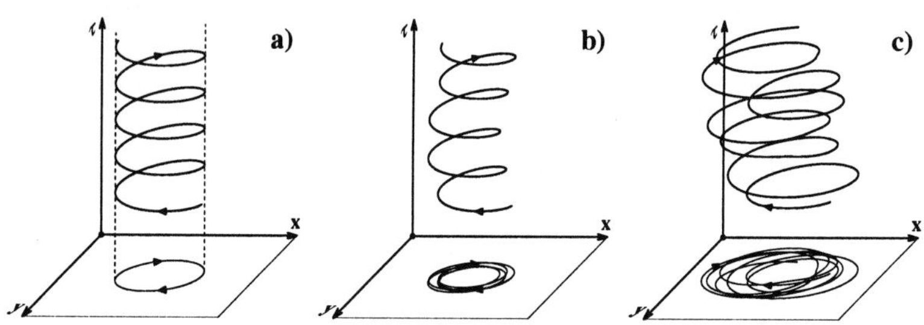

Fig. 5.12 (a) Periodic, (b) quasiperiodic, and (c) aperiodic trajectories in the extended phase space (x, y, τ) and their projections onto the phase plane (x, y).

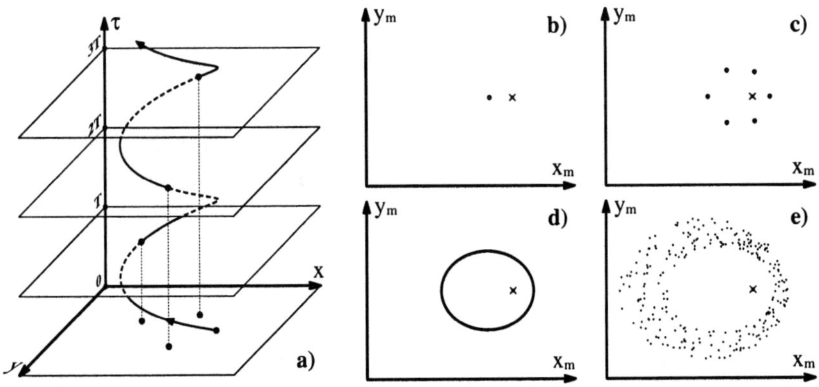

Fig. 5.13 (a) Trajectory in the extended phase space with the four first crossings at the time instants mT ($m = 0, 1, 2, \ldots$). The Poincaré sections (b) for a trajectory with the period T, (c) for a periodic trajectory with the period mT, (d) for a quasiperiodic trajectory, and (e) for aperiodic trajectory. The initial position of the corresponding trajectory is shown by the cross.

on the Poincaré section (Fig. 5.13(d)), whereas an aperiodic trajectory is represented by a cloud of points (Fig. 5.13(e)). It should be stressed that the pattern of the motion in a multi-dimensional system can be different for different surfaces of sections. Full information about the phase space topology can be obtained only by more detailed analysis of different map Poincaré sections.

Appendix A

Models of a Waveguide

In this Appendix we shall describe some models of an underwater sound channel. All those models, except for the first one, allow for a simple analytical derivation of the action-angle variables.

A.1 Canonical Munk Profile

The most famous model of an underwater sound channel is the so-called canonical Munk waveguide [Munk (1974)]. It is derived from a simple thermodynamic model based on the assumption of the exponential stratification. The fractional sound-speed gradient can be written as

$$\frac{1}{c}\frac{\partial c}{\partial z} = -\frac{\mu}{g}N^2(z) + \gamma_{\mathrm{a}}, \tag{A.1}$$

where μ is a dimensionless constant, $g = 9.8$ m/sec^2 is the gravitational acceleration, $N(z)$ is the Väisälä-Brunt frequency (or buoyancy frequency, see subsection 1.5.1), γ_{a} is the fractional sound-speed gradient in an adiabatic ocean. The exponential stratification implies

$$N = N_0 \exp\left(-\frac{z}{B}\right), \tag{A.2}$$

where B is a thermocline depth. The depth of the channel axis, being the solution of the equation $dc/dz = 0$, is

$$z_{\mathrm{a}} = \frac{B}{2}\ln\frac{\mu N_0^2}{g\gamma_{\mathrm{a}}}. \tag{A.3}$$

Let us introduce a dimensionless depth variable reckoned with respect to the channel axis

$$\eta = \frac{2}{B}(z - z_{\mathrm{a}}) \tag{A.4}$$

and set

$$c(z_a) = c_0. \tag{A.5}$$

Then the solution of (A.1) can be simply written as

$$c = c_0[1 + \epsilon(\eta + e^{-\eta} - 1)], \tag{A.6}$$

where

$$\epsilon = \frac{1}{2}B\gamma_a. \tag{A.7}$$

Formula (A.6) defines the *canonical Munk sound-speed profile* depicted in Fig. A.1(a).

The main geometrical properties of a wavefield in an USC can be studied within the ray approach. In the small-angle approximation the ray Hamiltonian is given by

$$H = \frac{p^2}{2} + U(z). \tag{A.8}$$

For the canonical Munk waveguide, the function $U(z)$ looks as follows:

$$U(z) = \epsilon(\eta + e^{-\eta} - 1). \tag{A.9}$$

All rays can be classified dependently on the value of H. Rays that refract within the water column without reflections either from the ocean surface or the bottom satisfy the inequality

$$0 \leq H < H_r. \tag{A.10}$$

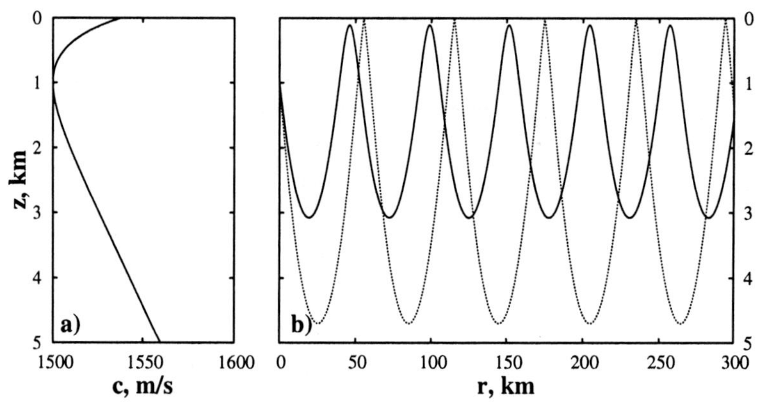

Fig. A.1 (a) The canonical Munk sound-speed profile with $c_0 = 1500$ m/s, $\epsilon = 0.0057$, $B = 1$ km. (b) Examples of ray trajectories in the canonical Munk waveguide. Solid line is a purely water ray, dotted line is a surface-bouncing ray.

where

$$H_{\mathrm{r}} = \epsilon \left(e^{2z_{\mathrm{a}}/B} - \frac{2z_{\mathrm{a}}}{B} - 1 \right). \tag{A.11}$$

Rays, reflecting from the ocean surface without reflections from the bottom at the depth h, satisfy the inequality

$$H_{\mathrm{r}} \leq H < H_{\mathrm{b}}, \tag{A.12}$$

where

$$H_{\mathrm{b}} = U(z = h). \tag{A.13}$$

Reflections from the ocean bottom are followed by significant losses of sound energy, therefore, bottom-reflected rays are irrelevant for long-range sound propagation and will not be considered in this Appendix. Examples of purely-water and surface-bouncing rays are shown in Fig. A.1(b). The rays with $E > E_{\mathrm{b}}$ exhibit reflections from the both ocean boundaries.

Unfortunately, the Munk canonical waveguide is not a very suitable model for the analytical investigation, and only some rough characteristics can be easily found. For instance, the double axial loop, being the cycle length for the axial ray and regarded as the distance between the convergence zones in a range-independent ocean, can be calculated in the following way. Expanding the exponent in Eq. (A.9) into the power series up to the second power, we can recast Eq. (A.8) as

$$H \simeq \frac{p^2}{2} + \frac{\omega_0^2 (z - z_{\mathrm{a}})^2}{2}, \tag{A.14}$$

where

$$\omega_0 = \frac{2\sqrt{\epsilon}}{B}. \tag{A.15}$$

The expression (A.14) is equivalent to the Hamiltonian of a linear oscillator with the frequency ω_0. In this scope, the double axial loop D_0 is an analogue of the oscillator's period and is related to ω_0 by the simple formula

$$D_0 = \frac{2\pi}{\omega_0}. \tag{A.16}$$

Thus, we find

$$D_0 = \pi B \epsilon^{-1/2}. \tag{A.17}$$

With $B = 1$ km and $\epsilon = 5.7 \cdot 10^{-3}$, the double axial loop is 41.6 km. The double axial loop increases with increasing the ray inclination with respect to the waveguide axis until rays become surface-bouncing. Then a local minimum appears.

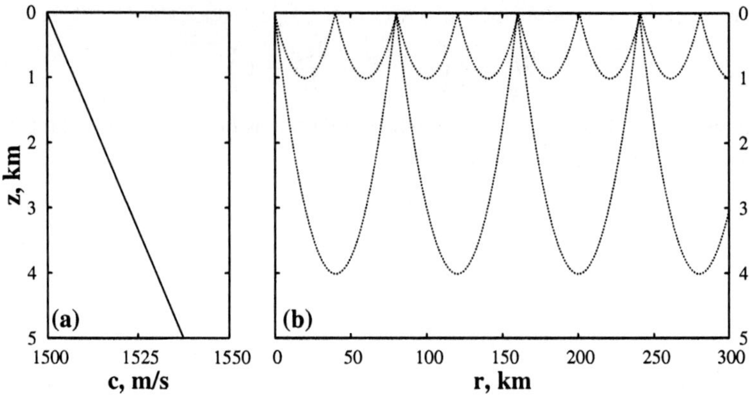

Fig. A.2 (a) The linear sound-speed profile with $a = 0.005\ \mathrm{km}^{-1}$. (b) Examples of ray trajectories.

A.2 Linear Profile

Here we consider the case when sound speed decreases linearly with increasing of depth (depicted in Fig. A.2(a))

$$c(z) = c_0(1 + az). \tag{A.18}$$

In this model all rays reflect from the ocean surface (see Fig. A.2(b)). The respective ray Hamiltonian is

$$H = \frac{p^2}{2} - az. \tag{A.19}$$

The ray action is the following function of H:

$$I(H) = \frac{1}{3a\pi}(2H)^{3/2}. \tag{A.20}$$

Inverting this formula, we express the Hamiltonian as a function of the action

$$H(I) = \frac{(3a\pi I)^{2/3}}{2}. \tag{A.21}$$

The frequency of spatial ray oscillations in a waveguide is given by

$$\omega(I) = \frac{dH}{dI} = (a\pi)^{2/3}(3I)^{-1/3}. \tag{A.22}$$

In this Appendix we shall assume that the angle ϑ varies within the interval $[-\pi : \pi]$. It leads to a more concise form of expressions. For the linear profile, the formula for the angle is

$$\vartheta = \pm\frac{\omega}{a}\left[\sqrt{2H} - \sqrt{2(H - az)}\right], \tag{A.23}$$

where the upper (lower) sign corresponds to positive (negative) values of the ray momentum p. The inverse transformation of the ray variables is given by the formulae

$$z(I, \vartheta) = -\frac{(3\pi aI)^{2/3}}{2a}\left(1 - \frac{\vartheta^2}{\pi^2},\right) \tag{A.24}$$

$$p(I, \vartheta) = (3\pi aI)^{1/3}\frac{\vartheta}{\pi}. \tag{A.25}$$

A.3 Bilinear Profile

The bilinear sound-speed profile is given by the following expression:

$$\frac{\Delta c}{c_0} = \begin{cases} -a(z - z_a), & z < z_a, \\ b(z - z_a), & z > z_a, \end{cases} \tag{A.26}$$

where a and b are positive constants. Typical ocean implies $a > b$, i.e., the vertical sound-speed gradient is larger in the upper ocean than in the lower one. A profile satisfying that condition is depicted in Fig. A.3(a). It allows for two kinds of rays propagating without reflections from the lossy bottom: purely-water rays with $H < az_a$ and surface-bouncing rays with $H \geq az_a$ (see Fig. A.3(b)).

For purely-water rays, the ray action is given by the formula

$$I(H) = \frac{2\sqrt{2}}{3\pi}\left(\frac{1}{a} + \frac{1}{b}\right)H^{3/2}. \tag{A.27}$$

The Hamiltonian as a function of the action is

$$H(I) = \frac{1}{2}\left(\frac{3\pi abI}{a+b}\right)^{2/3}. \tag{A.28}$$

The frequency of spatial ray oscillations in a waveguide is given by

$$\omega = \frac{\pi ab}{\sqrt{2H}(a+b)} = \left(\frac{\pi ab}{a+b}\right)^{2/3}(3I)^{-1/3}. \tag{A.29}$$

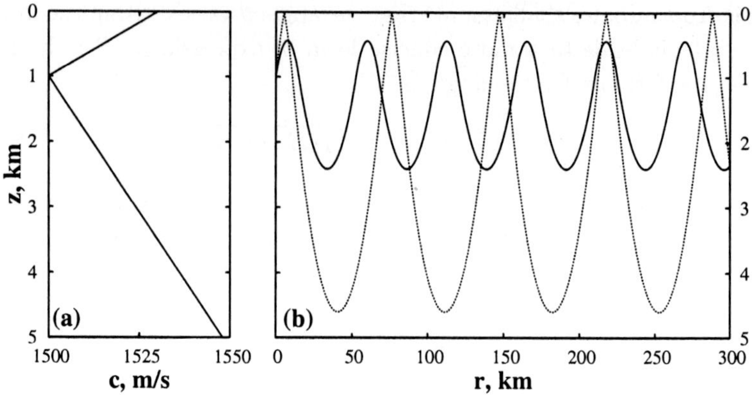

Fig. A.3 (a) Bilinear sound-speed profile with $z_a = 1$ km, $a = 0.02$ km^{-1}, $b = 0.008$ km^{-1}. (b) Examples of ray trajectories. A purely-water ray is plotted with a solid line and a surface-bouncing ray is plotted with a dotted line.

Above the waveguide axis ($z < z_a$), the angle variable is given by the formula

$$\vartheta = \pm\frac{\omega}{a}\sqrt{2(H - az_a + az)}. \tag{A.30}$$

The respective formula below the axis is

$$\vartheta = \pm\omega\sqrt{2H}\left(\frac{1}{a}+\frac{1}{b}\right) \mp \frac{\omega}{b}\sqrt{2(H + bz_a - bz)}. \tag{A.31}$$

In (A.30) and (A.31) the upper (lower) sign corresponds to positive (negative) values of the ray momentum p. For the purely-water rays, the depth variable is expressed by means of the action-angle variables in the following way:

$$z(I,\vartheta) = \begin{cases} \dfrac{a\vartheta^2}{2\omega^2} - \dfrac{H(I)}{a} + z_a, & |\vartheta| \le \vartheta_a, \\ \dfrac{1}{b}\left\{H(I) + bz_a - \left[\sqrt{H(I)}\left(\dfrac{b}{a}+1\right) \mp \dfrac{b\vartheta}{\sqrt{2}\omega}\right]^2\right\}, & |\vartheta| > \vartheta_a, \end{cases} \tag{A.32}$$

where

$$\vartheta_a = \frac{\omega}{a}\sqrt{2H}. \tag{A.33}$$

The respective expression for the ray momentum is

$$p(I,\vartheta)=\begin{cases}\dfrac{a\vartheta}{\omega}, & |\vartheta|\le\vartheta_{\rm a},\\[2ex] \pm\sqrt{2H}\left(\dfrac{b}{a}+1\right)+\dfrac{b\vartheta}{2\omega}, & |\vartheta|>\vartheta_{\rm a}.\end{cases}\tag{A.34}$$

For surface-bouncing rays, the action is given by the formula

$$I(H)=\frac{(2H)^{3/2}}{3\pi}\left[\frac{1}{a}-\frac{1}{a}\left(1-\frac{az_{\rm a}}{H}\right)^{3/2}+\frac{1}{b}\right].\tag{A.35}$$

The frequency of spatial oscillations now is

$$\omega=\frac{\pi ab}{(a+b)\sqrt{2H}-b\sqrt{2(H-az_{\rm a})}}.\tag{A.36}$$

Above the waveguide axis, the angle variable is given by

$$\vartheta=\pm\frac{\omega}{a}\left(\sqrt{2(H-az_{\rm a}+az)}-p_{\rm r}\right),\tag{A.37}$$

where $p_{\rm r}$ is the modulo of the ray momentum at the moment of a reflection from the surface. It is given by

$$p_{\rm r}=\sqrt{2H-2az_{\rm a}}.\tag{A.38}$$

The respective formula for the angle variable below the waveguide axis is

$$\vartheta=\pm\omega\sqrt{2E}\left(\frac{1}{a}+\frac{1}{b}\right)\mp\frac{\omega p_{\rm r}}{a}\mp\frac{\omega}{b}\sqrt{2(E+bz_{\rm a}-bz)}.\tag{A.39}$$

In Eqs. (A.37) and (A.39), the frequency ω is given by (A.36). Inverting (A.37) and (A.39), we find the formula for the ray depth as a function of ϑ and H

$$z=\begin{cases}\dfrac{a\vartheta^2}{2\omega^2}\pm\dfrac{\vartheta p_{\rm r}}{\omega}, & |\vartheta|\le\vartheta_{\rm a}^*,\\[2ex] \dfrac{1}{b}\left\{H(I)+bz_{\rm a}-\left[\dfrac{b\vartheta}{\sqrt{2}\omega}\mp\sqrt{H(I)}\left(\dfrac{b}{a}+1\right)\pm\dfrac{bp_{\rm r}}{\sqrt{2}a}\right]^2\right\}, & |\vartheta|>\vartheta_{\rm a},\end{cases}\tag{A.40}$$

where $\vartheta_{\rm a}$ is given by

$$\vartheta_{\rm a}^*=\frac{\omega}{a}\left(\sqrt{2H}-p_{\rm r}\right).\tag{A.41}$$

The respective formulae for the ray momentum of surface-bouncing rays are

$$p = \begin{cases} \dfrac{a\vartheta}{\omega} \pm p_{\mathrm{r}}, & |\vartheta| \le \vartheta_{\mathrm{a}}^*, \\ -\dfrac{b\vartheta}{\omega} \pm \sqrt{2H}\left(\dfrac{b}{a}+1\right) \mp \dfrac{bp_{\mathrm{r}}}{a}. & \end{cases} \tag{A.42}$$

A.4 Biexponential Profile

Here we consider the model originally introduced in Ref. [Makarov *et al.* (2002)].

$$c(z) = c_0\left[1 + \frac{b^2}{2}\left(e^{-az} - \eta\right)^2\right]. \tag{A.43}$$

We shall refer to the model (A.43) as the biexponential profile. With an appropriate choice of values of the parameters, this model is similar to the canonical Munk profile (see Fig. A.4(a)), but its analytical description is much simpler. The waveguide axis is placed at the depth

$$z_{\mathrm{a}} = \begin{cases} -\dfrac{1}{a}\ln\eta, & \eta < 1, \\ 0, & \eta \ge 1. \end{cases} \tag{A.44}$$

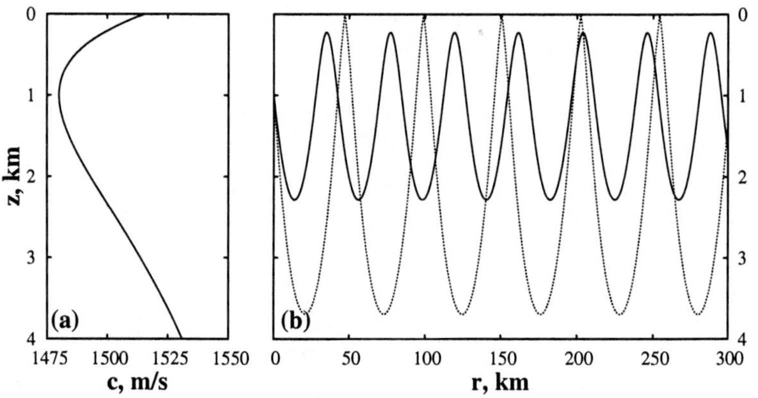

Fig. A.4 (a) The biexponential profile with the parameters $c_0 = 1480$ m/s, $\eta = 0.6065$, $a = 0.5$ km^{-1}, $b = 0.557$. (b) examples of ray trajectories. A purely-water ray is plotted with a solid line, a surface-bouncing ray is plotted with a dashed line.

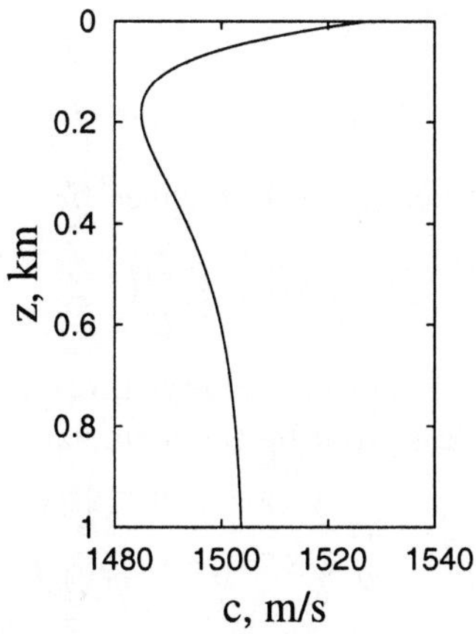

Fig. A.5 The biexponential profile with the parameters $c_0 = 1485$ m/s, $\eta = 0.4$, $a = 5$ km^{-1}, $b = 0.4$.

If the depth of the ocean bottom h is larger than h_c given by

$$h_c = \frac{1}{a} \ln \frac{1}{2\eta - 1}, \tag{A.45}$$

then the biexponential model allows for purely-water rays, surface-bouncing rays (without reflections from the bottom), and bottom-bouncing rays. This case is relevant for deep-water USCs and depicted in Fig. A.4(a). Throughout this book we shall refer to the sound-speed profile presented in this figure as the deep-water biexponential model.

The value of H, separating purely water rays and surface-bouncing ones, is given by

$$H_s = \frac{b^2}{2} (1 - \eta)^2 . \tag{A.46}$$

The examples of purely-water and surface-bouncing rays are shown in Fig. A.4(b). With $h < h_c$ the biexponential model doesn't allow for rays propagating with reflections from the surface but without reflections from the bottom. A respective profile is depicted in Fig. A.5. This situation is met in shallow seas.

The ray turning points inside the water column are given by the following formulae:

$$z_{\text{min}} = -\frac{1}{a} \ln(\eta + Q) \tag{A.47}$$

is the upper turning point, and

$$z_{\max} = -\frac{1}{a}\ln(\eta - Q) \qquad \text{(A.48)}$$

is the lower one. The function $Q(H)$ is given by

$$Q(H) = \frac{\sqrt{2H}}{b}. \qquad \text{(A.49)}$$

The action and angle variables for the rays propagating without reflections from the ocean surface are given by the formulae

$$I = \frac{\eta b}{a}\left(1 - \sqrt{1 - \frac{2H}{b^2\eta^2}}\right), \qquad \text{(A.50)}$$

$$\theta = \pm\frac{\pi}{2} \mp \theta_1(z,\, H), \qquad \text{(A.51)}$$

where θ_1 has the form

$$\theta_1(z,\, H) = \arcsin\frac{\eta + [Q^2(H) - \eta^2]e^{az}}{Q(H)}. \qquad \text{(A.52)}$$

The upper and the lower signs in (A.51) correspond to positive and negative values of the ray momentum, respectively. The inverse transformation for the rays, propagating without reflections from the surface, is expressed as follows:

$$z(I,\, \theta) = \frac{1}{a}\ln\frac{a^2b^2\,(\eta - Q\cos\theta)}{\omega^2}, \qquad \text{(A.53)}$$

$$p(I,\, \theta) = \frac{\omega\, Q\sin\theta}{a\,(\eta - Q\cos\theta)}, \qquad \text{(A.54)}$$

where ω is the spatial frequency of ray oscillations in a waveguide. It depends on H in the following way:

$$\omega = a\sqrt{\eta^2b^2 - 2H}. \qquad \text{(A.55)}$$

The action and the angle for the surface-bouncing rays are given by

$$I = \frac{b}{a}\left(\frac{\eta}{2} - \frac{\eta}{\pi}\arcsin\frac{\eta - 1}{Q} - \frac{\pi + 2\theta_2}{2\pi}\sqrt{\eta^2 - Q^2}\right) + \frac{|p(z=0)|}{\pi a}, \qquad \text{(A.56)}$$

$$\theta = \pm\pi\frac{\theta_2 - \theta_1(z,\, H)}{\theta_2 + \pi/2}. \qquad \text{(A.57)}$$

In (A.56) and (A.57) we used the notation

$$\theta_2 = \arcsin\left(\frac{\eta - \eta^2 + Q^2}{Q}\right). \tag{A.58}$$

Under reflections from the surface, the ray momentum is given by the formula

$$p(z=0) = \pm b\sqrt{Q^2 - 1 + \eta^2}. \tag{A.59}$$

The spatial frequency for the surface-bouncing rays is

$$\omega = \frac{2\pi a}{2\vartheta_2 + \pi}\sqrt{\eta^2 b^2 - 2H}. \tag{A.60}$$

The formulae for the ray depth and the ray momentum in terms of the angle variable for a surface-bouncing ray are the following:

$$z = \frac{1}{a}\ln\frac{\eta - Q\sin\vartheta'}{\eta^2 - Q^2}, \tag{A.61}$$

$$p = -\frac{\sqrt{\eta^2 b^2 - 2H}\cos\vartheta'}{\eta - Q\sin\vartheta'}, \tag{A.62}$$

where

$$\vartheta' = \vartheta_2 \mp \left(\frac{1}{2} + \frac{\vartheta_2}{\pi}\right)\vartheta. \tag{A.63}$$

Here the upper and lower signs correspond to positive and negative values of the ray momentum, respectively.

Appendix B

Derivation of Ray Equations

In this Appendix we briefly explain how the eikonal (Hamilton-Jacobi) equation reduces to a system of ordinary differential (Hamilton) equations. This is done using the method of characteristics for the ray theories constructed on the basis of the parabolic and Helmholtz equations.

Parabolic equation. Using Eqs. (1.28) and (1.30), we represent the eikonal equation (1.24) in the form

$$-H + \frac{p^2}{2} + U + V = 0. \tag{B.1}$$

In this formula H and p are the Hamiltonian and momentum, respectively, of a ray path arriving at the observation point (r, z). Both H and p are considered functions of r and z. Differentiating Eq. (B.1) with respect to z and taking into account that (according to Eqs. (1.28) and (1.30))

$$\frac{\partial H}{\partial z} = -\frac{\partial p}{\partial r}, \tag{B.2}$$

we obtain

$$\frac{\partial p}{\partial r} + p\frac{\partial p}{\partial z} + \frac{\partial (U + V)}{\partial z} = 0. \tag{B.3}$$

Introducing a function $\Psi(r, z, p) = p(r, z) - p$, we represent Eq. (B.3) as $\mathbf{N} \bigtriangledown \Psi = 0$, where $\mathbf{N} = (1, p, -\partial (U + V) / \partial z)$ and $\nabla = (\partial/\partial r, \partial/\partial z, \partial/\partial p)$ is the gradient in a 3-D space (r, z, p). Function $\Psi(r, z, p)$ takes constant values along the field lines of the vector field $\mathbf{N}$. These field lines (*characteristics*) are determined by a system of ordinary differential equations

$$\frac{dr}{1} = \frac{dz}{p} = -\frac{dp}{\partial (U + V) / \partial z} \tag{B.4}$$

equivalent to the Hamilton equations (1.33). Solutions of Eqs. (B.4) (and (1.33)), $z(r)$ and $p(r)$, determine a ray path. Take a path Γ connecting the points (r_0, z_0) and (r, z). Making use of Eqs. (1.28) and (1.30), we find

$$S(r,z) - S(r_0, z_0) = \int_\Gamma (p dz - H dr)\,. \tag{B.5}$$

Assuming $S(r_0, z_0) = 0$, we arrive at Eq. (1.37).

Helmholtz equation. In this case, Eq. (B.1) transforms to the equation

$$H^2 + p^2 - n^2 = 0, \tag{B.6}$$

obtained by substituting Eqs. (1.84) and (1.85) into Eq. (1.80). All three quantities H, p, and n in Eq. (B.6) are considered functions of r and z. Differentiating Eq. (B.6) with respect to z and taking into account that Eq. (B.2) remains valid, we have

$$-2H\frac{\partial p}{\partial r} + 2p\frac{\partial p}{\partial z} - \frac{\partial n^2}{\partial z} = 0. \tag{B.7}$$

Along the same lines as above we find a system of ordinary differential equations determining ray paths (characteristics),

$$\frac{dr}{-H} = \frac{dz}{p} = \frac{2dp}{\partial n^2/\partial z}, \tag{B.8}$$

equivalent to the Hamilton equations (1.89). According to the first equation in (B.8), the ray path is directed along the vector $\nabla S = (-H, p)$, which is perpendicular to the line $S = \text{const}$.

Appendix C

Solution of the Transport Equation

In this Appendix, the formulas, expressing the ray amplitude A through solutions of the Hamilton (ray) equations, are derived. We consider a family of ray trajectories labeled by a parameter α.

Parabolic approximation. Using the formula $p = \partial S/\partial z$ (Eq. (1.28)), we rewrite the transport equation (1.25) as

$$\frac{dA}{dr} + \frac{1}{2}A\frac{\partial p}{\partial z} = 0, \tag{C.1}$$

where

$$\frac{dA}{dr} = \frac{\partial A}{\partial r} + p\frac{\partial A}{\partial z} \tag{C.2}$$

is the derivative of the amplitude A along the ray path and $p = dz/dr$ is the momentum of a ray arriving at the observation point (r, z). In Eqs. (C.1) and (C.2), the momentum p is considered a function of r and z. The rays corresponding to different values of parameter α arrive at, generally, different points. Therefore,

$$\frac{\partial p}{\partial z} = \frac{\partial p/\partial \alpha}{\partial z/\partial \alpha} = \frac{1}{\partial z/\partial \alpha}\frac{\partial^2 z}{\partial r \partial \alpha}. \tag{C.3}$$

In the middle and right expressions, p and z are considered functions of r and α. Substituting Eqs. (C.2) and (C.3) into Eq. (C.1) yields an equation

$$\frac{1}{A}\frac{dA}{dr} + \frac{1}{2\partial z/\partial \alpha}\frac{\partial^2 z}{\partial r \partial \alpha} = 0 \tag{C.4}$$

whose solution is expressed by Eq. (1.39).

Helmholtz equation. Take the first two terms on the left-hand side of the transport equation (1.81). Using Eqs. (1.84) and (1.85), we rewrite their sum as

$$\frac{\partial A}{\partial r}\frac{\partial S}{\partial r} + \frac{\partial A}{\partial z}\frac{\partial S}{\partial z} = -H\left(\frac{\partial A}{\partial r} + \frac{dz}{dr}\frac{\partial A}{\partial z}\right)$$
$$= -H\frac{dA}{dr}, \tag{C.5}$$

where the symbol dA/dr again denotes the range derivative of the amplitude along the ray path.

It follows from Eq. (1.28) that

$$\frac{\partial^2 S}{\partial z^2} = \frac{\partial p}{\partial z} = \frac{\partial p/\partial \alpha}{\partial z/\partial \alpha}. \tag{C.6}$$

Using Eq. (1.30), we find

$$\frac{\partial^2 S}{\partial r^2} = -\frac{\delta H}{\delta r}, \tag{C.7}$$

where δH is the difference between Hamiltonians at two infinitesimally close points $(r + \delta r, z)$ and (r, z). The rays arriving at these points have close values of the parameter α labeling the paths. These values will be denoted by $\alpha + \delta\alpha$ and α. The connection between δa and δr is determined by the equation $z(r + \delta r, \alpha + \delta a) = z(r, \alpha)$. This yields

$$\frac{\partial z}{\partial r}\delta r = -\frac{\partial z}{\partial \alpha}\delta\alpha. \tag{C.8}$$

In this equation, $\partial z/\partial r$ is the same derivative that is denoted by dz/dr in Eq. (C.5). The symbol of partial derivative in Eq. (C.8) emphasizes that the ray path is considered a function of two arguments, r and α. The Hamiltonian H at the end point of a ray path can also be considered a function of r and α. Therefore, $\delta H = \partial H/\partial r\ \delta r + \partial H/\partial\alpha\ \delta a$. Combining this with Eqs. (C.8) and (C.7) yields

$$\frac{\partial^2 S}{\partial r^2} = -\frac{\partial H}{\partial r} + \frac{\partial H}{\partial \alpha}\frac{\partial z/\partial r}{\partial z/\partial \alpha}. \tag{C.9}$$

From Eq. (1.86) and the second of Eqs. (1.89) it follows that

$$\frac{\partial z}{\partial r} = -\frac{p}{H}. \tag{C.10}$$

Differentiating this equation with respect to α, we find

$$\frac{\partial^2 z}{\partial r \partial \alpha} = -\frac{1}{H}\frac{\partial p}{\partial \alpha} + \frac{p}{H^2}\frac{\partial H}{\partial \alpha}. \tag{C.11}$$

Substituting Eqs. (C.5), (C.6) and (C.9) into Eq. (1.81) and taking into account Eqs. (C.10) and (C.11), we obtain

$$\frac{1}{A}\frac{dA}{dr} + \frac{1}{2}\left(\frac{1}{\partial z/\partial \alpha}\frac{\partial^2 z}{\partial r \partial \alpha} + \frac{1}{H}\frac{\partial H}{\partial r}\right) = 0. \tag{C.12}$$

This equation is readily solved for A to yield Eq. (1.95).

Appendix D

Stationary Phase Technique

In this book several times we should evaluate integrals which have the form

$$J = \int_{-\infty}^{\infty} A(z) e^{ik_0 \Phi(z)} dz. \tag{D.1}$$

If $A(z)$ and $\Phi(z)$ are smooth enough functions, this integral may be approximately evaluated using the stationary phase technique. A detailed derivation of this method is given in many textbooks (e.g. [Brekhovskikh and Godin (1999); Felsen and Marcuvitz (1973)]). Here we only briefly explain the main idea of this approach on a "physical" level of rigor.

It is intuitively obvious that in the limit of large k_0, the main contribution to J comes from the neighborhood of the so-called stationary point $\bar{z}$ determined by the condition

$$\frac{d\Phi}{dz} = 0. \tag{D.2}$$

Contributions from points z close to $\bar{z}$ are summed up in phase. In contrast the contributions from points located far from $\bar{z}$ cancel each other. Approximating $\Phi(z)$ in the vicinity of the stationary point by an expression

$$\Phi(z) = \Phi(\bar{z}) + \frac{1}{2}\Phi''(\bar{z})(z - \bar{z})^2 \tag{D.3}$$

where

$$\Phi''(\bar{z}) = \left.\frac{d^2\Phi(z)}{dz^2}\right|_{z=\bar{z}},$$

and assuming that the variations of $A(z)$ at an interval of z providing the main contribution to J are negligible, we find

$$J = A(\bar{z})e^{ik_0\Phi(\bar{z})} \int_{-\infty}^{\infty} e^{i\frac{k_0}{2}\Phi''(\bar{z})(z-\bar{z})^2} dz$$

$$= \sqrt{\frac{2\pi}{k_0 |\Phi''(\bar{z})|}} A(\bar{z})e^{ik_0\Phi(\bar{z})+i\mathrm{sgn}(\Phi''(\bar{z}))\pi/4}. \qquad \text{(D.4)}$$

We have exploited the well-known formula for the Gaussian integral [Felsen and Marcuvitz (1973)].

In spite of the formally infinite integration limits in Eq. (D.4), the main contribution to J comes from an interval $(\bar{z} - \delta z, \bar{z} + \delta z)$, where

$$\delta z = \sqrt{\frac{2\pi}{k_0 |\Phi''(\bar{z})|}}. \qquad \text{(D.5)}$$

If there are a few stationary points (Eq. (D.2) has more than one solution) spaced apart by intervals exceeding $2\delta z$, J is expressed by a sum of terms expressed by Eq. (D.4) which represent contributions from different stationary points.

Bibliography

Abdullaev, S. S. and Zaslavsky G. M. (1981). Nonlinear dynamics of rays in inhomogeneous media, *Sov. Phys. JETP* **53**, pp. 265–270.

Abdullaev, S. S. and Zaslavsky G. M. (1983). Nonlinear and stochastic dynamics of rays in regular transversely inhomogeneous media, *Sov. Phys. JETP* **58**, pp. 915–921.

Abdullaev, S. S. and Zaslavsky G. M. (1984). Stochastic instability of rays and the speckle structure of the field in inhomogeneous media, *Sov. Phys. JETP* **60**, pp. 435–441.

Abdullaev, S. S. and Zaslavsky G. M. (1988). Fractals and ray dynamics in longitudinally inhomogeneous media, *Sov. Phys. Acoust.* **34**, pp. 334–336.

Abdullaev, S. and Zaslavsky, G. (1991). Classical nonlinear dynamics and chaos of rays in wave propagation problems in inhomogeneous media, *Usp. Fiz. Nauk* **161**(8), pp. 1–43.

Abdullaev, S. (1993). *Chaos and Dynamics of Rays in Waveguide Media, ed. by G. Zaslavsky* (Gordon and Breach, New York).

Abramovitz, M. and Stegun, I. (eds.) (1964). *Handbook of Mathematical Functions with Formulas, Graphs and Tables, Appl. Math. Ser. Vol. 55* (National Buro of Standards, Washington).

Afanasiev, V., Sagdeev, R. and Zaslavsky, G. (1991). Chaotic jets with multifractal space-time random walk, *Chaos* **1**, pp. 143–159.

Agam, O. and Fishman, S. (1994). Semiclassical criterion for scars in wave functions of chaotic systems, *Phys. Rev. Lett.* **73**, pp. 806–809.

Agam, O. and Brenner, N. (1995). Semiclassical wigner functions for quantum maps on a torus, *J. Phys. A, Math. Gen.* **28**, pp. 1345–1360.

Ahluwalia, D. and Keller, J. (1977). Exact and asymptotic representations of the sound field in a stratified ocean, in J. Keller and J. Papadakis (eds.), *Lecture Notes in Physics No. 70, Wave Propagation and Underwater Acoustics* (Springer-Verlag, Berlin).

Akulichev, V. A., Borodin, A. E., Burenin, A. V., Morgunov, Yu. N. and Strobykin, D. S. (2007). Application of complex acoustic signals in the long-range navigation of underwater objects, *Doklady Earth Science* **417**(5), pp. 693–696.

Anderson, P. W. (1958). Absence of diffusion in certain random lattices, *Phys. Rev.* **109**, pp. 1492–1505.

Anderson, P. W. (1978). Local moments and localized states, *Rev. Mod. Phys.* **50**, pp. 191–201.

Arnold, J. and Felsen, L. (1983). Rays and local modes in a wedge-shaped ocean, *J. Acoust. Soc. Am.* **73**(4), pp. 1105–1119.

Arnold, V. I. and Avez A. (1968). *Ergodic Problems of Classical Mechanics* (Benjamin, New York).

Arnold, V. I., Kozlov, V. V. and Neishtadt A. I. (1988). *Mathematical Aspects of Classical and Celestial Mechanics. Encyclopaedia of Mathematical Sciences, Vol. 3* (Springer-Verlag, Berlin).

Arnold, V. I. (1989). *Mathematical Methods of Classical Mechanics* (Springer, Berlin).

The ATOC Consortium (Baggeroer, A. B., Birdsall, T. G., Clark, C., Colosi, J. A., Cornuelle, B. D., Costa, D., Dushaw, B. D., Dzieciuch, M., Forbes, A. M. G., Hill, C., Howe, B. M., Marshall, J., Menemenlis, D., Mercer, J. A., Metzger, K., Munk, W., Spindel, R. C., Stammer, D., Worcester, P. F. and Wunsch, C.) (1998). Ocean climate change: comparison of acoustic tomography, satellite altimetry, and modeling, *Science* **281**, pp. 1327–1332.

Batorsky, D. and Felsen, L. (1971). Ray-optic calculation of modes excited by sources and scatterers in a weakly inhomogeneous duct, *Radio Science* **6**, pp. 911–923.

Bayfield, J. E. and Koch P. M. (1974). Multiphoton ionization of highly excited hydrogen atoms, *Phys. Rev. Lett.* **33**, pp. 258–261.

Berman, G. and Zaslavsky, G. (1978). Condition of stochasicity in quantum nonlinear systems, *Physica A* **91**, pp. 450–460.

Berman, G. and Zaslavsky, G. (1979). Condition of stochasicity in quantum nonlinear systems: II. Kinetic description of quantum K-systems, *Physica A* **97**, pp. 367–382.

Berman, G. and Kolovsky, A. (1992). Quantum chaos in interactions of multilevel quantum systems with a coherent radiation field, *Sov. Phys. Usp.* **35**(4), pp. 303–326.

Beron-Vera, F. J. and Brown, M. G. (2003). Ray stability in weakly range-dependent sound channels, *J. Acoust. Soc. Am.* **114**, pp. 123–130.

Beron-Vera, F. J., Brown, M. G., Colosi, J. A., Tomsovic, S., Virovlyansky, A. L., Wolfson, M. A. and Zaslavsky, G. M. (2003). Ray dynamics in a long-range acoustic propagation experiment, *J. Acoust. Soc. Am.* **114**, pp. 1226–1242.

Beron-Vera, F. J. and Brown, M. G. (2004). Travel time stability in weakly range-dependent sound channels, *J. Acoust. Soc. Am.* **115**, pp. 1068–1077.

Berry, M. (1977). Regular and irregular semiclassical wavefunctions, *J. Phys. A. Math. Gen.* **10**(12), pp. 2083–2091.

Berry, M. V. (1989). Quantum scars of classical closed orbits in phase space, *Proc. R. Soc. Lond. A* **423**, pp. 219–231.

Berry, M. V. and Keating, J. P. (1992). A new asymptotic representation for $\zeta(\frac{1}{2}+it)$ and quantum spectral determinants, *Proc. R. Soc. Lond. A* **437**, pp. 151–173.

Bezotvetnykh V. V., Burenin A. V., Morgunov Yu. N. and Polovinka Yu. A. (2009). Experimental studies of pulsed signal propagation from the shelf to deep sea, *Acoustical Physics* **55**, pp. 376–382.

Bies, W. E., Kaplan, L. and Heller, E. J. (2001). Scarring effects on tunneling in chaotic double-well potentials, *Phys. Rev. E* **64**, 016204.

Birdsall, T. G., Metzger, K. and Dzieciuch, M. A. (1994). Signals, signal processing, and general results, *J. Acoust. Soc. Am.* **96**, pp. 2343–2352.

Bogolyubov, N. N. and Mitropolsky, Yu. A. (1961) *Asymptotic Methods in the Theory of Nonlinear Oscillations* (Gordon and Breach, New York).

Bogomolny, E. B. (1988). Smoothed wave functions of chaotic quantum systems, *Physica D* **31**, pp. 169–189.

Born, M. and Wolf, E. (1968). *Principles of Optics* (Pergamon Press, Oxford).

Bowles, A., Smultea, M., De Master, B. D. P. and Palka, D. (1994). Does acoustic testing strand whales? *J. Acoust. Soc. Am.* **96**, pp. 2469–2484.

Boyles, L.M. (1984). *Acoustic Waveguides, Application to Ocean Science* (Wiley, New York).

Brekhovskikh, L. M. (1949). *Dokl. Akad. Nauk SSSR* **69**, p. 157.

Brekhovskikh, L. (1980). *Waves in Layered Media* (Academic Press, New York).

Brekhovskikh, L. and Lysanov, Y. (1991). *Fundamentals of Ocean Acoustics* (Springer-Verlag, Berlin).

Brekhovskih, L. M., Goncharov, V. V. and Kurtepov, V. M. (1995). Weakly divergent bundles of sound rays in the Arctic, *Atmos. Oceanic. Phys.* **31**, pp. 441–446.

Brekhovskikh, L. and Godin, O. (1999). *Acoustics of Layered Media. II: Point Sources and Bounded Beams* (Springer-Verlag, Berlin), 524 p.

Brown, M., Tappert, F. and Goñi, G. (1991). An investigation of sound ray dynamics in the ocean volume using an area preserving mapping, *Wave Motion* **14**, pp. 93–99.

Brown, M., Viechnicki, J. and Tappert, F. (1996). On the measurement of modal group time delays in the deep ocean, *J. Acoust. Soc. Am.* **100**(4), pp. 2093–2102.

Brown, M. and Viechnicki, J. (1998). Stochastic ray theory for long-range sound propagation in deep ocean environment, *J. Acoust. Soc. Am.* **104**, pp. 2090–2104.

Brown, M. G., Colosi J. A., Tomsovic, S., Virovlyansky, A. L. Wolfson, M. A. and Zaslavsky G. M. (2003). Ray dynamics in long-range deep ocean sound propagation, *J. Acoust. Soc. Am.* **113**, pp. 2533–2547.

Brown, M. G., Beron-Vera, F. J., Rypina, I. and Udovydchenkov, I. A. (2005). Rays, modes, wavefield structure, and wavefield stability, *J. Acoust. Soc. Am.* **117**(4), pp. 1607–1609.

Budyansky, M. V., Prants S. V. and Uleysky M. Yu. (2004). Chaotic scattering, transport, and fractals in a simple hydrodynamic flow, *J. Exper. Theor. Phys.* **99**, 1018-1027.

Budyansky, M. V. Uleysky M. Yu. and Prants, S. V. (2009). Detecting barriers to cross-jet Lagrangian transport and its destruction in a meandering flow, *Phys. Rev. E* **79**, 056215.

Burov, V. A., Sergeev, S. N. and Shmelev, A. A. (2007). The possibility of reconstructing the seasonal variability of the ocean using acoustic tomography methods, *Acoustical Physics* **53**, pp. 257–267.

Casati, G., Chirikov B. V., Ford J. and Izrailev F. M. (1979). Stochastic behavior of a quantum pendulum under a periodic perturbation, *Stochastic Behaviour in Classical and Quantum Hamiltonian Systems, Lecture Notes in Physics*, Ed. by Casati, J. and Ford, J. (Springer-Verlag, Berlin), **93**, pp. 334–352.

Cerruti, N. R. and Tomsovic, S. (2002). Sensitivity of wave field evolution and manifold stability in chaotic systems, *Phys. Rev. Lett.* **88**, 054103.

Chandre, C. and Jauslin, H. R. (2002). Renormalization-group analysis for the transition to chaos in Hamiltonian systems, *Physics Reports* **365**, pp. 1–64.

Chirikov, B. V. (1959). *Atomnaya Energiya* **6**, p. 630 (in Russian).

Chirikov, B. V. (1979). A universal instability of many-dimensional oscillator systems, *Physics Reports* **52**, pp. 265–379.

Codona, J., Creamer, D., Flatté, S., Frehlich, R. and Henyey, F. (1985). Average arrival time of wave pulses through continuous random media, *Phys. Rev. Lett.* **55**(1), pp. 9–12.

Colosi, J. A., Flatté S. M. and Bracher, C. (1994). Internal-wave effects on 1000-km oceanic acoustic pulse propagation: Simulation and comparison with experiment, *J. Acoust. Soc. Am.* **96**, pp. 452–468.

Colosi, J. and Flatté, S. (1996). Mode coupling by internal waves for multimegameter acoustic propagation in the ocean, *J. Acoust. Soc. Am.* **100**, pp. 3607–3620.

Colosi, J. A. and Brown, M. G. (1998). Efficient numerical simulation of stochastic internal-wave-induced sound-speed perturbation fields, *J. Acoust. Soc. Am.* **103**, pp. 2232–2235.

Colosi, J. A., Scheer, E. K., Flatté, S., Cornuelle, B. D., Dzieciuch, M. A., Munk, W. H., Worcester, P. F., Howe, B. M., Mercer, J. A., Spindel, R. C., Colosi, J. A., Metzger, K., Birdsall, T. G. and Baggeroer, A. B. (1999). Comparisons of measured and predicted acoustic fluctuations for a 3250-km propagation experiment in the eastern North Pacific Ocean, *J. Acoust. Soc. Am.* **105**, pp. 3202–3218.

Colosi, J. A., Tappert, F. and Dzieciuch, M. (2001). Further analysis of intensity fluctuations from a 3252-km acoustic propagation experiment in the eastern North Pacific Ocean, *J. Acoust. Soc. Am.* **110**, pp. 163–169.

Colosi, J.A., Baggeroer, A. B., Cornuelle, B. D., Dzieciuch, M. A., Munk, W. H., Worcester, P. F., Dushaw, B. D., Howe, B. M., Mercer, J. A., Spindel, R. C., Birdsall, T. G., Metzger, K. and Forbes, A. M. G. (2005). Analysis of multipath acoustic field variability and coherence in the finale of broadband basin-scale transmissions in the North Pacific Ocean, *J. Acoust. Soc. Am.* **117**, pp. 1538–1564.

Cornuelle, B., Wunsch, C., Behringer, D., Birdsall, T., Brown, M., Heinmeller, R., Knox, R., Metzger, K., Munk, W., Spiesberger, J., Spindel, R., Webb, D. and Worcester, P. (1985). Tomographic Maps of the Ocean Mesoscale. Part 1: Pure Acoustics, *J. Phys. Oceanogr.* **15**, pp. 133–152.

De Santo, J. A. (1979) (ed.). *Ocean Acoustics* (Springer-Verlag, Berlin).

Dosso, S. E. and Chapman, N. R. (1987). Measurement and modeling of downslope acoustic propagation loss over a continental slope, *J. Acoust. Soc. Am.* **81**, pp. 258–268.

Dozier, L. and Tappert, F. (1978). Statistics of normal mode amplitudes in a random ocean. I. Theory, *J. Acoust. Soc. Am.* **63**(2), pp. 353–365.

Dushaw, B. (1999). Inversion of multimegameter-range acoustic data for ocean temperature, *IEEE Journal of Oceanic Engineering* **24**(2), pp. 215–223.

Dyson, F., Munk, W. and Zetler, B. (1976). Interpretation of multipath scintillations Eleuthera to Bermuda in terms of internal waves and tides, *J. Acoust. Soc. Am.* **59**, pp. 1121–1133.

Ewing, M., Worcel, J. L. and Pekeris, C. L. (1948). *Propagation of Sound in the Ocean, Geological Society of America Memoir* **27**, New York.

Felsen, L. B. and Marcuvitz, N. (1973). *Radiation and Scattering of Waves. Vol. 1* (Prentice-Hall, Englewood Cliffs, New Jersey).

Felsen, L. B. (1981). Hybrid ray-mode fields in inhomogeneous waveguides and ducts, *J. Acoust. Soc. Am.* **69**(2), pp. 352–361.

Feynman, R. P. (1948). Space-time approach to non-relativistic quantum mechanics, *Rev. Mod. Phys.* **20**(2), pp. 367–387.

Fishman, S., Grempel R. E. and Prange R. E. (1982). Chaos, quantum recurrences and Anderson localization, *Phys. Rev. Lett.* **49**, pp. 509–512.

Flatté, S., Dashen, R., Munk, W., Watson, K. and Zachariasen, F. (1979). *Sound Transmission Through a Fluctuating Ocean* (Cambridge University Press, London).

Gao, T. and Shang, E.-C. (1982). The transformation between the mode representation and the generalized ray representation of a sound field, *J. Sound. Vib.* **80**(1), pp. 105–115.

Gardiner, C. (1985). *Handbook of Stochastic Methods for Physics, Chemistry and the Natural Sciences* (Springer-Verlag, New York).

Garrett, C. R. and Munk, W. H. (1972). Space-time scales of internal waves, *Geophys. Fluid Dyn.* **2**, pp. 225–264.

Gaspard, P. (1998). *Chaos, Scattering and Statistical Mechanics* (Cambridge University Press, Cambridge).

Godin, O. A. (2007). Restless rays, steady wave fronts, *J. Acoust. Soc. Am.* **122**(6), pp. 3353–3363.

Goncharov, V. V. and Kurtepov, V. M. (1994). Formation and propagation of weakly diverging bundles of rays in a horizontally inhomogeneous ocean, *Acoustical Physics* **40**, pp. 685–692.

Gorodetskaya, E., Malekhanov, A., Sazontov, A. and Vdovicheva, N. (1999). Deep water acoustic coherence at long ranges: Theoretical prediction and effects on large array signal processing, *IEEE J. Ocean Eng.* **24**(2), pp. 156–171.

Grempel, D., Prange, R. and Fishman, S. (1984). Quantum dynamics of a non-integrable system, *Phys. Rev. A* **29**, pp. 1639–1647.

Guckenheimer, J. and Holmes, P. (1983). *Nonlinear Oscillations, Dynamical Systems, and Bifurcations of Vector Fields* (Springer, New York).

Gutzwiller, M. (1967). Phase-integral approximation in momentum space and the bound states of an atom, *J. Math. Phys.* **8**, pp. 1979–2000.

Gutzwiller, M. (1990). *Chaos in Classical and Quantum Mechanics* (Springer-Verlag, New York), 432 p.

Haake, F. (2001). *Quantum Signatures of Chaos* (Springer-Verlag, Berlin).

Hegewisch, K. C., Cerruti, N. R. and Tomsovic, S. (2005). Ocean acoustic wave propagation and ray method correspondence: Internal wave fine structure, *J. Acoust. Soc. Am.* **117**, pp. 1582–1594.

Heller, E. J. (1984). Bound-state eigenfunctions of classically chaotic Hamiltonian systems: Scars of periodic orbits, *Phys. Rev. Lett.* **53**, pp. 1515–1518.

Hufnagel, L., Ketzmerick, R., Otto, M.-F. and Schanz, H. (2002). Eigenstates ignoring regular and chaotic phase-space structures, *Phys. Rev. Lett.* **89**(15), 154101.

Iomin, A. and Bliokh, Yu. (2003). Wave localization as a manifestation of ray chaos in underwater acoustics, *Commun. Nonlin. Sci. Numer. Simul.* **8**, pp. 389–399.

Itin, A. P., Neishtadt, A. I. and Vasiliev, A. A. (2000). Captures into resonance and scattering on resonance in dynamics of a charged relativistic particle in magnetic field and electrostatic wave, *Physica D* **141**, pp. 281–296.

Jalabert, R. A. and Pastawski, H. M. (2001). Environment-independent decoherence rate in classically chaotic systems, *Phys. Rev. Lett.* **86**, pp. 2490–2493.

Jensen, F., Kuperman, W., Porter, M. and Schmidt, H. (1994). *Computational Ocean Acoustics* (AIP, Woodbury, New York).

Jones, R. M., Shang, E. C. and Georges, T. M. (1993). Nonperturbative modal tomography inversion. Part I. Theory, *J. Acoust. Soc. Am.* **94**, pp. 2296–2302.

Kamel, A. and Felsen, L. (1982). On the ray equivalent of a group of modes, *J. Acoust. Soc. Am.* **71**(6), pp. 1445–1452.

Kohler, W. and Papanicolaou, G. (1977). Sound propagation in a randomly inhomogeneous ocean, in J. Keller and J. Papadakis (eds.), *Lecture Notes in Physics No. 70, Wave Propagation and Underwater Acoustics* (Springer-Verlag, Berlin).

Kon'kov, L.'E., Makarov, D. V., Sosedko, E. V. and Uleysky M. Yu. (2007). Wave chaos and ghost orbits in an underwater sound channel, *Phys. Rev. E* **76**, 056212.

Koshel, K. V. and Prants, S. V. (2008). *Chaotic Advection in the Ocean* (Institute for Computer Science, Moscow-Izhevsk) [in Russian].

Kravtsov, Y. and Orlov, Y. (1990). *Geometrical Optics of Inhomogeneous Media*, Springer Ser. Wave Phen., Vol. 6 edn. (Springer, Berlin, Heidelberg).

Landau, L. and Lifshitz, E. (1976). *Course of Theoretical Physics*, Vol. 1: *Mechanics*, 3rd edn. (Pergamon Press, Oxford).

Landau, L. D. and Lifshitz E. M. (1977). *Course of Theoretical Physics*, Vol. 3: *Quantum Mechanics: Non-Relativistic Theory* (Pergamon, New York).

Leontovich, M. (1944). On a method for solving the problem of electromagnetic wave propogation along the Earth surface, *Izv. Akad. Nauk. SSSR. Ser. Fiz.* **8**, pp. 16–22.

Leontovich, M. and Fock, V. (1946). Solution of the problem of radio wave propagation over the Earth surface by the parabolic equation method, *J. Phys. USSR* **10**, pp. 13–24.

Levstein, P. R., Usaj G. and Pastawski, H. M. (1998). Attenuation of polarization echoes in nuclear magnetic resonance: A study of the emergence of dynamical irreversibility in many-body quantum systems, *J. Chem. Phys.* **108**, pp. 2718–2724.

Lichtenberg, A. J. and Liberman, M. A. (1983). *Regular and Chaotic Dynamics* (Springer-Verlag, New York).

Lichtenberg, A. and Lieberman, M. (1992). *Regular and Chaotic Dynamics* (*Applied Mathematical Sciences, Vol 38*) (Springer Verlag, New York).

MacKay, R. S., Meiss, J. D. and Percival, I. C. (1984). Transport in Hamiltonian systems, *Physica D* **13**, pp. 55–81.

Makarov, D. V., Prants, S. V. and Uleysky, M. Yu. (2002). Structure of spatial nonlinear resonance of rays in an inhomogeneous underwater sound channel, *Doklady Earth Science* **382**, pp. 394–396.

Makarov, D. V., Uleysky, M. Yu. and Prants, S. V. (2004). Ray chaos and ray clustering in an ocean waveguide, *Chaos* **14**, pp. 79–95.

Makarov, D. V. and Uleysky, M. Yu. (2006). Specific Poincaré map for a randomly-perturbed nonlinear oscillator, *J. Phys. A: Math. and Gen.* **39**, pp. 489–497.

Makarov, D. V., Uleysky, M. Yu., Budyansky M. V. and Prants, S. V. (2006). Clustering in randomly driven Hamiltonian systems, *Phys. Rev. E* **73**, 066210.

Makarov, D. V. and Uleysky, M. Yu. (2007). Giant acceleration in slow-fast space-periodic Hamiltonian systems, *Phys. Rev. E* **75**, 065201(R).

Makarov, D. V. and Uleysky, M. Yu. (2007). Ray escape from a range-dependent underwater sound channel, *Acoustical Physics* **53**, pp. 495–502.

Makarov, D. V. and Kon'kov, L. E. (2007). Chaotic diffusion at sound Propagation in a range-Dependent underwater sound channel, *Rus. J. Nonlin. Dyn.* **3**, pp. 157–174.

Makarov, D. V., Kon'kov, L. E. and Uleysky, M. Yu. (2008). The ray-wave correspondence and the suppression of chaos in long-range sound propagation in the ocean, *Acoust. Phys.* **54**, pp. 382–391.

Makarov, D. V. and Uleysky, M. Yu. (2008). Local chaos induced by spatial oscillations of a perturbation, *Commun. Nonl. Sci. Numer. Simul.* **13**, pp. 400–406.

Mandelbrot, B. (1982). *The Fractal Geometry of Nature* (New York, Freeman).

Melnikov, V. K. (1963). On the stability of the center for time periodic perturbations, *Trans. Moscow Math. Soc.* **12**, pp. 1–57.

Moore, F. L., Robinson, J. C., Bharucha, C., Williams P. E. and Raizen M. G. (1994). Observation of dynamical localization in atomic momentum transfer: A new testing ground for quantum chaos, *Phys. Rev. Lett.* **73**, pp. 2974–2977.

Morozov, A. K. and Colosi, J. A. (2005). Entropy and scintillation analysis of acoustic beam propagation through ocean internal waves, *J. Acoust. Soc. Am.* **117**(3), pp. 1611–1623.

Morozov, A. and Colosi, J. (2007). Stochastic differential equation analysis for sound scattering by random internal waves in the ocean, *Acoustical Physics* **53**(3), pp. 335–347.

Mouchet, A., Eltschka, C. and Schlagheck, P. (2006). Influence of classical resonances on chaotic tunneling, *Phys. Rev. E* **74**, 026211.

Munk, W. (1974). Sound channel in an exponentially stratified ocean with application to SOFAR, *J. Acoust. Soc. Am.* **55**, pp. 220–226.

Munk, W. and Wunsch, C. (1979). Ocean acoustic tomography: A scheme for large scale monitoring, *Deep-Sea Res.* **26**, pp. 123–161.

Munk, W. and Wunsch, C. (1983). Ocean acoustic tomography: Rays and modes, *Rev. Geophys. and Space. Phys.* **21**, pp. 777–793.

Munk, W. H., O' Reilly, W. C. and Reid, J. L. (1988). Australia-Bermuda Sound Transmission Experiment (1960) Revisited, *J. Phys. Oceanograph.* **18**, pp. 1876–1898.

Munk, W. H., Spindel, R. C., Baggeroer, A. and Birdsall, T. G. (1994). The Heard Island Feasibility Test, *J. Acoust. Soc. Am.* **96**, pp. 2330–2342.

Munk, W., Worcester, P. F. and Wunsch, C. (1995). *Ocean Acoustic Tomography* (Cambridge University Press, Cambridge).

Neishtadt, A. and Vasiliev, A. (2005) Phase change between separatrix crossings in slow-fast Hamiltonian systems, *Nonlinearity* **18**, pp. 1393–1406.

Neishtadt, A. and Vasiliev, A. (2007) On the absence of stable periodic orbits in domains of separatrix crossings in nonsymmetric slow-fast Hamiltonian systems, *Chaos* **17**, 043104.

Ott, E. (1993) *Chaos in Dynamical Systems* (Cambridge University Press, New York).

Ottino, J. M. (1989). *The Kinematics of Mixing: Stretching, Chaos, and Transport* (Cambridge University Press, Cambridge).

Palmer, D., Brown, M., Tappert, F. and Bezdek, H. (1988). Classical chaos in nonseparable wave propagation problems, *Geophys. Res. Lett.* **15**, pp. 569–572.

Palmer, D., Georges, T. and Jones, R. (1991). Classical chaos and the sensitivity of the acoustic field to small-scale ocean structure, *Comput. Phys. Commun.* **65**, pp. 219–223.

Pedlosky, J. (1987). *Geophysical Fluid Dymamics* (Springer, New York).

Percival, I. C. (1973). Regular and irregular spectra, *J. Phys. B* **6**, pp. L229–L232.

Percival, I. C. (1979). *AIP Conference Proceedings* **57**, pp. 302–310.

Peres, A. (1984). Stability of quantum motion in chaotic and regular systems, *Phys. Rev. A* **30**, pp. 1610–1615.

Petukhov, Yu. V. (1994). A sound beam with minimal wavefront divergence in a stratified ocean waveguide, *Acoustical Physics* **40**, pp. 97–105.

Petukhov, Yu. V. and Khil'ko, A. A. (2008). Formation, propagation, and focusing of acoustic beams in oceanic waveguides. The quasi-optics approximation, *Acoustical Physics* **54**(5), pp. 689–697.

Reichl, L. (1992). *The Transition to Chaos in Conservative Classical Systems: Quantum Manifestations* (Springer-Verlag, New York), 551 p.

Rosenberg, L. D. (1949). *Dokl. Akad. Nauk SSSR* **69**, p. 175.

Rypina, I. and Brown, M. (2007). On the width of a ray, *J. Acoust. Soc. Am.* **122**(3), pp. 1440–1448.

Rytov, S., Kravtsov, Y. and Tatarsky, V. (1978). *Introduction to Statistical Radiophysics. Part II* (Nauka, Moscow).

Sagdeev, R., Usikov, D. and Zaslavsky, G. (1988). *Nonlinear Physics: From the Pendulum to Turbulence and Chaos* (Harwood Academic, Switzerland).

Sepulveda, M. A., Tomsovic, S. and Heller, E. J. (1992). Semiclassical propagation: How long can it last?, *Phys. Rev. Lett.* **69**, pp. 402–405.

Simmen, J., Flatte, S. and Wang, G.-Y. (1997). Wavefront folding, chaos, and diffraction for sound propagation through ocean internal waves, *J. Acoust. Soc. Am.* **102**, pp. 239–255.

Smale, S. (1967). Differentiable dynamical systems, *Bull. Amer. Math. Soc.* **73**, pp. 747–817.

Smirnov, I., Virovlyansky, A. L. and Zaslavsky, G. (2001). Theory and applications of ray chaos to underwater acoustics, *Phys. Rev. E* **64**, 036221.

Smirnov, I., Virovlyansky, A. L. and Zaslavsky, G. (2002). Sensitivity of ray travel times, *Chaos* **12**(3), pp. 617–635.

Smirnov, I., Virovlyansky, A. L. and Zaslavsky, G. (2004). Wave chaos and mode-medium resonances at long range sound propagation in the ocean, *Chaos* **14**(2), pp. 317–322.

Smirnov, I., Virovlyansky, A. L., Edelman, M. and Zaslavsky, G. (2005a). Chaos-induced intensification of wave scattering, *Phys. Rev. E* **72**, 026206.

Smirnov, I., Virovlyansky, A. L. and Zaslavsky, G. (2005b). Ray chaos, travel time modulation, and sensitivity to the initial conditions, *J. Acoust. Soc. Am.* **117**(3), pp. 1595–1606.

Smith, K., Brown, M. and Tappert, F. (1992). Ray chaos in underwater acoustics, *J. Acoust. Soc. Am.* **91**, pp. 1939–1949.

Soskin, S. M., Yevtushenko, O. M. and Mannella, R. (2005). Divergence of the chaotic layer width and strong acceleration of the spatial chaotic transport in periodic systems driven by an adiabatic ac force, *Phys. Rev. Lett.* **95**, 224101.

Spiesberger, J. L. (1985). Ocean acoustic tomography: Travel time biases, *J. Acoust. Soc. Am.* **77**, pp. 83–100.

Spiesberger, J. and Metzger., K. (1991). Basin-scale tomography: A new tool for studying weather and climate, *J. Geophys. Res.* **96**, pp. 4869–4889.

Spiesberger, J. L. and Tappert, F. D. (1996). Kaneohe acoustic thermometer further validated with rays over 3700 km and the demise of the idea of axially trapped energy, *J. Acoust. Soc. Am.* **99**, pp. 173–184.

Spiesberger, J. L. (2004). Acoustic identification of a single transmission at 3115 km from a bottom-mounted source at Kauai, *J. Acoust. Soc. Am.* **115**, pp. 1497–1504.

Spindel, R. C., Na, J., Dahl, P. H., Oh, S., Eggen, C., Kim, Y. G. Akulichev, V. A. and Morgunov, Yu. N. (2003). Acoustic tomography for monitoring the Sea of Japan: A pilot experiment, *IEEE Journ. Ocean. Eng.* **28**, pp. 297–302.

Stockmann, H. J. (1999). *Quantum Chaos: An Introduction* (Cambridge University Press, Cambridge).

Sundaram, B. and Zaslavsky, G. (1999). Wave analisis of ray chaos in underwater acoustics, *Chaos* **9**, pp. 483–492.

Takagaki, Y. and Ploog, K. H. (2004). Imaging of the scarred wave functions in chaotic quantum cavities, *Phys. Rev.B* **70**, 073304.

Tappert, F. (1977). The parabolic approximation method, in J. Keller and J. Papadakis (eds.), *Lecture Notes in Physics No. 70, Wave Propagation and Underwater Acoustics* (Springer-Verlag, Berlin), pp. 224–284.

Tappert, F. and Tang, X. (1996). Ray chaos and eigenrays, *J. Acoust. Soc. Am.* **99**, pp. 185–195.

Tappert, F. D., Spiesberger, J. L. and Wolfson, M. A. (2002). Study of a novel range-dependent propagation effect with application to the axial injection of signals from the Kaneohe source, *J. Acoust. Soc. Am.* **111**, pp. 757–762.

Tappert, F. D. (2003). Theory of explosive beam spreading due to ray chaos, *J. Acoust. Soc. Am.* **114**, pp. 2775–2781.

Tatarskii, V. (1983). The wigner representation of quantum mechanics, *Sov. Phys. Uspekhi* **26**(4), pp. 311–327 [Usp. Fiz. Nauk, v. 139, No. 4, pp. 587–619 (1983)].

Tikhonov, A. N. and Arsenin V. Y. (1977). *Solutions of Ill-Posed Problems* (Winston Sons, Washington).

Tikhonov, V. and Mironov, M. (1977). *Markov Processes* (Sovetskoe Radio, Moscow).

Timofeev, A. V. (1978). On the constancy of an adiabatic invariant when the nature of the motion changes, *Sov. Phys. JETP* **48**, pp. 656–659.

Tindle C. T. and Guthrie K. M. (1974). Rays as interfering modes in underwater acoustics, *J. Sound. Vib.* **34**, pp. 291–295.

Titchmarsh, E. (1948). *Introduction to the Theory of Fourier Integrals* (Oxford University Press, London).

Tolstoy, A., Diachok, O. and Frazer, L. N. (1991). Acoustic tomography via matched field processing, *J. Acoust. Soc. Am.* **89**, pp. 1119–1127.

Tomsovic, S. and Heller E. J. (1991). Semiclassical dynamics of chaotic motion: Unexpected long-time accuracy, *Phys. Rev. Lett.* **67**, pp. 664–667.

Tomsovic, S. and Heller E. J. (1993). Long-time semiclassical dynamics of chaos: The stadium billiard, *Phys. Rev. E* **47**, pp. 282–299.

Tomsovic, S. and Heller, E. J. (1993b). Semiclassical construction of chaotic eigenstates, *Phys. Rev. Lett.* **70**(10), pp. 1405–1408.

Tomsovic, S. and Ullmo D. (1994). Chaos-assisted tunneling, *Phys. Rev. E* **50**, pp. 282–162.

Udovidchenkov, I. and Brown, M. (2008). Modal group time spreads in weakly range-dependent deep ocean environments, *J. Acoust. Soc. Am.* **123**(1), pp. 41–50.

Vainchtein, D. L., Neishtadt, A. I. and Mezic, I. (2006). On passage through resonances in volume-preserving systems, *Chaos* **16**, 043123.

Van der Weele, J. P. and Valkering, T. P. (1990) The birth process of periodic orbits in non-twist maps, *Physica A* **169**, pp. 42–72.

Van Trees, H. L. (1968). *Detection, Estimation, and Modulation Theory* (Wiley, New York).

Vera, M. D., Heaney, K. D. and the NPAL Group [Colosi, J. A., Cornuelle, B. D., Dushaw, B. D., Dzieciuch, M. A., Howe, B. M., Mercer, J. A., Munk, W. H., Spindel, R. C. and Worcester, P. F.] (2005). The effect of bottom interaction on transmissions from the North Pacific Acoustic Laboratory Kauai source, *J. Acoust. Soc. Am.* **117**(2), pp. 1624–1634.

Virovlyanskii, A. L. (1985). Travel times of acoustic pulses in the ocean, *Sov. Phys. Acoust.* **31**(5), pp. 399–401.

Virovlyanskii, A. L. (1985a). Temporal structure of a pulse signal in an underwater sound channel, *Sov. Phys. Acoust.* **31**(6), pp. 480–481.

Virovlyansky, A. L. (1995). On general properties of ray arrival sequences in oceanic acoustic waveguides, *J. Acoust. Soc. Am.* **97**, pp. 3180–3183.

Virovlyansky, A. L., Kurin, V. V., Pronchatov-Rubtsov, N. V. and Simdyankin, S. I. (1997). Fresnel zones for modes, *J. Acoust. Soc. Am.* **101**, pp. 163–173.

Virovlyansky, A. L. and Zaslavsky, G. M. (1999). Wave chaos in terms of normal modes, *Phys. Rev. E* **59**(2), pp. 1656–1668.

Virovlyansky, A. L. (2000). Manifestation of ray stochastic behavior in a modal structure of the wave field, *J. Acoust. Soc. Am.* **108**(1), pp. 84–95.

Virovlyansky, A. L. and Zaslavsky, G. M. (2000). Evaluation of the smoothed interference pattern under conditions of ray chaos, *Chaos* **10**(1), pp. 211–223.

Virovlyansky, A. L. (2001). Ray travel times in range-dependent acoustic waveguides, *nlin.CD/0012015*.

Virovlyansky, A. L. (2003a). Ray chaos in the long-range propagation of sound in the ocean, *Radiophysics and Quantum Electronics* **46**(7), pp. 502–516.

Virovlyansky, A. L. (2003). Ray travel times at long ranges in acoustic waveguides, *J. Acoust. Soc. Am.* **113**, pp. 2523–2532.

Virovlyansky, A. L. (2005a). Signal travel times along chaotic rays in long-range sound propagation in the ocean, *Acoustical Physics* **51**(3), pp. 271–281.

Virovlyansky, A. L. (2005b). Statistical description of ray chaos in an underwater acoustic waveguide, *Acoustical Physics* **51**(1), pp. 71–80.

Virovlyansky, A. L., Kazarova, A. Yu. and Lyubavin, L. Ya. (2005). Ray-based description of normal mode amplitudes in a range-dependent waveguide, *Wave Motion* **42**(4), pp. 317–334.

Virovlyansky, A. L. and Zaslavsky G. M. (2005). Manifestation of scarring in a driven system with wave chaos, *Chaos* **15**, 023301.

Virovlyansky, A. L. (2006). *Ray Theory of Long-Range Sound Propagation in the Ocean* (Institute of Applied Physics, Nizhny Novgorod).

Virovlyansky, A. L., Kazarova, A. Yu. and Lyubavin, L. Ya. (2006). *Ocean Acoustics, Proceedings of the 11th L.M. Brekhovskikh's Conference*, chap. Modal structure of the field under conditions of wave chaos (GEOS, Moscow), pp. 40–43.

Virovlyansky, A. L., Kazarova, A. Yu. and Lyubavin, L. Ya. (2007). Statistical description of chaotic rays in a deep water acoustic waveguide, *J. Acoust. Soc. Am.* **121**, pp. 2542–2552.

Virovlyansky, A. L., Kazarova, A. Yu. and Lyubavin, L. Ya. (2007). Reconstruction of the average ocean temperature from the measured travel times of sound pulses, *Acoustical Physics* **53**(2), pp. 181189.

Virovlyansky, A. L., Kazarova, A. Yu. and Lyubavin, L. Ya. (2008a). The possibilty of using a vertical array for estimating the delays of sound pulses at multimegameter ranges, *Acoustical Physics* **54**(4), pp. 486–494.

Virovlyansky, A. L., Kazarova, A. Yu, Lyubavin, L. Ya. and Nefedova, D. V. (2008b). Ray description of the field of a distributed source in a waveguide, *Acoustical Physics* **54**(5), pp. 654–663.

Virovlyansky, A. L., Kazarova, A. Yu. and Lyubavin, L. Ya. (2009). Ray-based description of normal modes in a deep ocean acoustic waveguide, *J. Acoust. Soc. Am.* **125**(3), pp. 1362–1373.

Virovlyansky, A. L. and Nefedova D. V. (2009). Wave beam with minimum divergence in a smoothly inhomogeneous medium, *Radiophysics and Quantum Electronics* **52**(1), pp. 41–49.

Voronovich, A. and Ostashev, V. (2006). Low-frequency sound scattering by internal waves in the ocean, *J. Acoust. Soc. Am.* **119**(3), pp. 1406–1419.

Wage, K. E., Baggeroer, A. and Preisig, J. (2003). Modal analysis of broadband acoustic receptions at 3515-km range in the north pacific using short-time fourier techniques, *J. Acoust. Soc. Am.* **113**(2), pp. 801–817.

Wage, K. E., Dzieciuch, M. A., Munk, W. H., Worcester, P. F., Howe, B. M. and Mercer, J. A. (2005). Mode coherence at megameter ranges in the North Pacific Ocean, *J. Acoust. Soc. Am.* **117**, pp. 1565–1581.

Watson, J. G., Siegman, W. L. and Jacobsen, M. J. (1981). Refracted ray characteristics from the geometrical spreading equation for depth-dependent media, *J. Acoust. Soc. Am.* **70**, pp. 583–590.

Wolfson, M. A. and Tappert F. D. (2000). Study of horizontal multipath and ray chaos due to ocean mesoscale structure, *J. Acoust. Soc. Am.* **107**, pp. 154–162.

Wolfson, M. A. and Tomsovic S. (2001). On the stability of long-range sound propagation through a structured ocean, *J. Acoust. Soc. Am.* **109**, pp. 2693–2703.

Worcester, P. F., Cornuelle B. D., Dzieciuch M. A., Munk, W. H. Howe, B. M., Mercer, J. A., Spindel, R. C., Colosi, J. A. Metzger, K., Birdsall, T. G. and Baggeroer, A. B. (1999). A test of basin-scale acoustic thermometry using a large-aperture vertical array at 3250-km range in the eastern North Pacific Ocean, *J. Acoust. Soc. Am.* **105**, pp. 3185–3201.

Worcester, P. F. and Spindel, R. C. (2005). North pacific acoustic laboratory, *J. Acoust. Soc. Am.* **117**(3), pp. 1499–1510.

Zaslavsky, G. (1981). Stochasticity in Quantum Systems, *Physics Reports* **80**, pp. 157–250.

Zaslavsky, G. (1984). *Stochasticity in Dynamical Systems* (Nauka, Moscow).

Zaslavsky, G. M., Edelman M. and Niyazov B. A. (2001). Self-similarity, renormalization, and phase space nonuniformity of Hamiltonian chaotic dynamics, *Chaos* **7**, pp. 159–181.

Zaslavsky, G. M. and Abdullaev, S. S. (1997). Chaotic transmission of waves and "cooling" of signals, *Chaos* **7**, pp. 182–186.
Zaslavsky, G. (1998). *Physics of Chaos in Hamiltonian Systems* (Imperial College Press, Singapore).
Zaslavsky, G. M. (2002) Chaos, fractional kinetics, and anomalous transport, *Physics Reports*, **371**, pp. 461–580.
Zaslavsky, G. M. (2002). Dynamical traps, *Physica D*, **168**, pp. 292–304.
Zaslavsky, G. M. (2005). *Hamiltonian Chaos and Fractional Dynamics* (University Press, Oxford).

Index

acoustic tomography 42
action-angle variables 15, 19, 25, 351
  in a range-dependent waveguide 15, 19
angle variable 16
  wrapped 16
  unwrapped 16
averaging over ray starting parameters 234

Bernoulli shift 319
bifurcation 111, 112, 320
  pitchfork bifurcation 111
  saddle-center 111
Brillouin waves 152, 162

canonical transformation 15, 19, 20, 26
cantorus 321
capturing into resonance 107
caustic 14
chaotic sea 83, 90
Chirikov criterion 90
coupled mode equations 158, 159
  slow and strong range dependence 158
  weak range dependence 159

degeneracy 91, 94, 95
diffusivity 225
dynamical localization 147
dynamical trap 77, 322

eddy 41, 306
  synoptic 41, 306
Ehrenfest 141, 142
  time 141
  break distance 142
eigenray 14
eikonal 11, 13, 18, 21, 22
eikonal equation 11, 22, 363
elliptic point 70, 87, 341–343
empirical orthogonal functions 44
ergodic motion 75

fidelity 145
Floquet modes 168, 177, 181, 198
"focusing" of ray travel times 130
Fokker-Planck equation for action 222
fractal 322
fuzzy segment 123, 251
  spread and bias of 262

gap between travel times of chaotic and regular rays 126
generating function 15, 26
geometrical optics approximation 10, 22
  for the Helmholtz equation 22
  for the parabolic equation 11

Hamilton equations 12, 15, 16, 20, 68, 326
Hamiltonian 12, 15, 23, 83, 101, 108, 242

chaos 67, 72, 324
dynamics 326
formalism 10, 22
of nonlinear resonance 86, 246
Helmholtz equation 7, 8, 22, 160, 162, 168, 363, 364
heteroclinic structure 72, 327, 328
homoclinic structure 72, 327, 328
horseshoe map 329
hyperbolic point 70, 72, 327, 340
Husimi distribution function 182, 239

index of refraction 5
instantaneous frequency 275
arrival time of 274
internal waves 37
invariant torus 332
island of stability 333

Liouville theorem 16, 323, 326
Liouville equation 73, 237, 238
classical 238
quantum 237
Lyapunov exponent 76, 80, 141, 142, 185, 217, 320, 332, 334, 335
global 74, 80, 82, 95
finite-range 59, 60, 80–83
maximal 70–71, 145, 146, 334, 335

manifold
Lagrangian 96–98, 115–117, 145, 208
stable 72, 320, 335–337
unstable 72, 320, 335–337
Maslov index 14
matrix elements 153, 195
measure 74–77, 337, 338
Melnikov's method 338
mixing 76, 218, 324
modal structure 269, 274, 301
of the monochromatic wave field 170, 269
of the transient wave field 274
ray-based description 163
statistical description 269, 278
mode *see* normal mode
mode-medium resonance 171, 177
mode coupling 158, 170, 194, 269
equations, *see also* coupled-mode equations 158, 160
in the presence of a fine structure 194
under chaotic conditions 170
mode pulse 269, 274
in a range-independent waveguide 276
in the presence of internal waves 278
overlapping 277
smoothed 278
spread and bias of 282
mode ray 165
momentum 12, 23
momentum-position variables 12, 23
mudslide effect 52

nonlinear resonance 86, 246, 341
normal mode 150, 160
eigenfunction 150, 160
eigenvalue 150, 158, 160
local 158

overlap of wave fields 145

parabolic equation 8
phase space 68
Poincaré map 71, 344
for a randomly inhomogeneous waveguide 241
point source 8, 9
starting field 9, 159, 162
modal representation 152, 154, 162, 165, 168
pulse signal 25, 125, 274, 307
smoothed intensity of 134
probability density function (PDF) 73, 218, 230, 232
in the Wiener process approximation 227, 228, 230, 235
of coordinate and momentum 218, 236

of action and angle variables 219, 220, 228, 230, 235
of ray parameters 232

quantization rule 151, 270

ray 12
amplitude 14, 25, 365
identifier xiv, 28
equations 12, 15, 20, 21, 84
derivation 363
escaping 95
momentum 12, 23
pulse 14
trajectory (path) 12
travel time 1, 14, 18, 26
as a function of starting momentum (launch angle) 33, 267
statistical description 259
turning points 15
ray-based description 163, 237, 288
of a wave beam 288
of mode amplitudes 164
Helmholtz equation 168
parabolic equation 164
of Wigner function 237
ray chaos 67
in a random medium 217
statistical description 218
ray-mode relations 154, 163
ray-medium resonance 83
reference waveguide 20

scarring 186
scattering on resonance 105
semiclassical propagator 142
shadowing theorem 81
single mode source 166, 168, 270
smoothed mode intensity 271, 303
Snell's law 13, 24
specific Poincaré map 241
splitting of separatrices 88, 339–341
stability 345, 346
stability matrix 79
stationary phase technique 155, 164, 275, 291, 369
applicability conditions 167, 293
stationary points 88, 109, 275, 292, 327, 346, 369
stickiness xiv, 132, 137, 184, 321
stochastic layer 71, 325, 341
Sturm-Liouville eigenvalue problem 38, 150, 160
Helmholtz equation 150
parabolic equation 160

timefront 27, 117, 122, 251, 307
in a range-independent waveguide 27
in a waveguide with periodic range dependence 117, 122
in a randomly inhomogeneous waveguide 251
in the presence of synoptic eddy 307
variations in the presence of a synoptic eddy 307
segment 28, 123, 251
spread and bias of 262
"shelves" 34, 125, 267
trajectory
periodic 347–349
quasiperiodic 347–349
chaotic 347–3049
transient wave field 8, 9, 14, 23, 25
ray representation 14, 23, 25
modal representation 274
transport equation 11, 22
solution 14, 25, 365

underwater sound channel (USC) 3
axis 3
double 3
near-bottom 4
near-surface 4

vertical resonance 98

wave beam 287
in a free space 288

modal structure in the presence of random internal waves 301
ray-based description 290
weakly divergent 297
wave chaos 139
wavefront 96, 144
Wiener process 227
approximation 227
Wigner function 181, 237
coarse-grained 237
ray-based description 237
WKB approximation 150, 161

Advisory Board

Valentin Afraimovich
San Luis Potosi University, IICO-UASLP
Av. Karakorum 1470, Lomas 4a Seccion
San Luis Potosi, SLP 78210, Mexico
Email: valentin@cactus.iico.uaslp.mx

Maurice Courbage
Universite Paris 7-Denis Diderot/L.P.T.M.C.
Tour 24-14.5eme etage, 4, Place Jussieu
75251, Paris Cedex 05, France
Email: courbage@ccr.jussieu.fr

Ben-Jacob Eshel
School of Physics and Astronomy,
Tel Aviv University, 69978 Tel Aviv, Israel
Email: eshel@tamar.tau.ac.il

Bernold Fiedler
Freie Universität Berlin, Institut für
Mathematik I, Arnimallee 2-6
14195 Berlin, Germany
E-mail: fiedler@math.fu-berlin.de

James A. Glazier
Biocomplexity Institute
Department of Physics
Indiana University, Bloomington
Bloomington, IN 47405-7105, USA
Email: glazier@indiana.edu

Nail Ibragimov
Department of Mathematics, IHN,
Blekinge Institute of Technology,
S-371 79 Karlskrona, Sweden
Email: nib@bth.se

Anatoly Neishtadt
Space Research Institute
Russian Academy of Sciences
Profsoyuznaya 84/32
Moscow 117997, Russia
E-mail: aneishta@iki.rssi.ru

Leonid Shilnikov
Research Institute for Applied Mathematics &
Cybernetics
10 Ul'yanov Street
Nizhny Novgorod 603005, Russia
Email: diffequ@unn.ac.ru

Michael Shlesinger
Office of Naval Research, Physics and
Chemistry Department
800 North Quincy Str, Room 503
Arlington, VA 22217-5660, USA
Email: shlesim@onr.navy.mil

Dietrich Stauffer
Institute for Theoretical Physics
University of Cologne, Zülpicher Straße 77
D-50937 Köln, Germany
Email: stauffer@thp.uni-koeln.de

Jian Qiao Sun
Department of Mechanical Engineering,
University of Delaware, Newark
DE 19716, USA
Email: sun@me.udel.edu

Dimitry Treschev
Department of Mechanics and Mathematics
Moscow State University, Leninskie Gogy
Moscow 119899, Russia
E-mail: dtresch@mech.math.msu.su

Vladimir V. Uchaikin
Ulyanovsk State University
L.Tolstoy Str., 42, Ulyanovsk 432700
Russia
Email: uchaikin@sv.uven.ru

Angelo Vulpiani
Department of Physics,
University La Sapinza, P.le A.Moro 2
00185 Roma, Italy
E-mail: Angelo.Vulpiani@roma1.infn.it

Pei Yu
Department of Applied Mathematics
The University of Western Ontario, London
Ontario N6A 5B7, Canada
Email: pyu@pyu1.apmaths.uwo.ca